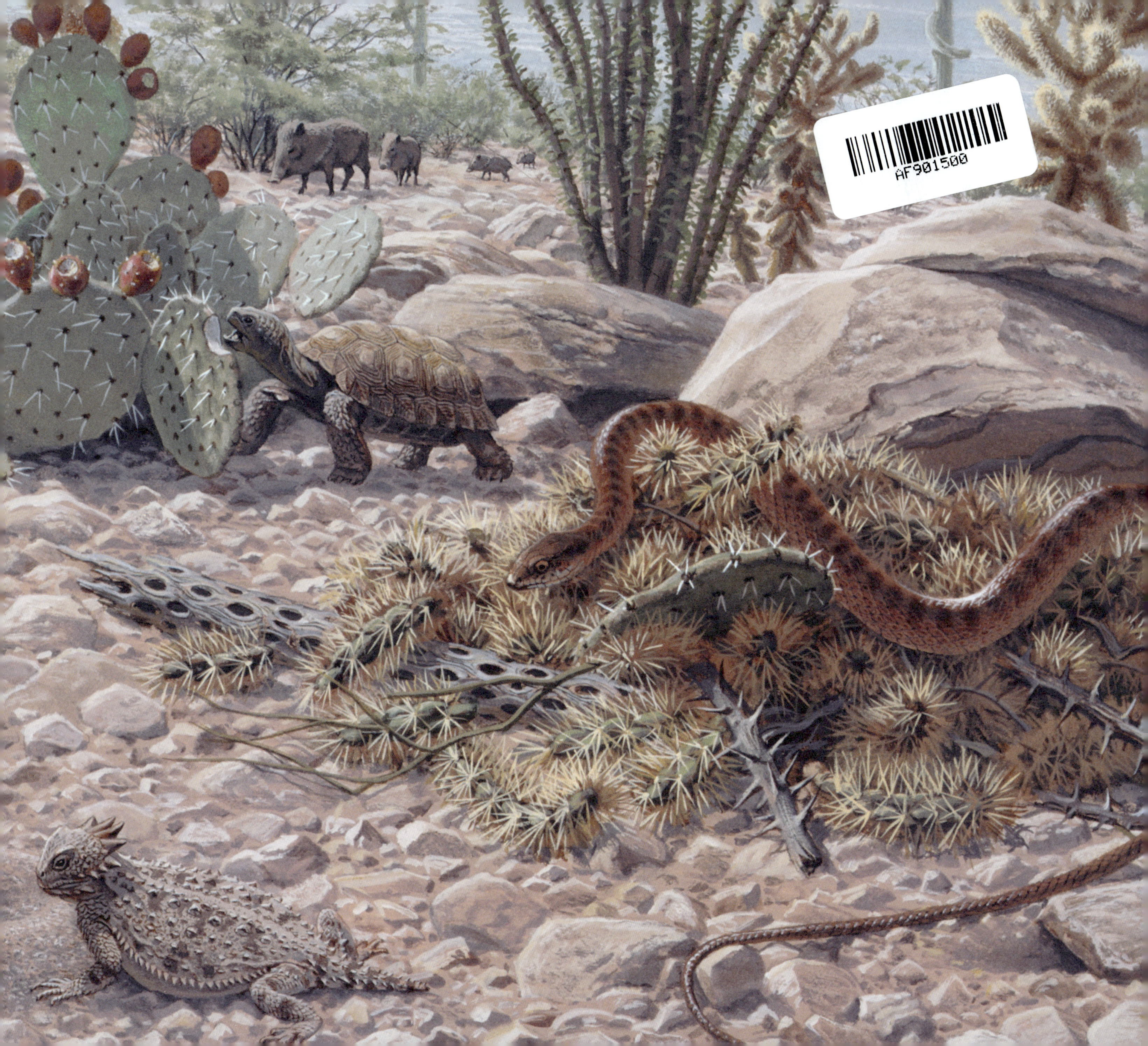

JAY MATTERNES

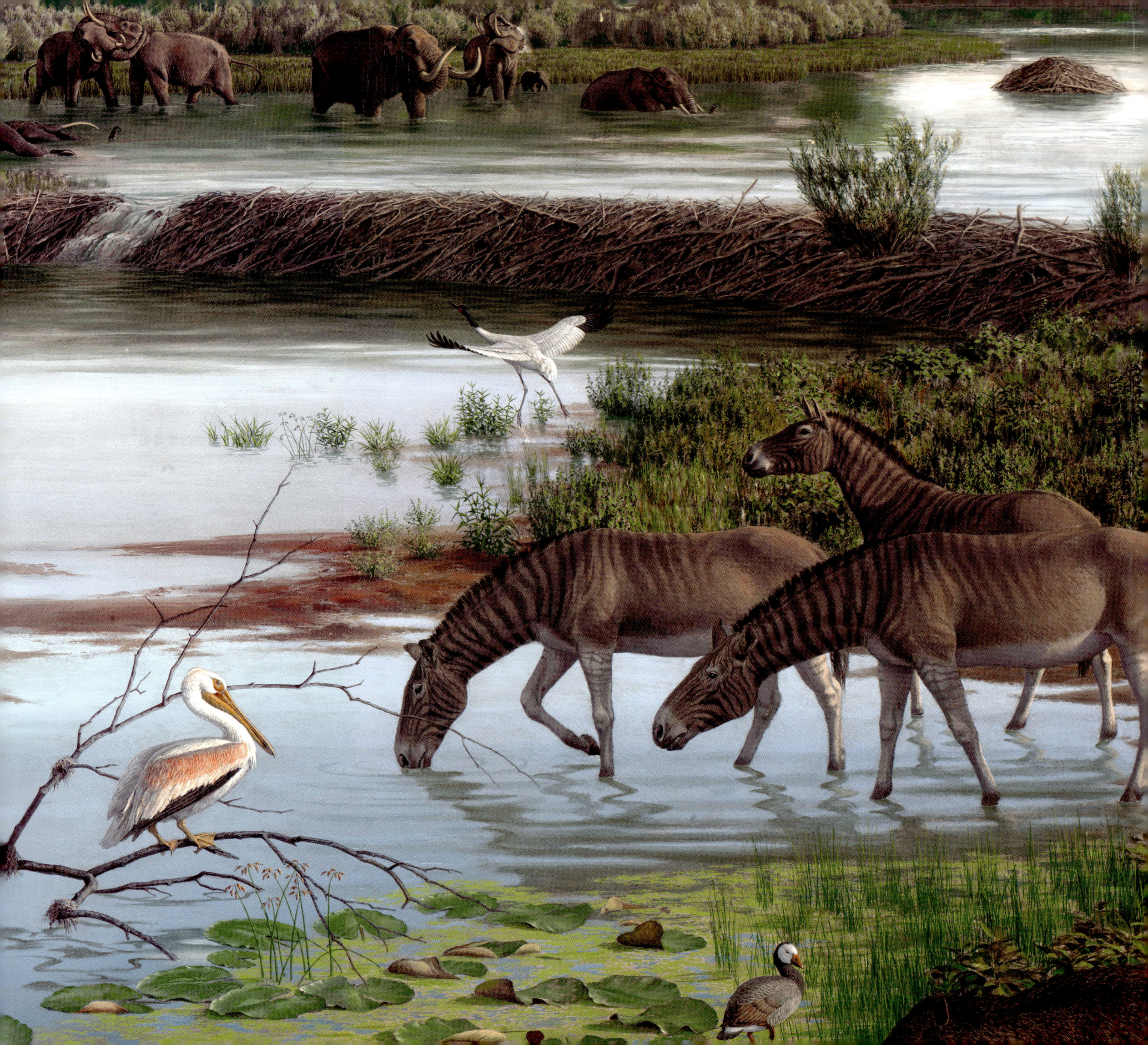

JAY MATTERNES

PALEOARTIST AND WILDLIFE PAINTER

BY RICHARD MILNER
WITH IAN TATTERSALL

FOREWORD BY
MAURICIO ANTÓN

ABBEVILLE PRESS
New York London

CONTENTS

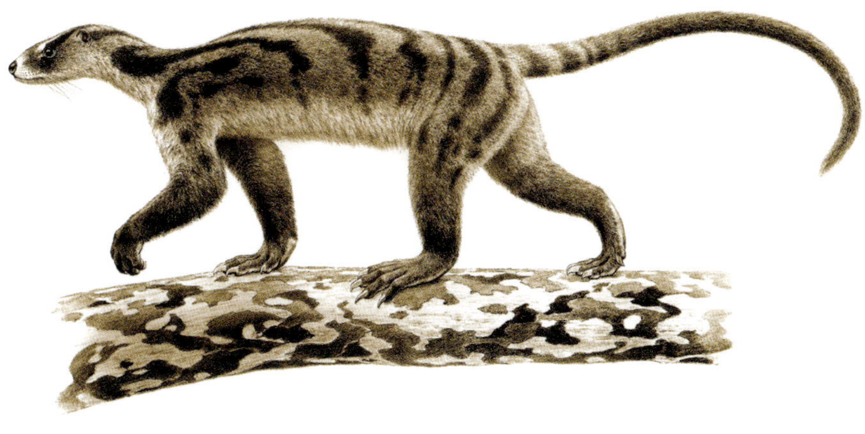

PALEOART'S RENAISSANCE MAN

AN APPRECIATIVE FOREWORD BY MAURICIO ANTÓN

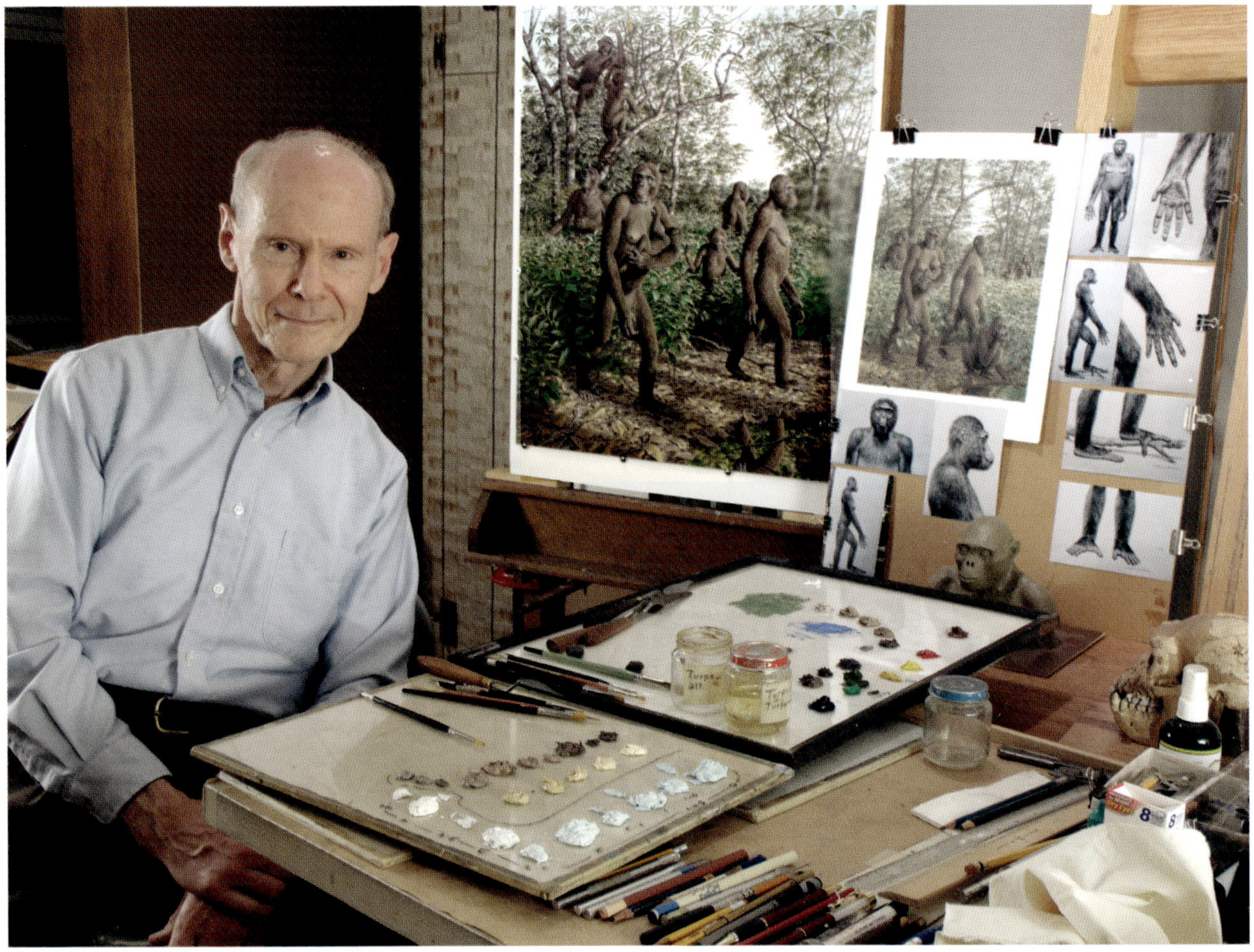

Jay Matternes in his Fairfax, Virginia, studio working on a reconstruction of a family group of the early hominin *Ardipithecus*, 2012.

I remember clearly the first time I saw the art of Jay Matternes. I was perhaps 15 years old, growing up in Caracas, Venezuela, and was browsing through my school library for anything that might have prehistoric animals in it. Then I came across a Spanish translation of the 1964 Time-Life book *The Land and Wildlife of South America*, which included a delightful seven-page section on extinct mammals, illustrated with Matternes' gorgeous depictions of such exotic creatures as the giant ground sloth *Megatherium*, the hippo-like *Toxodon,* and *Macrauchenia*, a strange long-snouted ungulate. But the highlight for me was a two-page spread showing a collection of sketches that summed up the methodology used by the artist to reconstruct the life appearance of the "marsupial sabertooth," *Thylacosmilus*. Seeing those drawings was a revelation: although I was already keen on paleontology and paleoart, I felt then as if for the first time I could stick my head out of a window and look into a clear three-dimensional world that previously I had only seen through the blur of darkly tinted glass.

Back then I was fascinated by sabertooth cats and had learned a little about *Smilodon* or the formidable catlike *Hoplophoneus*, but I had never heard of such a thing as a "marsupial sabertooth," so I was impressed by the combination of the

"Love of Nature and Honesty: these are the two great passions of geniuses."

—Auguste Rodin

familiar and the strange in this new creature as defined by Matternes' drawings. But it was also the format that made this set of images completely different. Previously, each time I learned about a new prehistoric beast it would be in the form of a finished life restoration featured in a book or a collectible card, but what relation that restoration had to the actual fossils was a big unknown. Those sketches by Matternes showed not only the final result, but the whole process. The fossils of *Thylacosmilus* were drawn in painstaking detail, differentiating the preserved sections of the skeleton from the reconstructed ones, and then the artist added the muscles and other soft tissues following a strict anatomical methodology. There were even comparisons with related species that served as reference for reconstructing missing parts. And then, once the hypothetical appearance of the animal had been established, Matternes went on to reconstruct its likely locomotion and predatory behavior through a series of dynamic but incredibly accurate sketches.

So, this was much more than a collection of beautiful drawings: it was a lesson in making paleobiological inferences from the fossils, which for me had all the excitement of scientific investigation. But the role of art was also essential. The aesthetic excellence and depth of those drawings kept me coming back to look at the book time and again. I would take pencil and paper with me and copy the sketches, trying to absorb an indefinable something: perhaps a combination of the elegance of the artwork and the precision of each deceivingly easy pencil stroke. In a way, those *Thylacosmilus* sketches established for me what it meant to be a paleoartist, and afterward I would never be completely satisfied with just borrowing somebody else's interpretation of an extinct species: only by studying the fossils and the anatomy of the animals would the task of reconstruction acquire full meaning. And of course, when shown as transparently as Matternes did, reconstruction revealed itself as an open process, always subject to revision as new evidence becomes available.

After that initial discovery I badly needed to see more of Matternes' art, and over the years I tried to track down every old issue of *National Geographic* with his illustrations and even travelled to Washington, DC, to admire his murals at the National Museum of Natural History (Smithsonian), an experience no less transcendent to me than seeing the work of Renaissance masters first hand. In fact, I see Matternes as a Renaissance man. Like Leonardo da Vinci before him, he never degrades his subjects to mere excuses for painted beauty; his art is a celebration of Nature, inseparable from the acquisition of knowledge.

Such a humble approach to art and science, based on a sense of reverence, is something that we desperately need if we are ever to reverse the destruction of the living world around us. Today science too often becomes a tool of dominion over nature (and art a tool of dominion over viewers). Knowledge no longer seems to be enough, and we are proving ourselves entirely capable of destroying what we think we understand. After all, understanding how something works can be the first step toward appreciating the natural world and wanting to preserve it, but it can also foster an arrogance that results in destruction. However, one fervently hopes that we will hesitate before destroying what we respect and admire. From his 1964 *Thylacosmilus* to his 2009 reconstruction of the hominin *Ardipithecus*, Matternes' creations reflect the scientific understanding of his time, but just as importantly they show his deep respect for his subjects—and for Nature. To me that is a mark of true, honest genius.

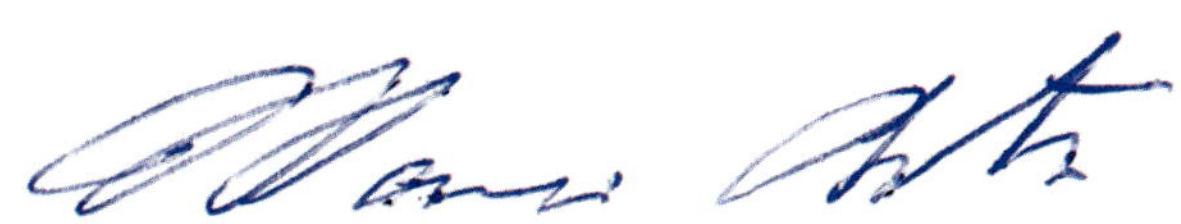

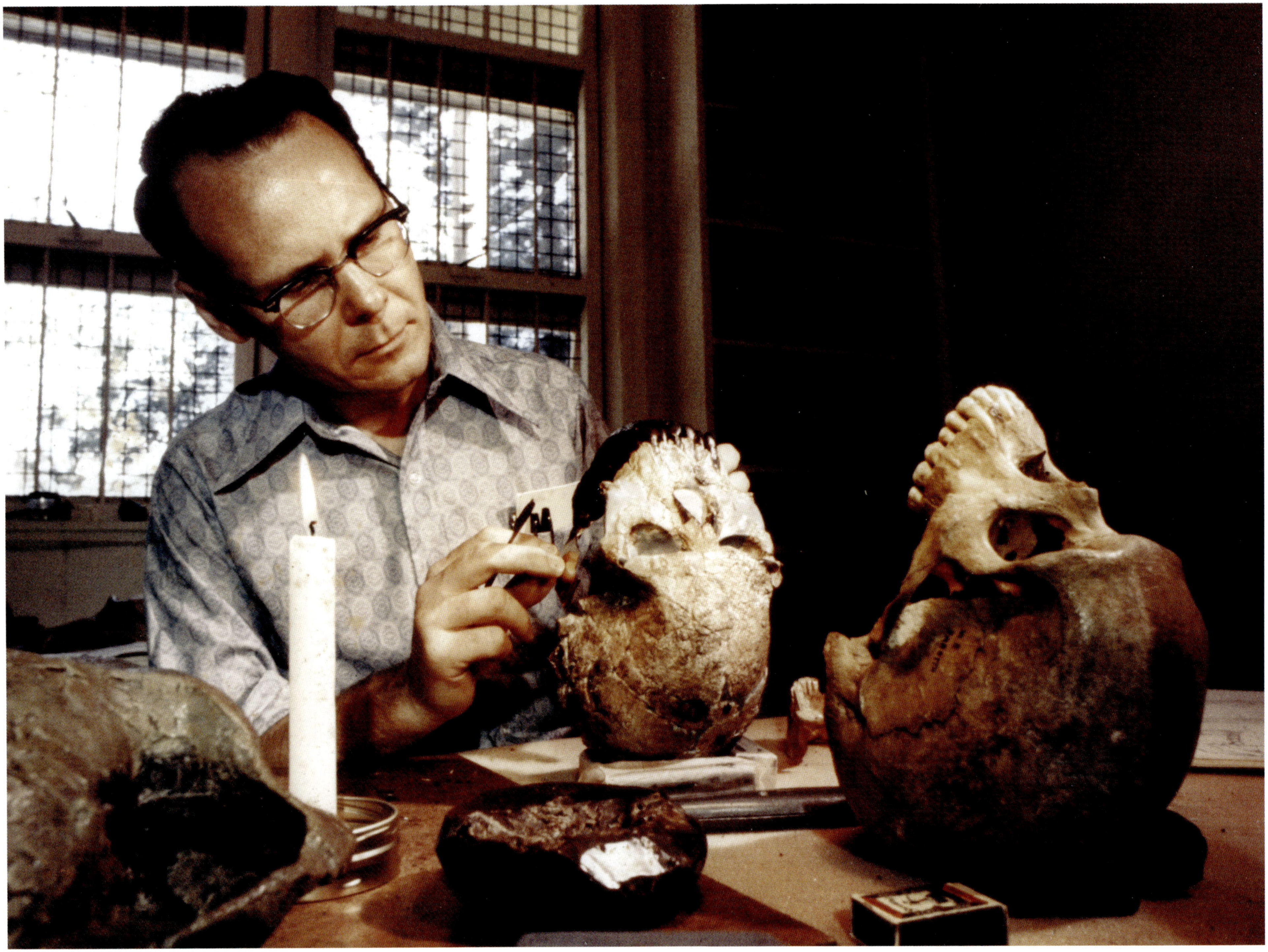

Jay sculpts a missing mandible in wax to complement the KNMT-ER 1470 skull, a 2-million-year-old hominin from Koobi Fora, at the Louis Leakey Center in Nairobi, 1973.

INTRODUCTION

THE ART OF JAY MATTERNES

Jay Matternes is a master of "paleoart"—the collaboration of wildlife artists and evolutionary biologists to create evidence-based restorations of animals and plants that vanished from the Earth millions of years ago.

This book contains a retrospective of Matternes' past seven decades of artwork, during which he helped to establish a gold standard of beauty and scientific accuracy within this "Time Machine" genre.

Generations of visitors to the Smithsonian's National Museum of Natural History in Washington, DC, have viewed his six classic murals on the Cenozoic era (Age of Mammals), painted between 1961 and 1975. Millions have been enthralled, educated, and entertained by their parade of astonishing animals that lived between 50 million and 14,000 years ago, including saber-toothed cats, ponderous brontotheres, and tiny three-toed horses. (The Smithsonian murals are lavishly displayed in *Visions of Lost Worlds: The Paleoart of Jay Matternes* by Matthew Carrano and Kirk Johnson, Smithsonian Books, Washington, DC, 2019.)

Matternes masterworks also adorn the American Museum of Natural History in New York, the Florida Museum of Natural History (Gainesville), and Japan's Gunma Prefecture Museum. Over the years, his work has appeared in *National Geographic*, *Nature*, *Scientific American*, and numerous books. Dozens of films and television programs have featured digital animations of his images.

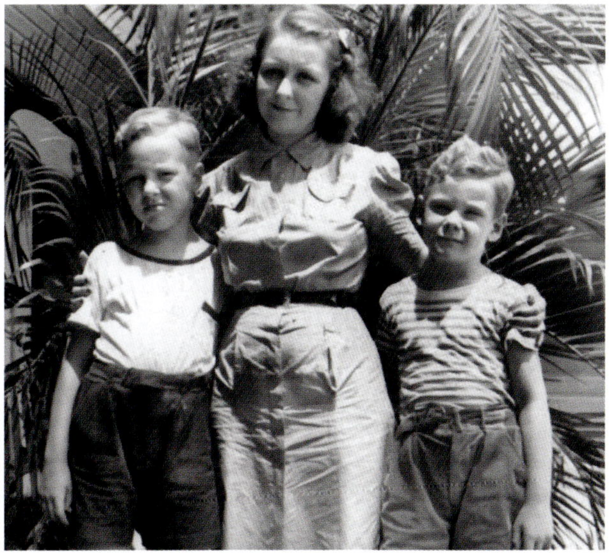
Young Jay (right) poses with his brother James and mother Janet for a snapshot taken at Ancon Hill, Panama City Canal Zone, 1941.

Matternes has particularly contributed to paleoanthropology—focusing on the quest for an accurate picture of our own remote ancestors. Working closely with the world's leading prehistorians, fossil hunters, anatomists, and primate experts, including Louis, Mary, and Richard Leakey, Tim White, Jane Goodall, and Dian Fossey, he has spent his life visiting sites of important fossil discoveries, and has absorbed libraries of research, dissected apes, measured fossil skeletons, painted living animals, and imaginatively navigated a vast geologic timescale spanning hundreds of millions of years.

Born in 1933 into a military family stationed on Corregidor Island, Philippines, the precocious youngster spent hours sprawled on the floor with his boxes of crayons, absorbed in sketching cowboys and Indians, World War II airplanes, and Hollywood-inspired jungles. Like many boys, he often filled in the blank spaces with flying bullets, bombs, and arrows as he narrated the action scenes and sound effects to himself (pp. 16–17).

As a boy, he became fascinated with the work of Charles R. Knight (1874–1953), the wildlife

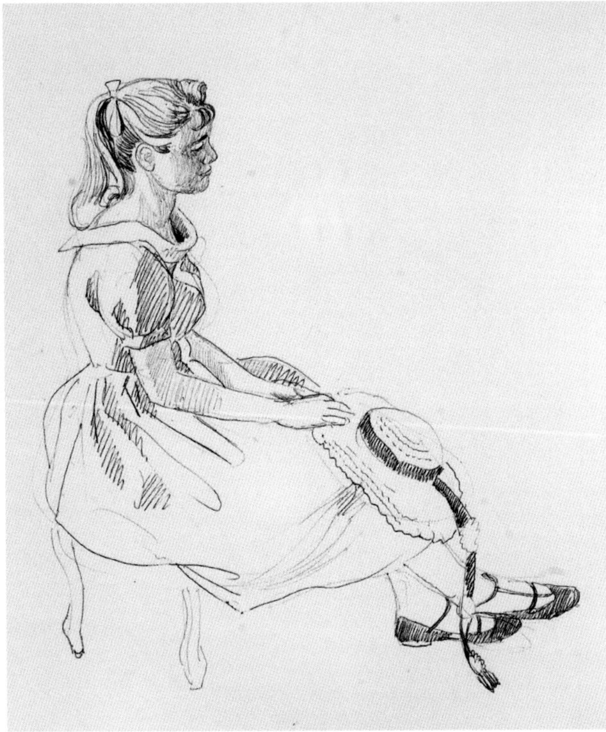

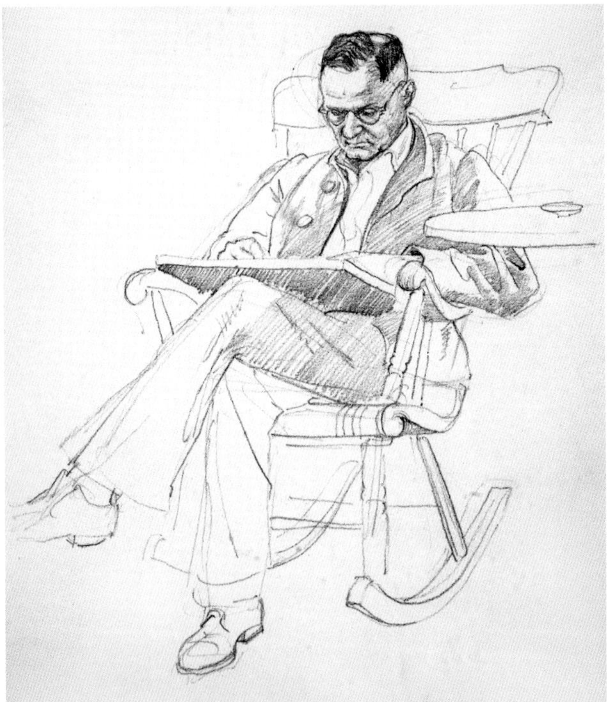

Top: Matternes sketch of his sister Anne, made as a young man at Governors Island, New York, where his father was stationed at an Army hospital. Bottom: Companion sketch of Jay's father, Lawrence Matternes, a physician and surgeon, made around the same time.

painter and "father of paleoart" at the American Museum of Natural History in New York, who created and dominated the field from the 1880s through the 1950s. Knight's lifelike reconstructions of monstrous dinosaurs, woolly mammoths, and "cavemen" were endlessly reproduced in movies, paleontology books, toys, and comic books, and became well known to children all over the world.

After being shunted among various army bases to which his surgeon father was assigned, the family settled in Pennsylvania, where he made frequent "sketching visits" to the Philadelphia Zoo and the Academy of Natural Sciences, and became obsessed with drawing animals. He recalls being particularly inspired when, in 1946, his father took him on a rare weekend outing to New York City's American Museum of Natural History and the Bronx Zoo. He always looked back on that visit as having been pivotal in determining the course of his life.

By his teens, Matternes was determined to become a professional wildlife artist, and balked at his family's insistence that he pursue medicine, like his father and both grandfathers before him. But when he won a scholarship to pursue his chosen career at the Carnegie Institute of Technology's College of Fine Arts in Pittsburgh, the family relented and his path was set.

From 1958 to 1960 he was drafted by the army, and assigned to work with other artists on various government projects. Shortly after leaving the service, he was engaged by the Smithsonian to create the murals of Cenozoic mammals that would occupy him for a full career at the National Museum of Natural History. While there he seized the opportunity to perform dissections of gorillas, chimps, and orangutans he had procured from various sources, including "zoo casualties." By developing meticulous anatomical drawings from his dissections over several years, he became an expert comparative anatomist, which proved invaluable in illustrating living animals as well as extinct species. Eventually, he became adept at understanding the construction of animals "from the inside out."

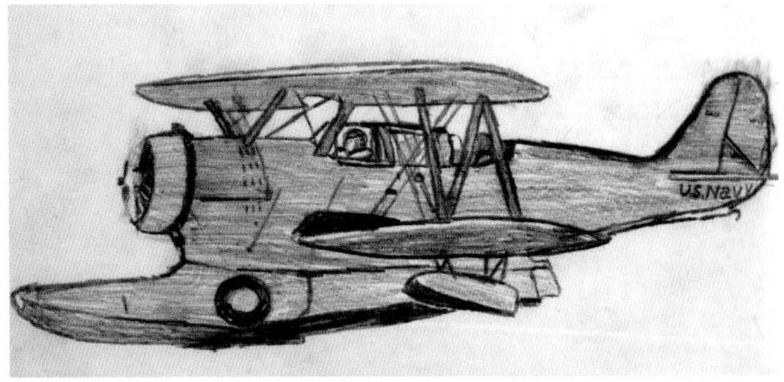

A US Navy amphibious biplane sketched by Matternes from newspaper photos in 1940, when he was seven years old. Used for reconnaissance, it featured both wheels and pontoons for water landings.

Matternes also studied the works of British artist Edwin Landseer (1802–1873), who painted mainly dogs, horses, and deer, and won fame for his Romantic portrait of a magnificent stag in the Scottish Highlands (*Monarch of the Glen*) and the heroic lion sculptures that guard the base of Nelson's Column in Trafalgar Square.

Another influence was the Swede Bruno Liljefors (1860–1939), whose wildlife paintings often dramatized the "survival of the fittest" by focusing on such familiar subjects as cats and foxes on the hunt. Matternes staged some of his own scenes of mortal combat on a much grander scale, such as lions attacking a Cape buffalo while dark storm

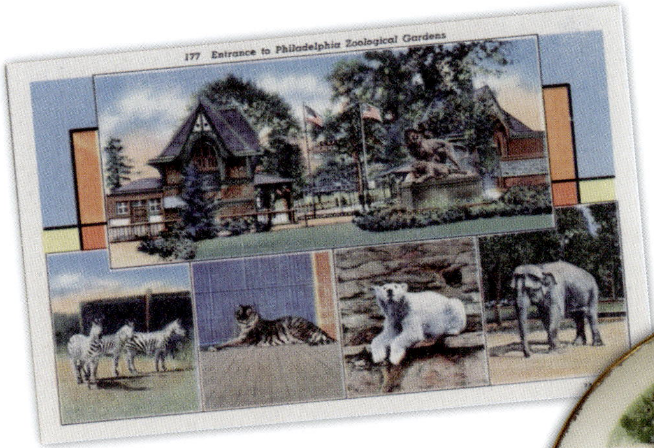

Vintage postcard from the Philadelphia Zoo, a haven for young Matternes, with his sketchbook always at hand.

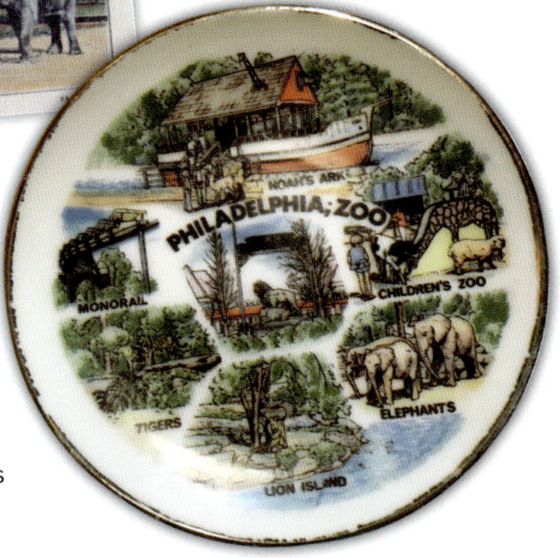

Souvenir plate from the Philadelphia Zoo, where teenage Jay spent weekends sketching animals from life.

Massa, a 400-pound western lowland gorilla, the young artist's favorite denizen of the Philadelphia Zoo. He held the longevity record for captive gorillas when he died in 1984 at the age of 54 years.

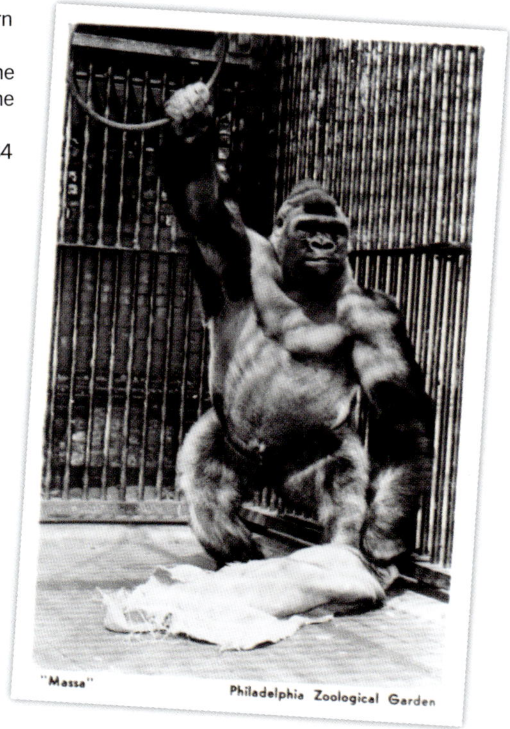

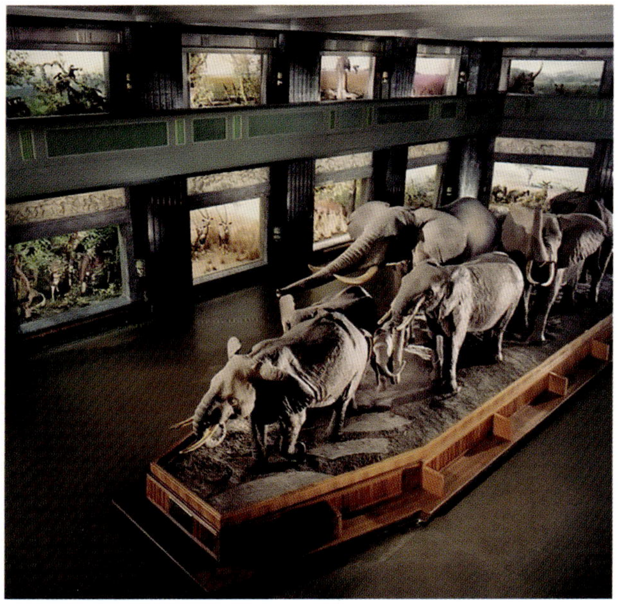

Akeley Hall of African Mammals at the American Museum of Natural History. In 1946, the young Jay traveled to New York City to meet his father for a weekend at the museum and the Bronx Zoo. He later wrote, "I never got over it. I think that museum visit more than anything else formed my career and aspirations."

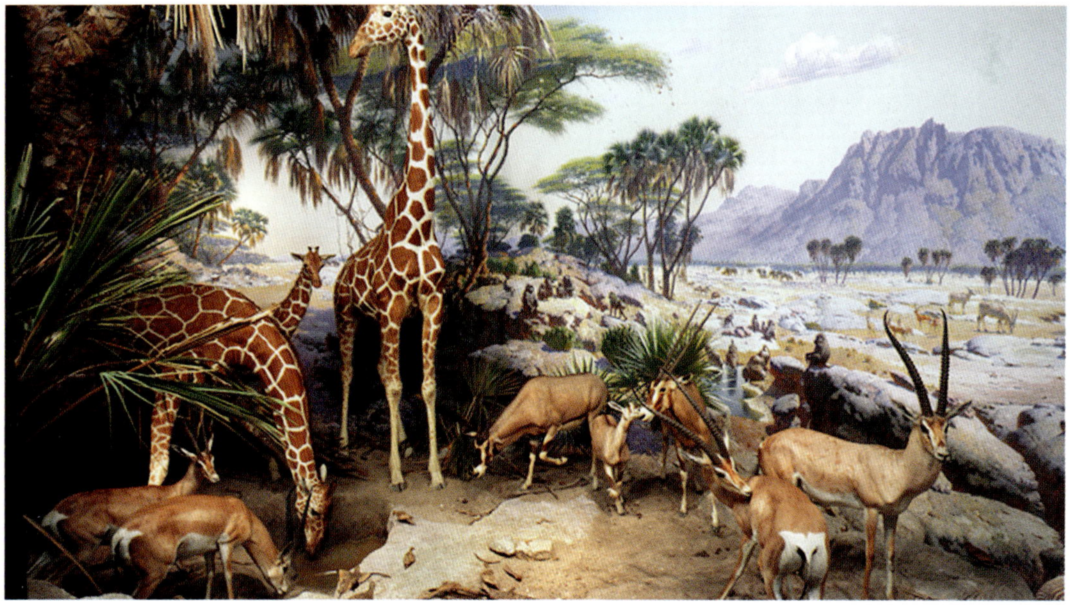

The waterhole group at the AMNH, showcasing giraffes, antelopes, and warthogs against a beautifully painted backdrop by William R. Leigh and J. Perry Wilson. Begun in 1928, it is the first and oldest diorama in Akeley African Hall, which opened in 1936, only a decade before Matternes and his father visited there.

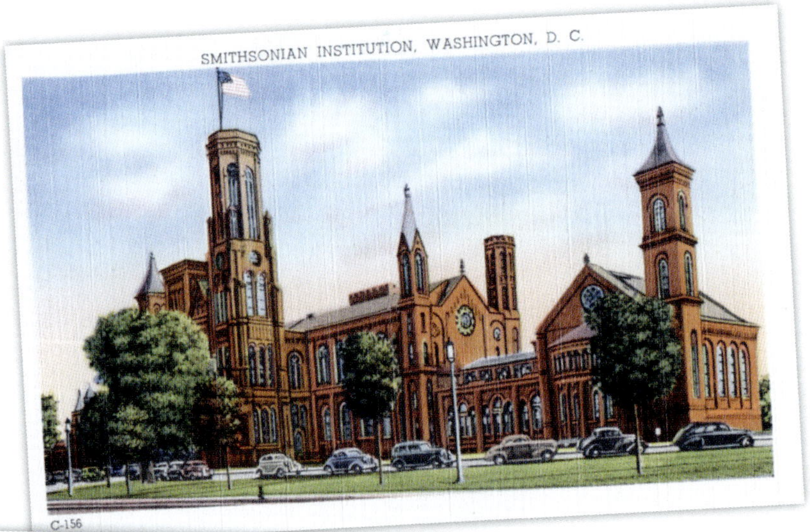

Postcard from the 1940s of the Smithsonian Institution, where Matternes spent much of his professional life. A snapshot in time, it shows a lineup of visitors' parked cars of the period.

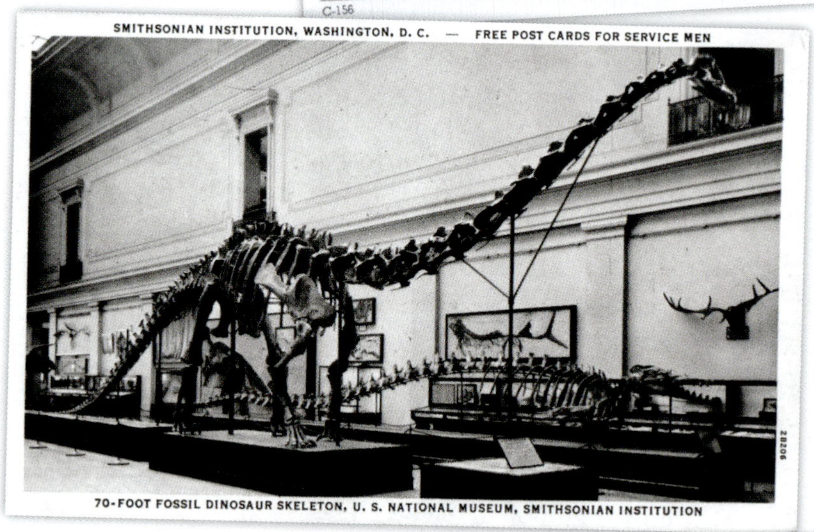

Diplodocus replica skeleton shown on Smithsonian postcard during WWII era. It proclaims "70-foot Fossil Dinosaur Skeleton" and "Free Post Cards for Service Men."

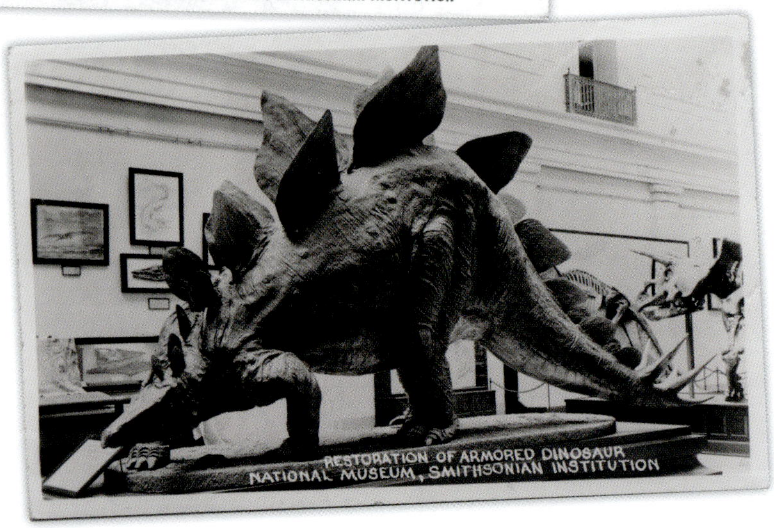

Stegosaurus as a full-sized sculpture on a Smithsonian souvenir postcard. Restorations of extinct animals at the Washington, DC, institution were already an important part of the paleontology exhibits.

clouds gather overhead (pp. 24–25) and the cinematic "Buffalo Jump," an action scene of Plains Indians stampeding a herd of bison over a cliff (pp. 180–81).

He also admired the pioneer experimental photographer Eadweard Muybridge (1830–1904), who used multiple camera setups to capture split-second photos of animals walking and running—including horses and camels galloping. (Muybridge famously won a bet to prove that during a horse's gallop, all four hooves simultaneously leave the ground.) But although Matternes used photographs for reference, he never copied directly from them and always preferred drawing from life.

As a young freelancer at the Smithsonian, Matternes still considered himself primarily a wildlife artist who loved painting birds of prey, hunting dogs, big cats, African antelopes, and primates, especially the great apes (pp. 18–76). During his early days at the museum, he recalls, it never entered his mind that a primary focus would become painting near-humans that no one had seen for a million years. His career took a sharp turn in that direction, however, when he was invited by the editors of Time-Life Books to create illustrations for their slim, mass-marketed volume *Early Man* (1965). Coordinated by F. Clark Howell, a leading American paleoanthropologist of the day, it became an instant classic.

In that book, which revolutionized the level of photorealism used in illustrating early human relatives, Matternes created a striking scene of hominins stealing a hyena's kill (pp. 156–57), and another of a faceoff between hostile gangs of two different hominin species (pp. 160–61). These images merged

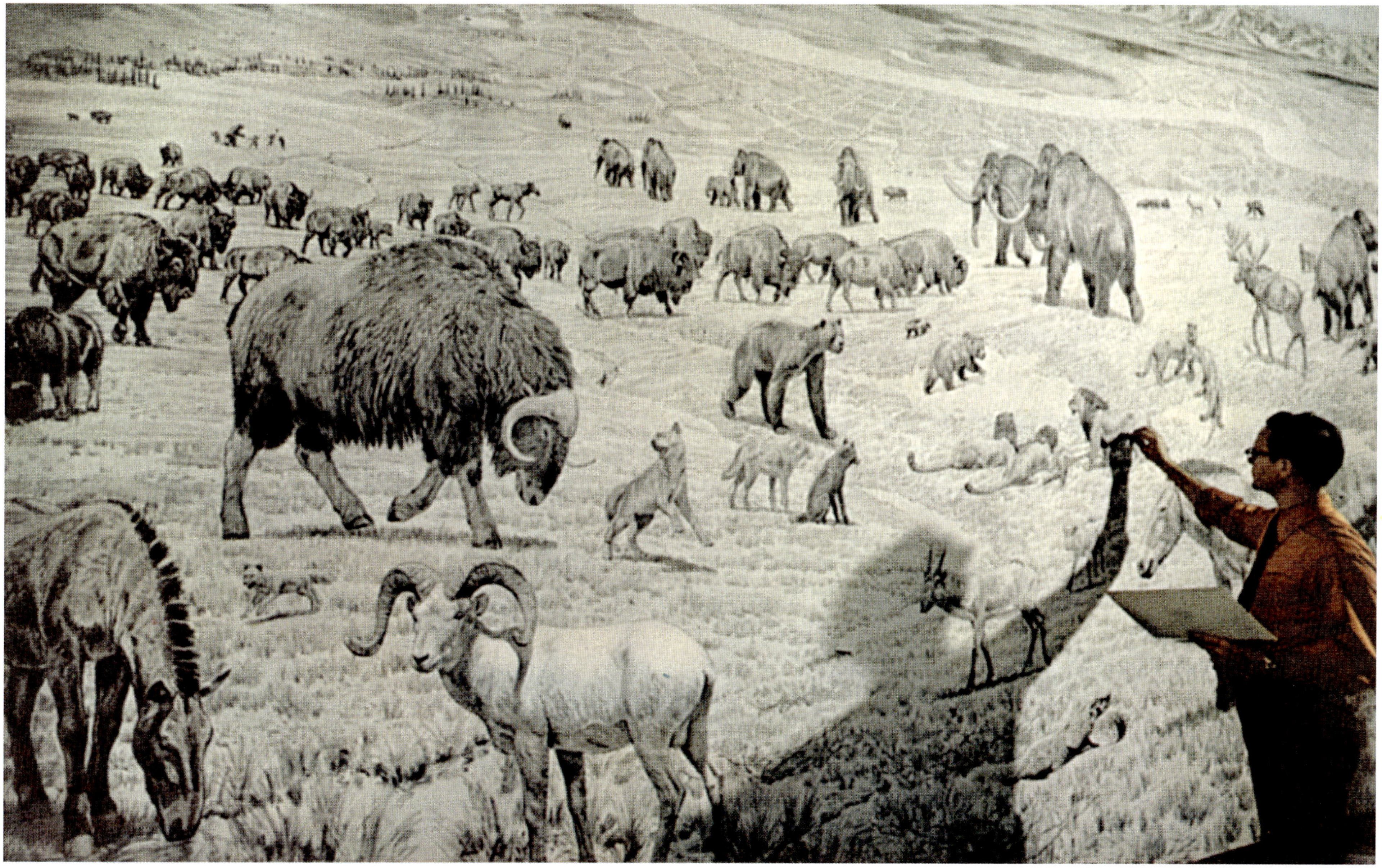

Young Jay Matternes completes his charcoal sketch of the Arctic (Fairbanks) mural at the Smithsonian, circa 1971.

a sense of drama with an attention to scientific detail that equaled and sometimes even surpassed those of his boyhood hero Charles R. Knight.

Then and now, Matternes' restoration technique involves several stages of preparation. First, working with casts of the fossils, he draws all bones and fragments of the specimen, sometimes actual size as measured with calipers. Often, he must restore partial and distorted elements or extrapolate missing parts. Working with both drawings and clay sculptures, he next assembles an individual's skull and skeleton, overlays soft tissues, then finishes with surface features and textures (pp. 88–91, 130–33). During this laborious task the series of detailed, accurate drawings transforms an incomplete collection of fossilized bone fragments into a convincing portrait of a living, breathing being.

When the legendary Anglo-African fossil hunter Louis Leakey, his archaeologist wife Mary, and their sponsors at the National Geographic Society saw the *Early Man* illustrations, their admiration led to further commissions. Jay created hominin reconstructions for the Leakeys' magazine articles and books, not only of their East African fossil

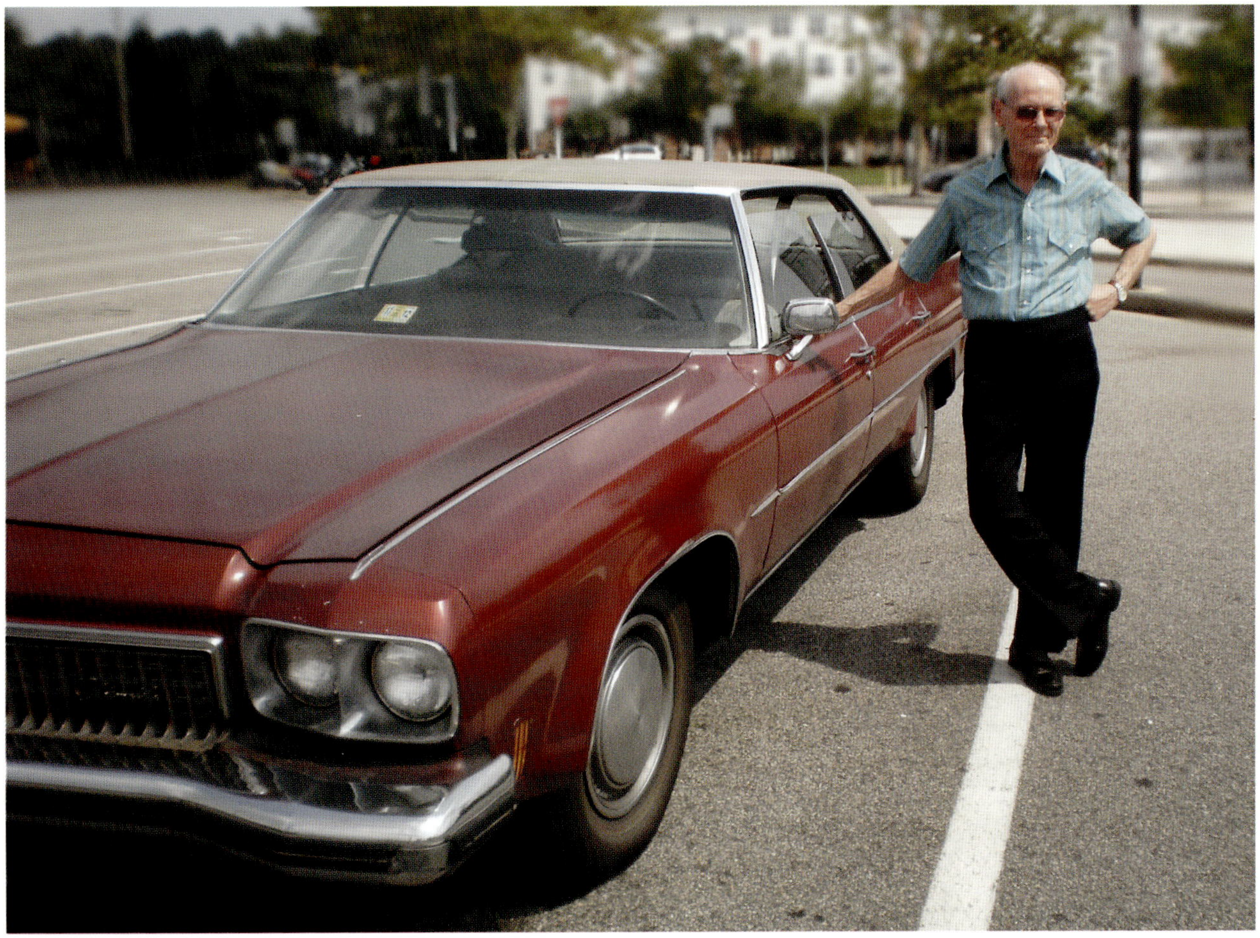

This red Oldsmobile 98 hardtop (1973) was the artist's pride and joy for thirty years. Fairfax, Virginia, 2018.

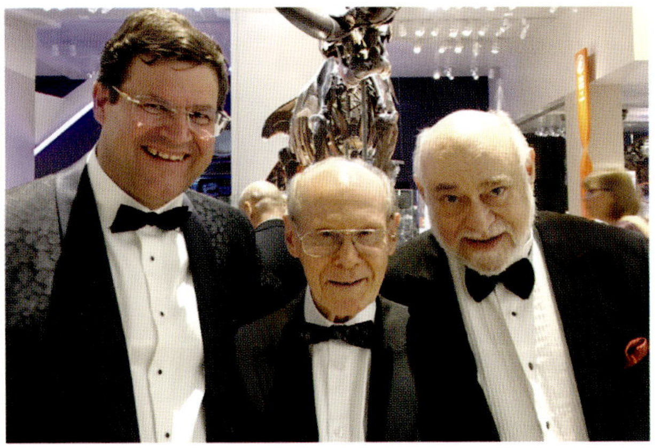

Kirk Johnson, Director of the Smithsonian's National Museum of Natural History, joined Matternes and author Milner at the gala opening of the museum's refurbished "Deep Time" Hall on June 5, 2019. Two of Jay's classic murals were digitally archived and installed there.

discoveries *Paranthropus* and *Homo habilis*, but also of the South African australopiths as well (pp. 138–47, 150–51).

Matternes went on to work with other distinguished anthropologists, and in addition enjoyed a twenty-year-long friendship with legendary gorilla ethologist Dian Fossey until her tragic murder in 1985. He was among the first to photograph Fossey during the early days of her fieldwork in the Rwandan forest, where he sketched the behavior of her free-living gorilla groups (pp. 56, 62–76). He also illustrated Jane Goodall's popular book on the chimpanzees of the Gombe Stream Reserve in Tanzania, published in 1971 (pp. 48–51).

Jay met his wife, Delphine Vermish, during the Early Miocene epoch—that is, while he was painting the mural. A native of Melbourne, Australia, she was working at the World Bank in Washington, DC, in 1961, when mutual friends from nearby Alexandria, Virginia, brought them together. They married four years later, and remained inseparable until her passing in 2011. "Marrying Del," he says, "was one of the wisest things I ever did."

In 2018, in his eighties, Matternes expressed his lifelong interest and affection for Native American culture by painting a bustling Lakota Sioux village in the 1890s, complete with dogs, horses, tepees, and an old, gesticulating storyteller (pp. 184–89). It is a happy but poignant morning frozen in time, just before an invasion of land-hungry settlers and fortune-seeking gold miners was to sweep away their ancestral mode of life forever.

Because of his obsession with scientific accuracy, Matternes' drawings and paintings provide a unique history of the progress of paleoanthropological knowledge. Fossil hunters no longer trumpet each new discovery to the press as a "missing link" in human evolution, as they once did. Instead, tracing human evolution has become an analytical, multidisciplinary science in which no one person or fossil has all the answers. Our present quest for understanding is now recognized as inevitably provisional—interim snapshots along the path to knowledge, vulnerable to errors of judgment and cultural fashions. Jay invites his audience, whether of scientists or the general public, to understand exactly how he transforms dry fossil bones into vibrant visions of vanished worlds, and welcomes discussions of the interpretations he presents.

Matternes' reconstructions of hominins have become a chronicle of the recent path to increasing knowledge about our distant relatives and ancestors. As a result, this book is much more than simply a retrospective gallery of dazzling art and rare virtuosity; it opens a portal to visualizing, vividly, humankind's remarkable evolutionary emergence.

Charles Darwin famously ended his 1859 masterpiece on evolution, the *Origin of Species*, by concluding that "there is grandeur in this view of life." To which Jay Matternes has added, "and wondrous beauty as well!"

—*Richard Milner and Ian Tattersall*

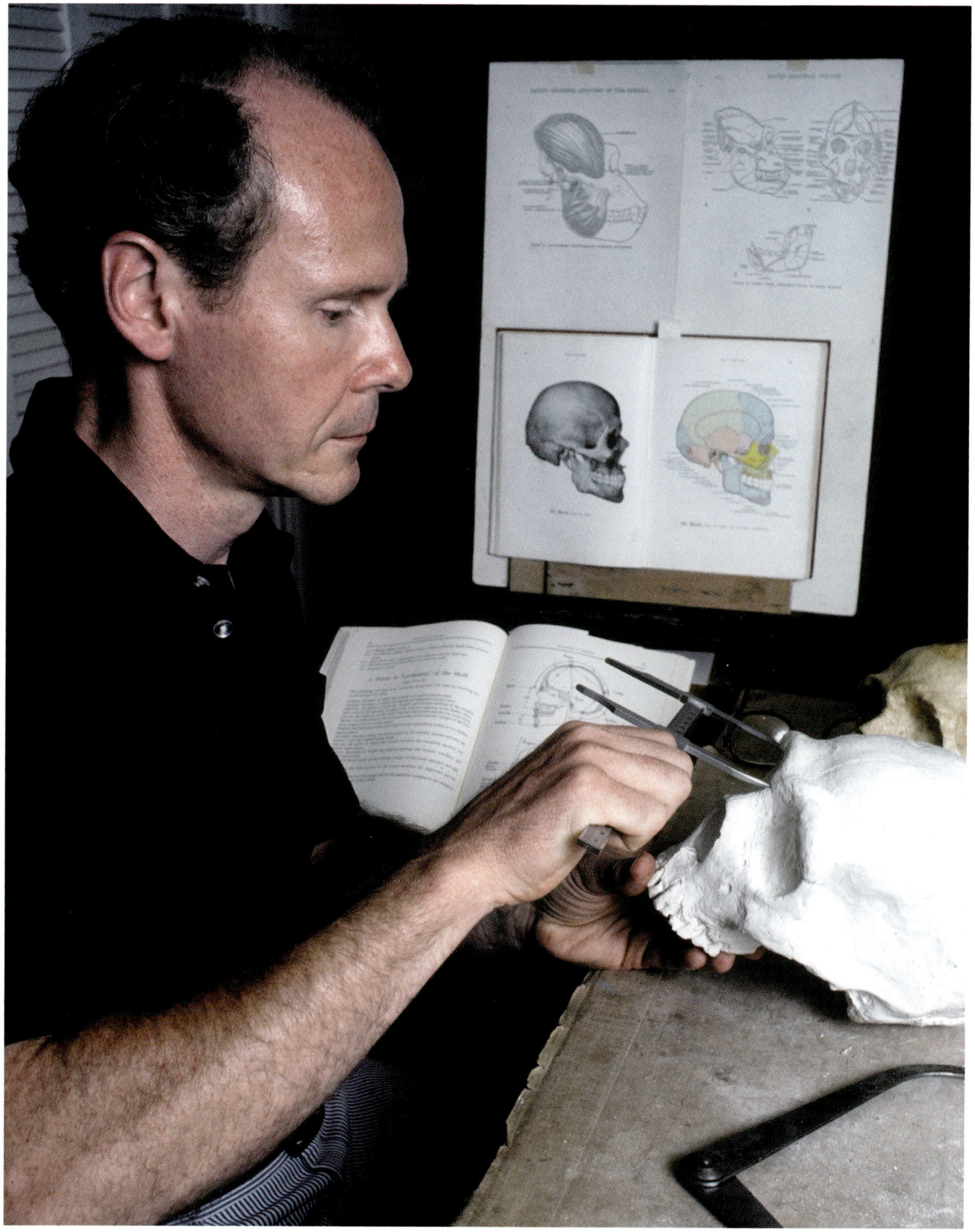

Using calipers, Jay measures a cast of the Shanidar I Neanderthal skull in his studio, circa 1980.

CHILDHOOD DRAWINGS

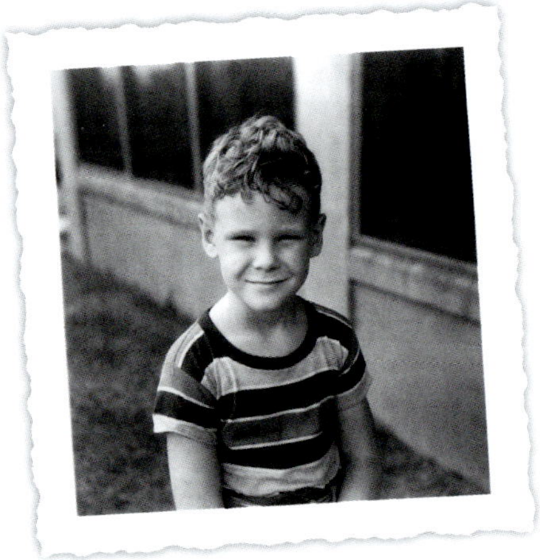

The artist at seven years old, Ancon Hill, Panama Canal Zone, 1940

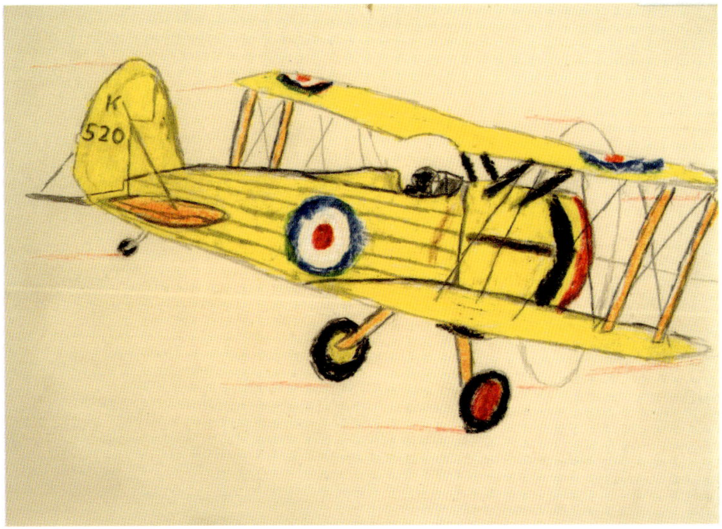

During World War II, seven-year-old Jay liked to sketch warplanes from newspapers. ("I was captivated by the Battle of Britain, which my family was following with great apprehension.") As the air war gathered momentum, the Kaydet 1425-5, an old-style biplane, was used by the U.S. Navy to train new pilots. The aircraft were often painted bright yellow to warn others of the "new driver" in their midst.

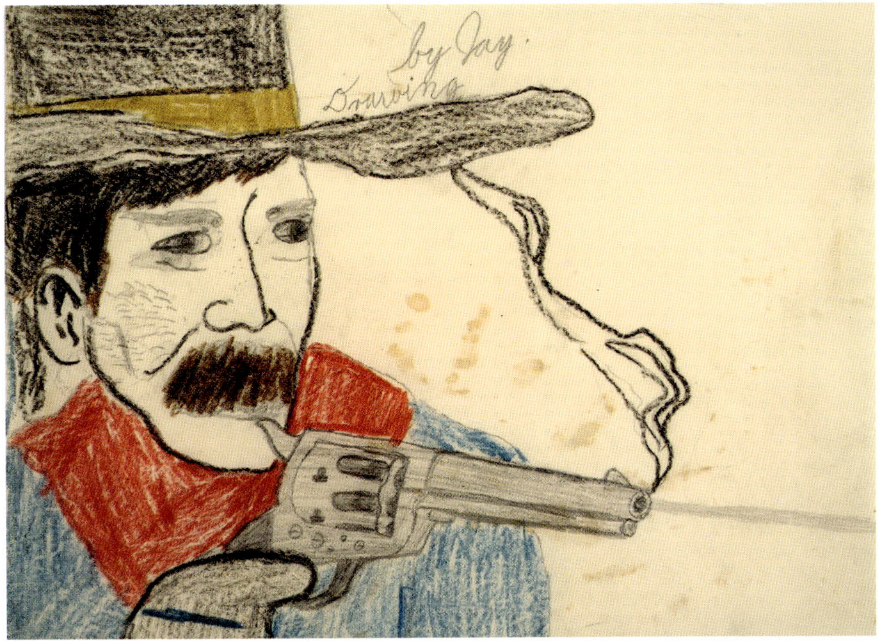

Matternes recalled that "film cowboys were my heroes, although I had no interest in the grueling, boring work of real cowpunchers. I liked to draw horses from an early age, although I had never ridden one until 1946 or so. I think that everyone at any age loves horses. As a child, I discovered quickly that when you get on a horse's back, he knows immediately whether or not you know what you're doing."

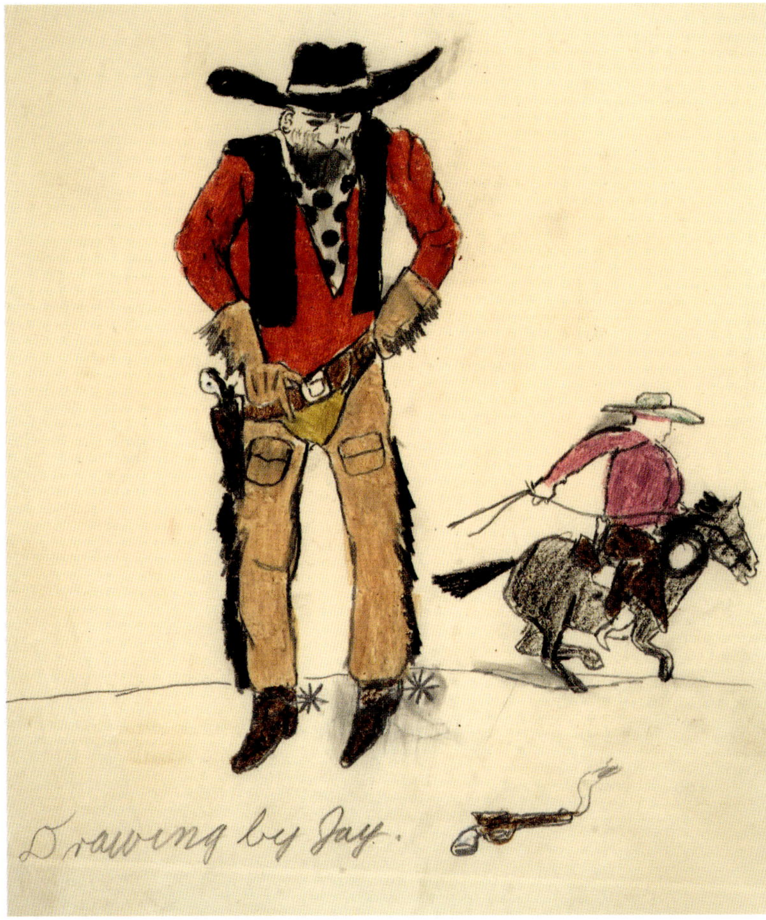

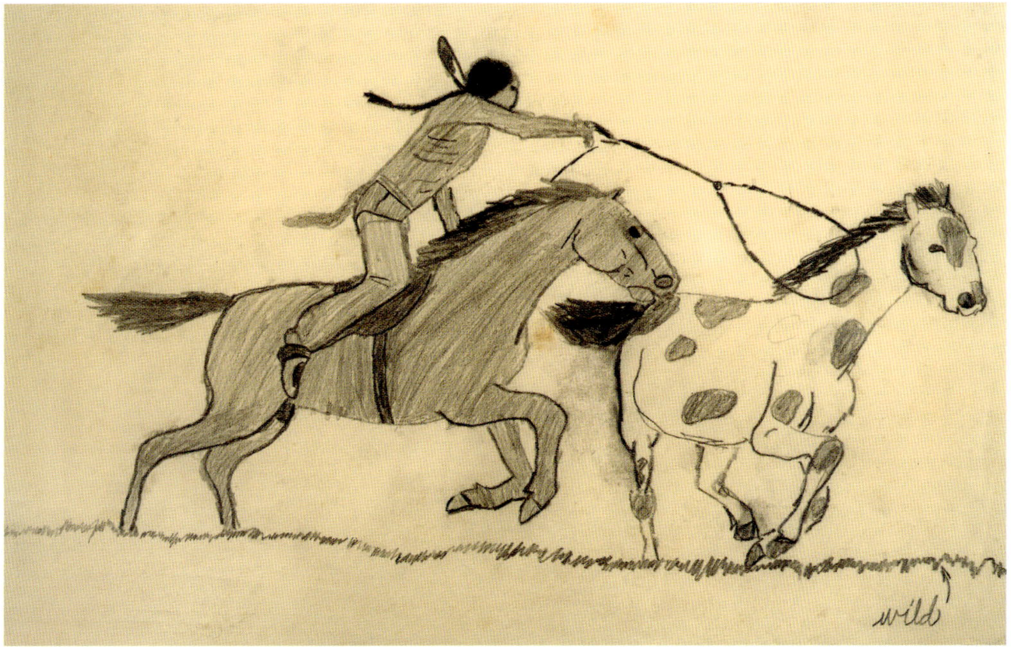

“The so-called Indians in films of the 1940s were depicted as villainous savages, but they, along with the cowboys, formed an important part of my romance with the Old West. In recent years, we’ve come to acknowledge how badly settlers treated the indigenous peoples: routinely broke treaties, stole their land, exterminated the bison on which they relied, and systematically attempted to dismantle their ancient cultures.”

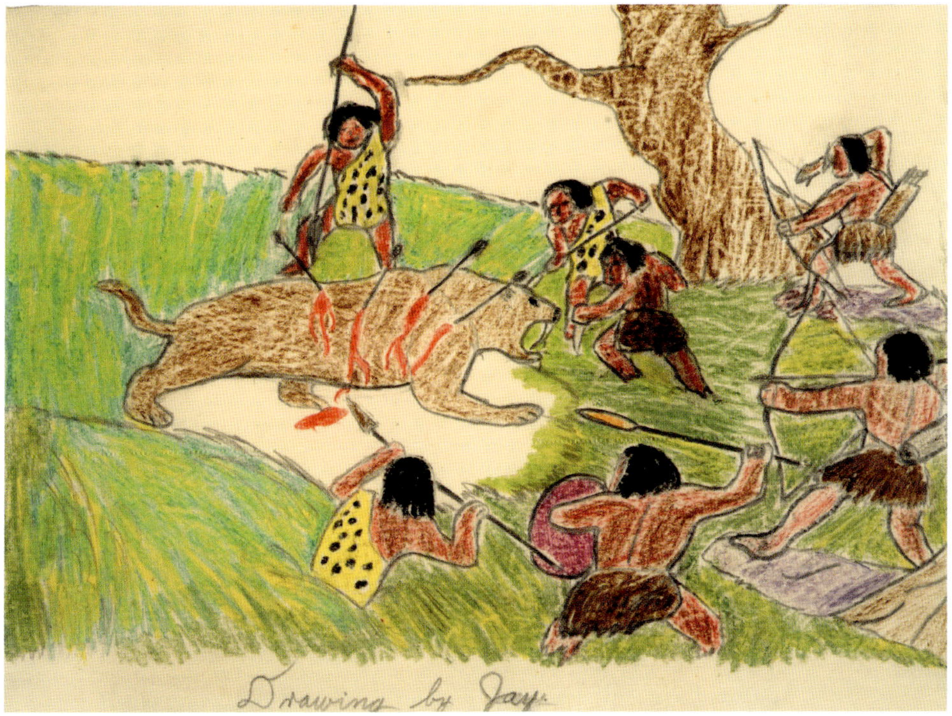

Another early crayoned drama shows prehistoric tribal hunters wearing leopard skins attacking a sabertooth cat. “I don’t know what inspired that particular scene,” Jay says, “but it’s going to be bad for the feline.” The tribal hunters were imagined through the trope of Hollywood *Tarzan* movies.

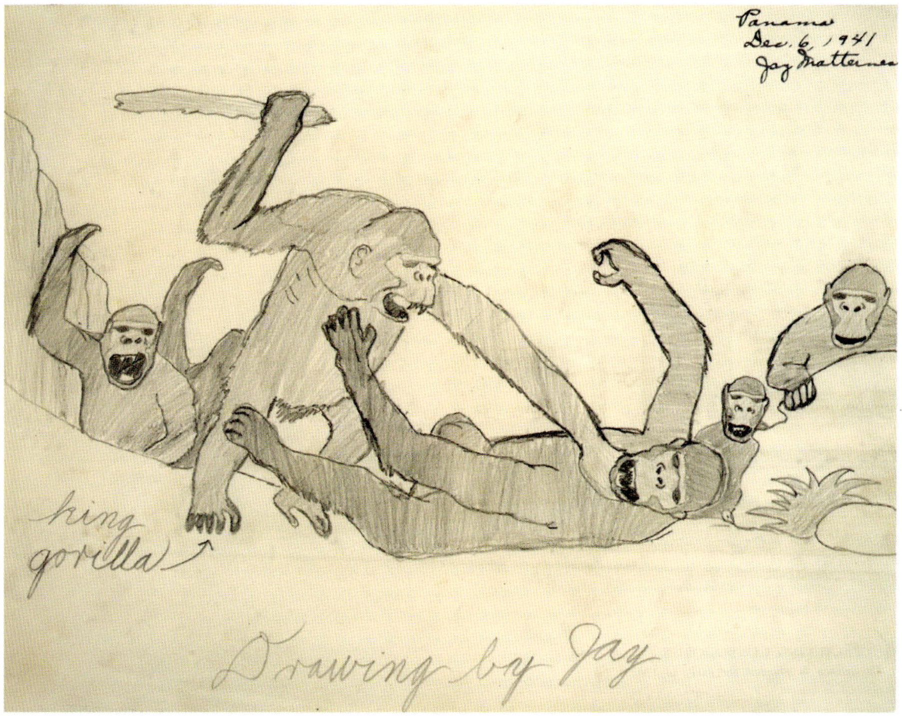

The brutal ape fight now appears to be more chimp-like than gorilla-like. “Of course, I knew nothing about the behavior of these ‘chimporillas,’ as I called them, for that was before any field studies had been conducted. Chimpanzees are now known to pick up stout clubs with which to bludgeon opponents, as in my childhood drawing, and violent fights between male gorillas do indeed break out.”

ELEGANT UNGULATES

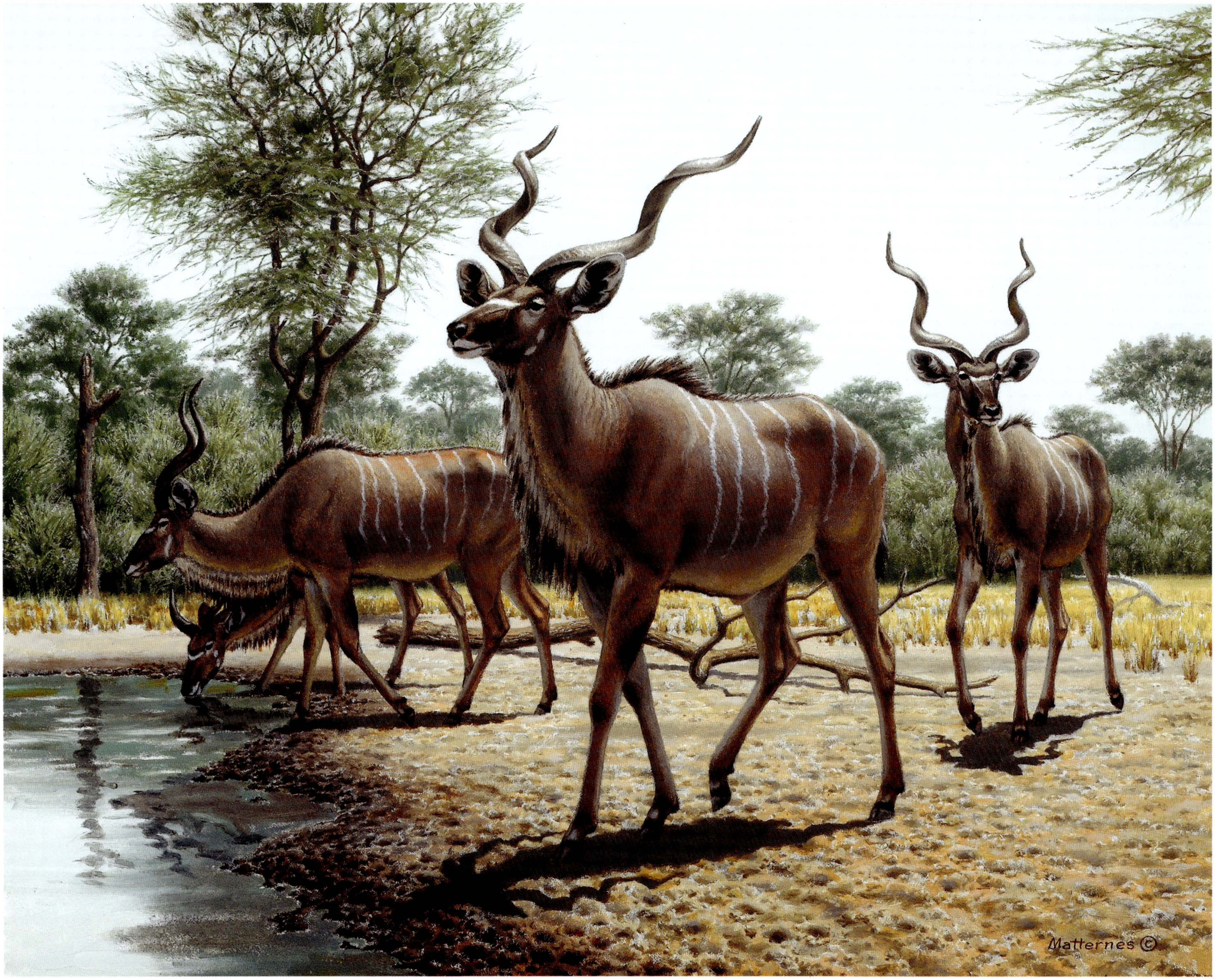

Greater Kudus at Waterhole

"Africa's larger antelopes are graced with magnificent forms and coloration, and none more so than the tall, stately greater kudu. I never saw one in the wild, but glimpsed its smaller and no less beautiful cousin, the lesser kudu, in a brushy area of Tsavo National Park, Kenya. The painting was inspired by a splendid television documentary that showed greater kudu near a river, their underbellies highlighted by reflected sunlight from the water. For kudu anatomy, I relied on my extensive clippings file and sketches from museum mounts. I envisioned how thirsty males might cautiously approach a riverbank, alert to danger from lions concealed near the water, as well as submerged crocodiles, poised to surface suddenly and seize any creatures that come to drink."

Bighorn Sheep

A trio of Rocky Mountain bighorn rams have just become aware of a disturbing sound ("The First Alert")— a possible danger to the community. Bighorn males hang out in same-sex groups in the summer alpine meadows, only descending to lower, warmer terrain with the approach of winter. Ewes and lambs also stick together. During the mating season, mature rams that were previously companions become fierce rivals. Two or more will often charge each other with horrendous impact, and then recoil, temporarily stunned, only to repeat the mutual assault again and again. The sharp cracking sound of these head-butting contests resounds throughout the mountains, and only the strongest males will win mating rights.

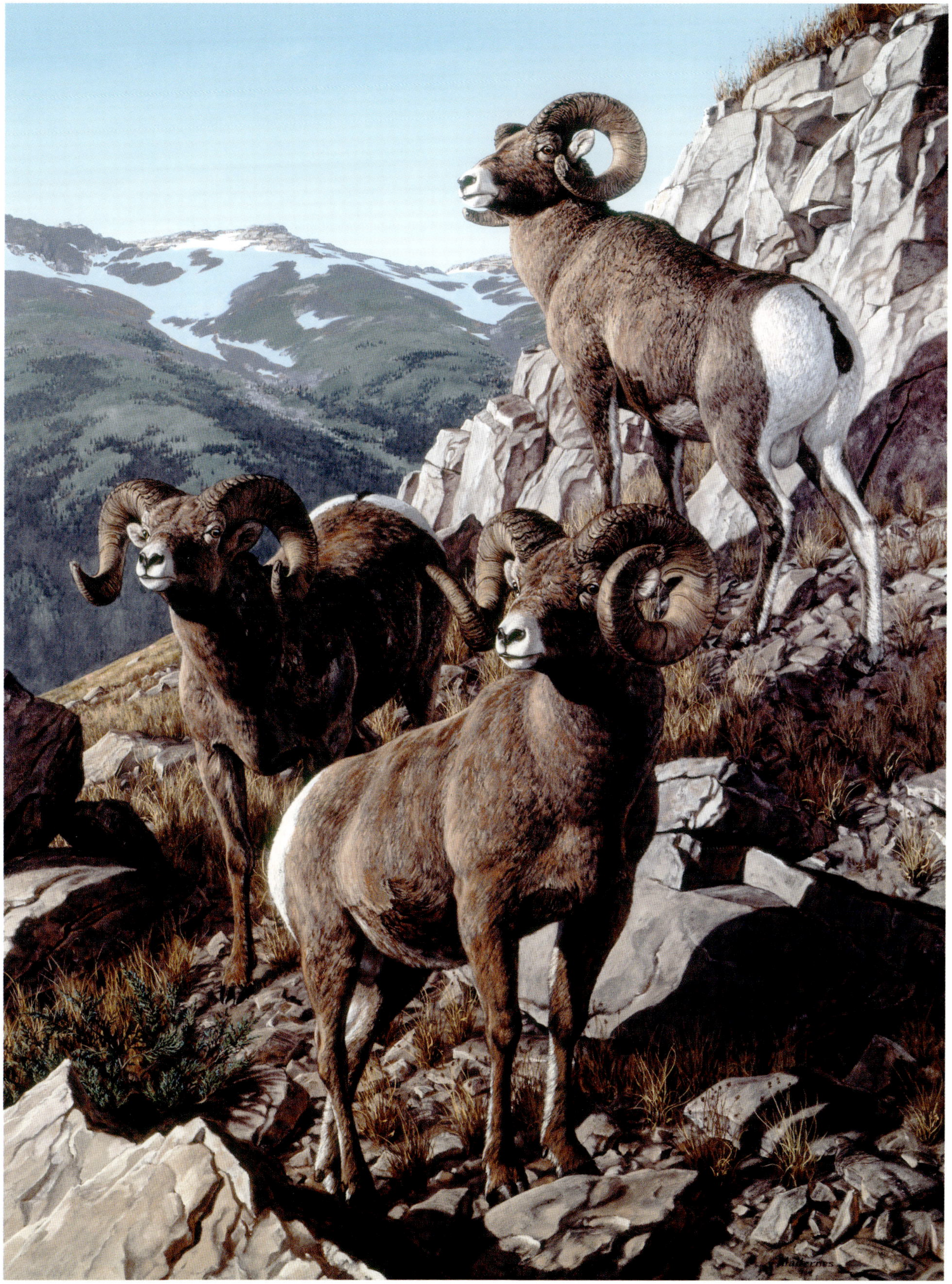

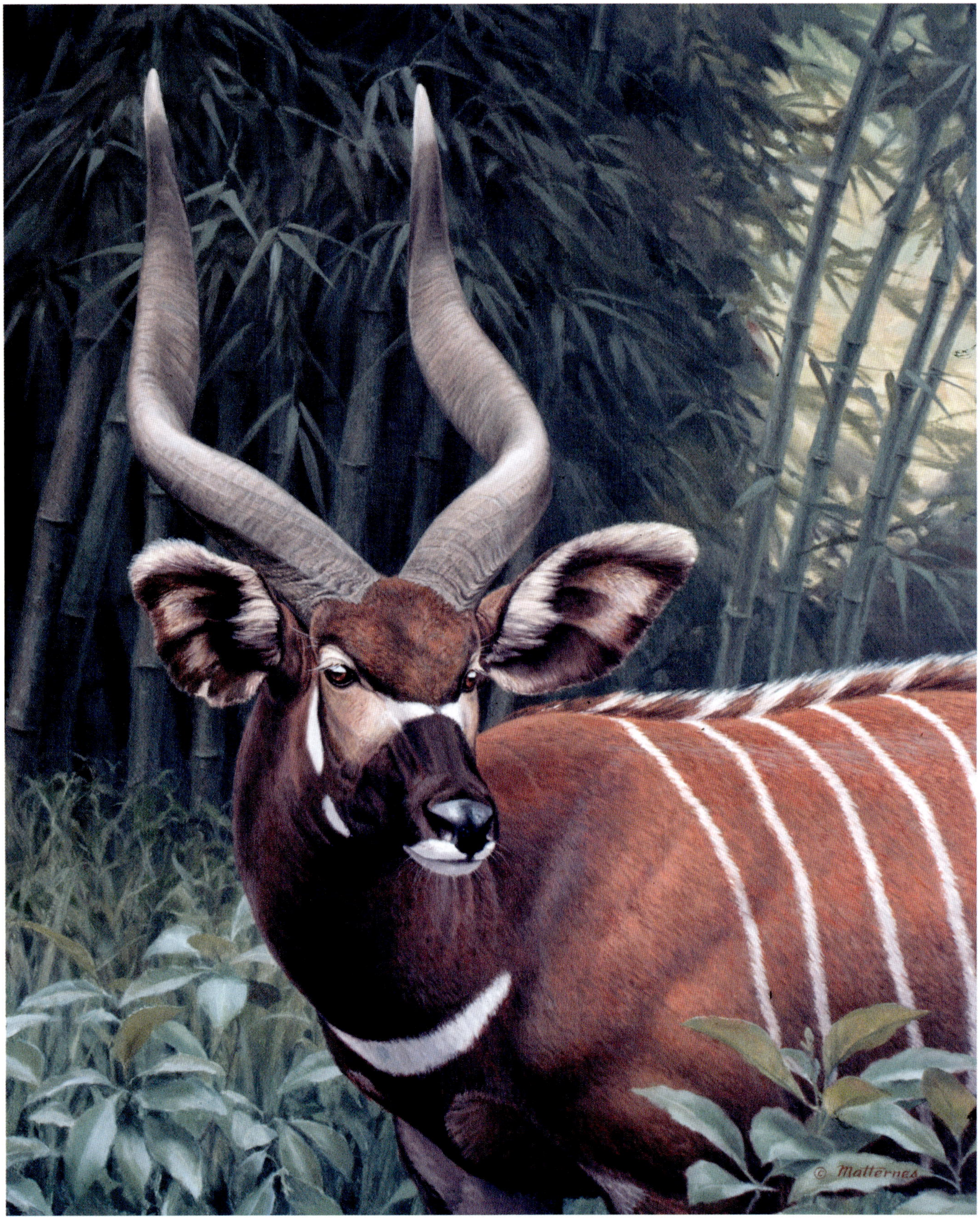

Bongo

This male eastern, or mountain, bongo is one of many species of East African antelope Matternes has painted. It is a critically endangered creature, now found in the wild only in a remote mountainous region of Kenya.

Sable Antelope on Beach

In 1968, while on safari in Kenya's Shimba Hills, Jay spotted some sable antelope, which he had "always regarded as among the handsomest of Africa's hoofed animals. The glossy, rich dark brown to black coats of the large 'horsey' males are a curiosity for a tropical animal, since dark colors absorb solar heat, and it's not clear how they remain cool. Dropping to their 'knees' (actually, their wrist joints), male sables often engage in impromptu sparring contests as shown. The hitchhiking birds are oxpeckers, distant relatives of pigeons, which help keep their hosts free of ticks and insect pests. I chose a beach, lakeside setting and a side view to best show the proportions of these statuesque creatures, with their heavily ringed, gracefully arched horns." When attacked or pursued, they can be powerful adversaries; those beautifully curved horns are not just for show!

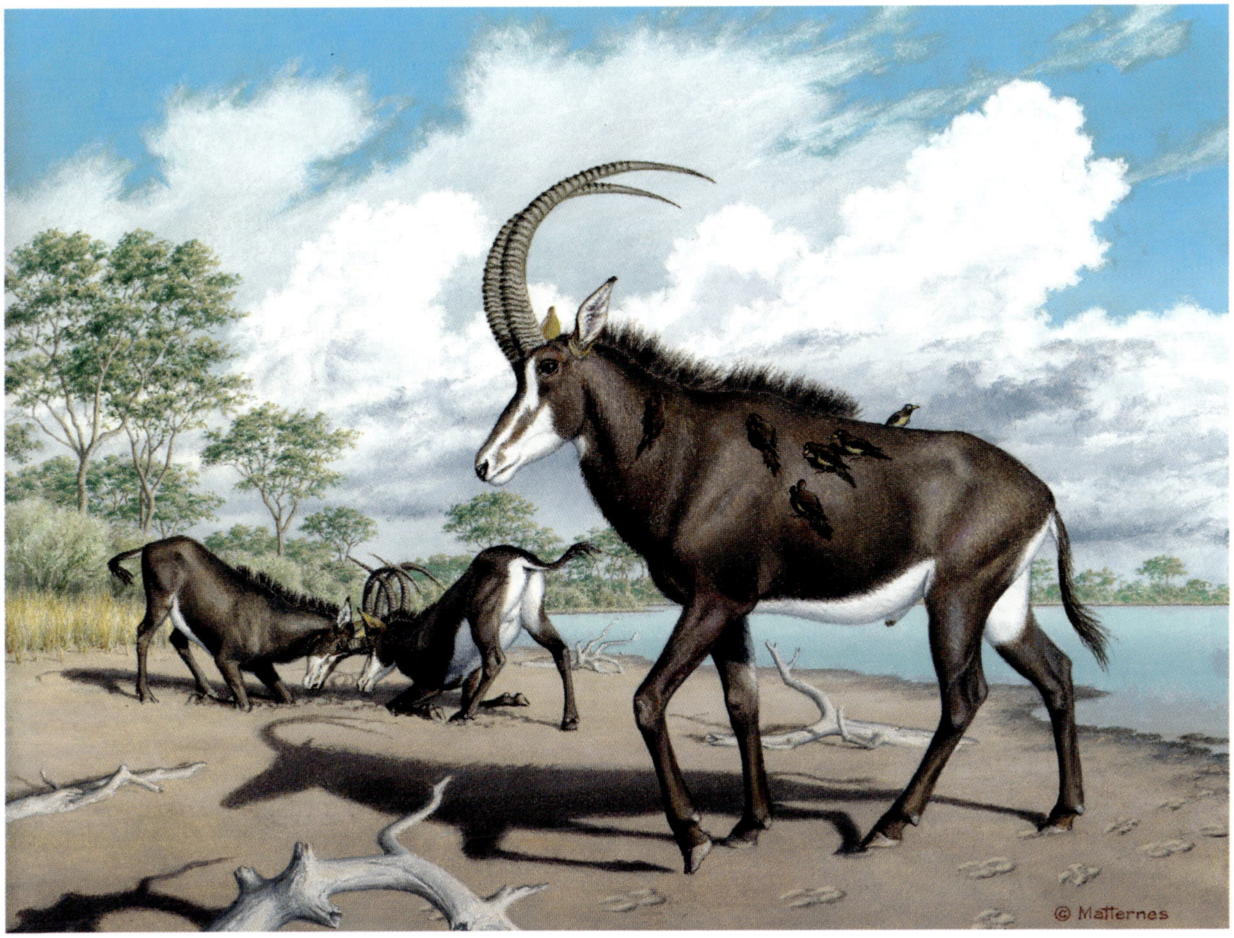

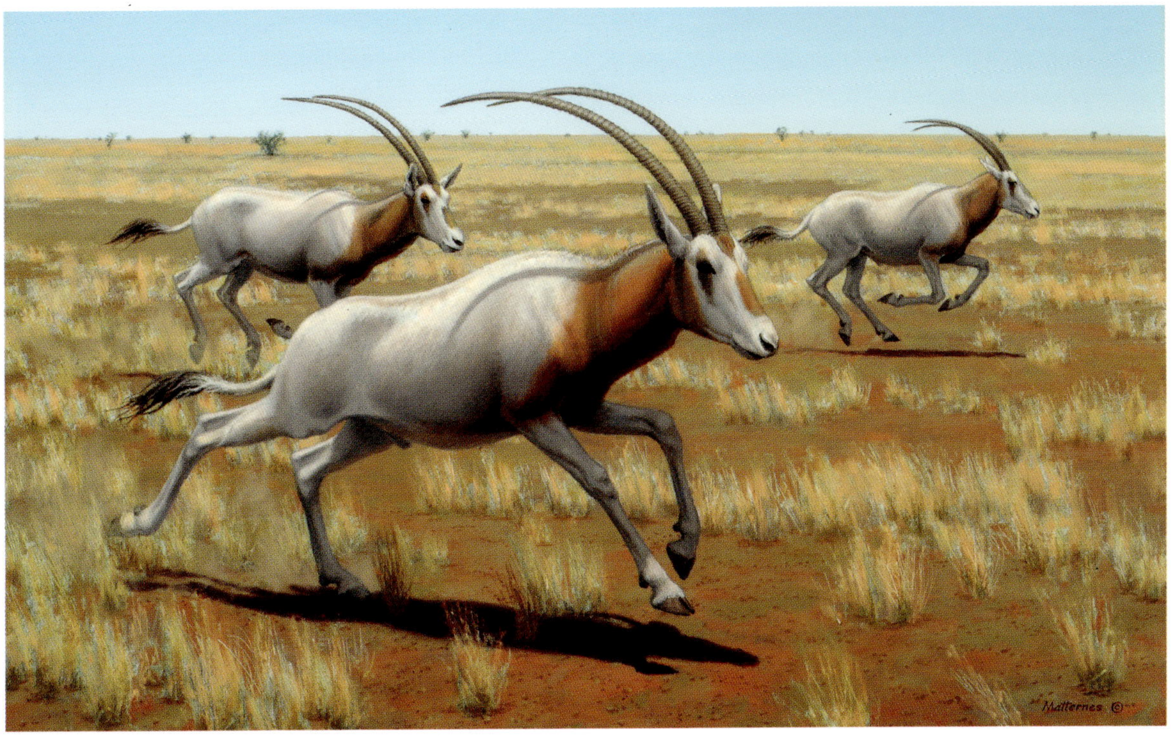

Scimitar-Horned Oryx Antelope

"A small herd of oryxes at the National Zoo in Washington, DC, drew my interest in painting the species. I already knew that they were denizens of deserts, and that their primary habitat is the Sahel, a vast arid band that stretches across Africa's entire sub-Saharan region. Having never visited this immense bleak territory, I relied on published sources about its extreme environment. I first made several 'thumbnail' sketches, moving my subjects around on paper in various attitudes and behaviors, since I had no clear vision of the final composition. (A moving pencil generates ideas.) All oryx sport long, sharp headgear, but this is the only species whose horns are gracefully curved. They have a nearly white, reflective coat, as there is little shade to shield them from the relentless desert sun."

ETERNAL ADVERSARIES

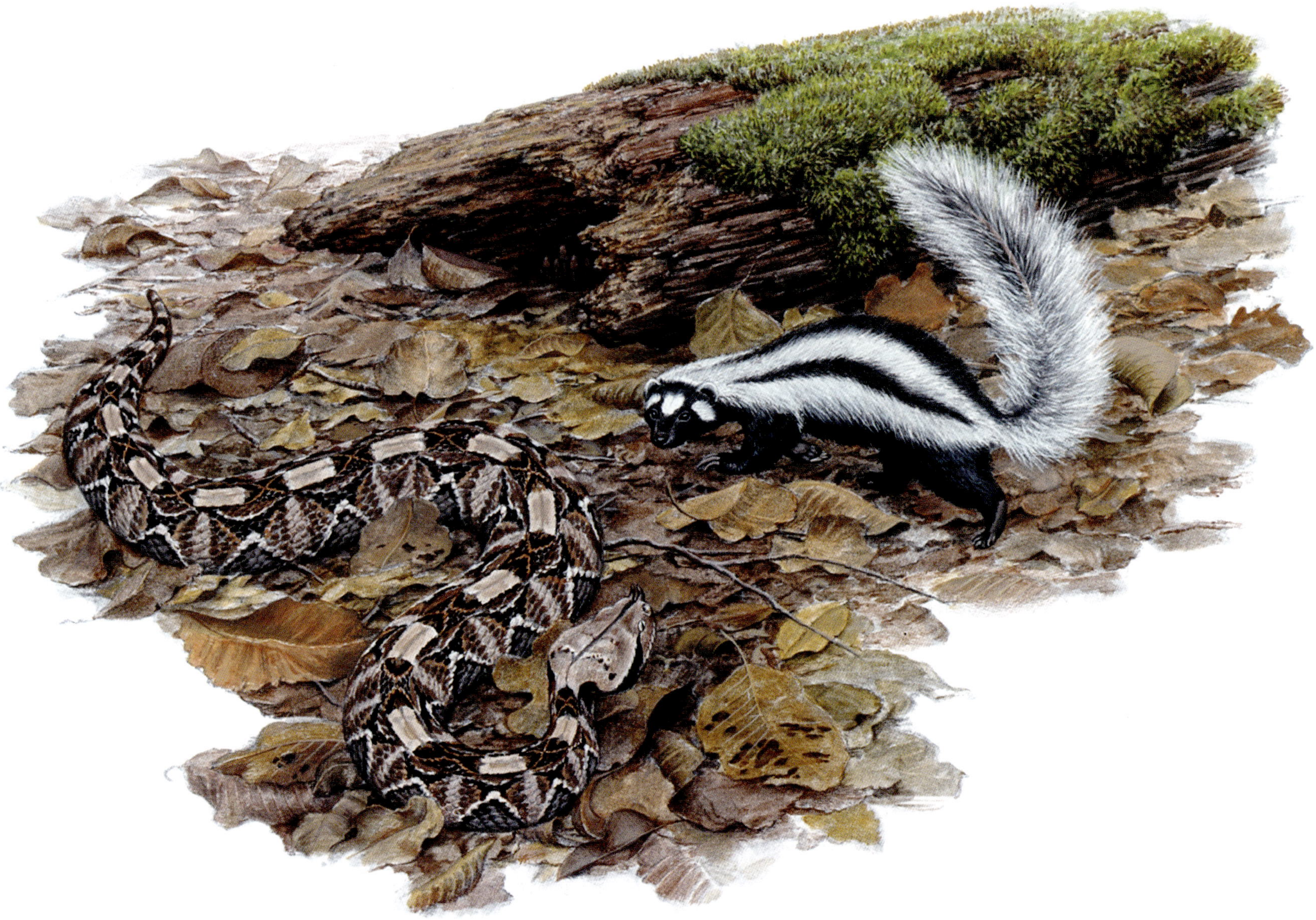

Zorilla vs. Gaboon Viper

The zorilla, also called striped polecat (*Ictonyx striatus*) or African skunk, is a member of the weasel family that lives in arid savannahs and open country. Like other mustelids, it is a carnivore that preys on rodents, snakes, birds, amphibians, and insects. A cranky, solitary animal, it sleeps most of the day and actively hunts at night.

Zorillas are aggressive and mark their territories with feces and a noxious anal spray, which also serves as a defense by temporarily blinding and disorienting their adversaries.

They have no fear of deadly poisonous snakes, such as this formidable Gaboon viper, which often loses or flees such confrontations. So meticulously has Matternes rendered the reptile's complex pattern of camouflage that if you hold this book at a slight distance, the boldly patterned snake will seem to entirely disappear among the dead leaves.

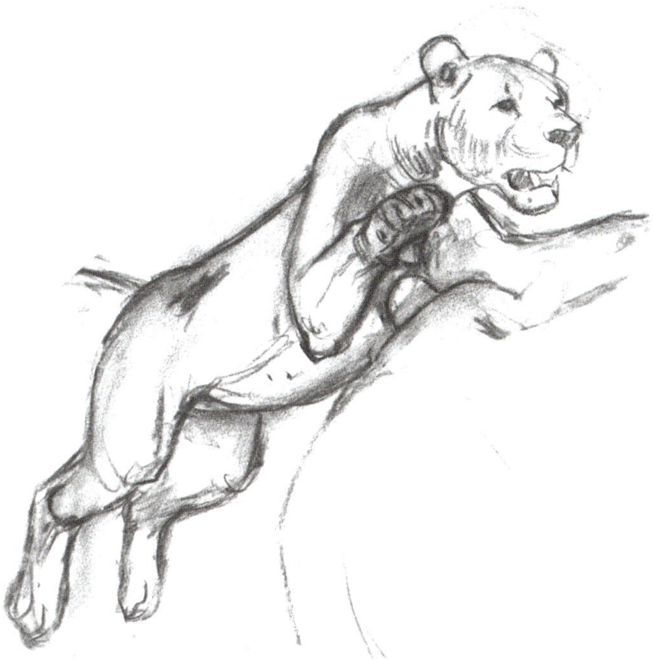

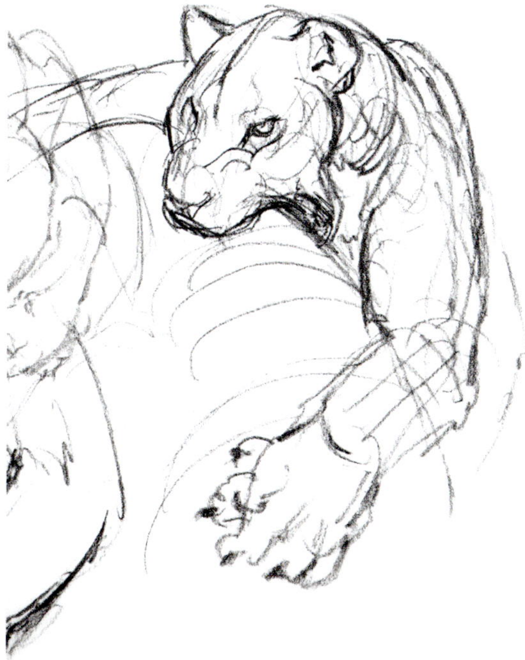

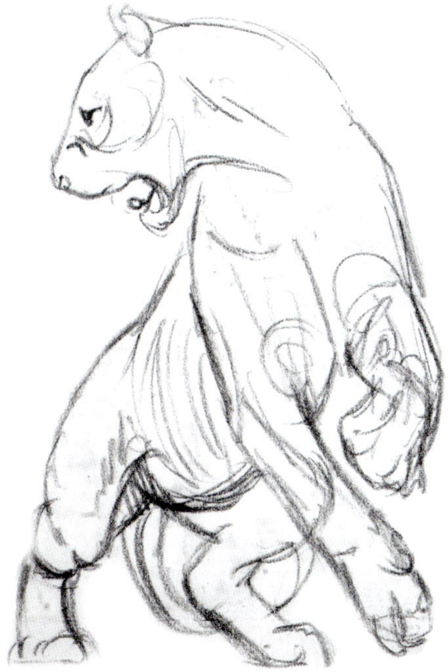

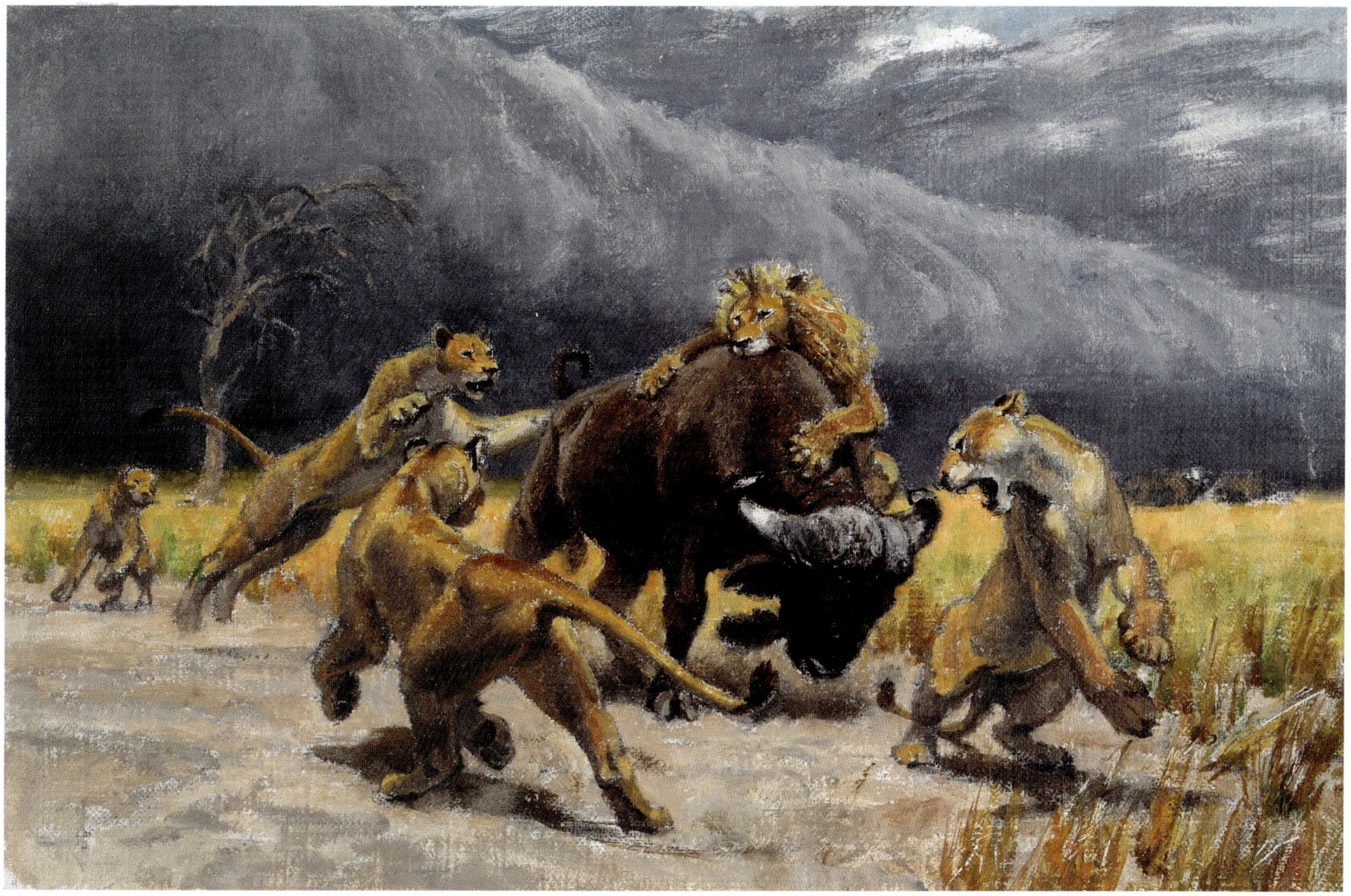

LIFE OR DEATH MOMENT, frozen in time, overleaf, was built up in stages. Matternes first produced fluid pencil sketches of the attacking lions, above, then made a small color sketch in oils, left, to define the colors and composition.

In retrospect, Jay likes the composition in the preliminary color sketch better than in the final painting. He admits, "Sometimes it turns out that way."

The Fight on the Pan

On a dried-up lake bed ("pan") in East Africa, a pride of lions attacks a formidable but aged Cape buffalo bull. Matternes' depiction of their desperate combat is based on his long studies of animal anatomy and behavior, and was not copied from photographs. The cracked mud is the result of a long period of drought, but the low, fast-moving "roll cloud" will soon bring torrents of rain.

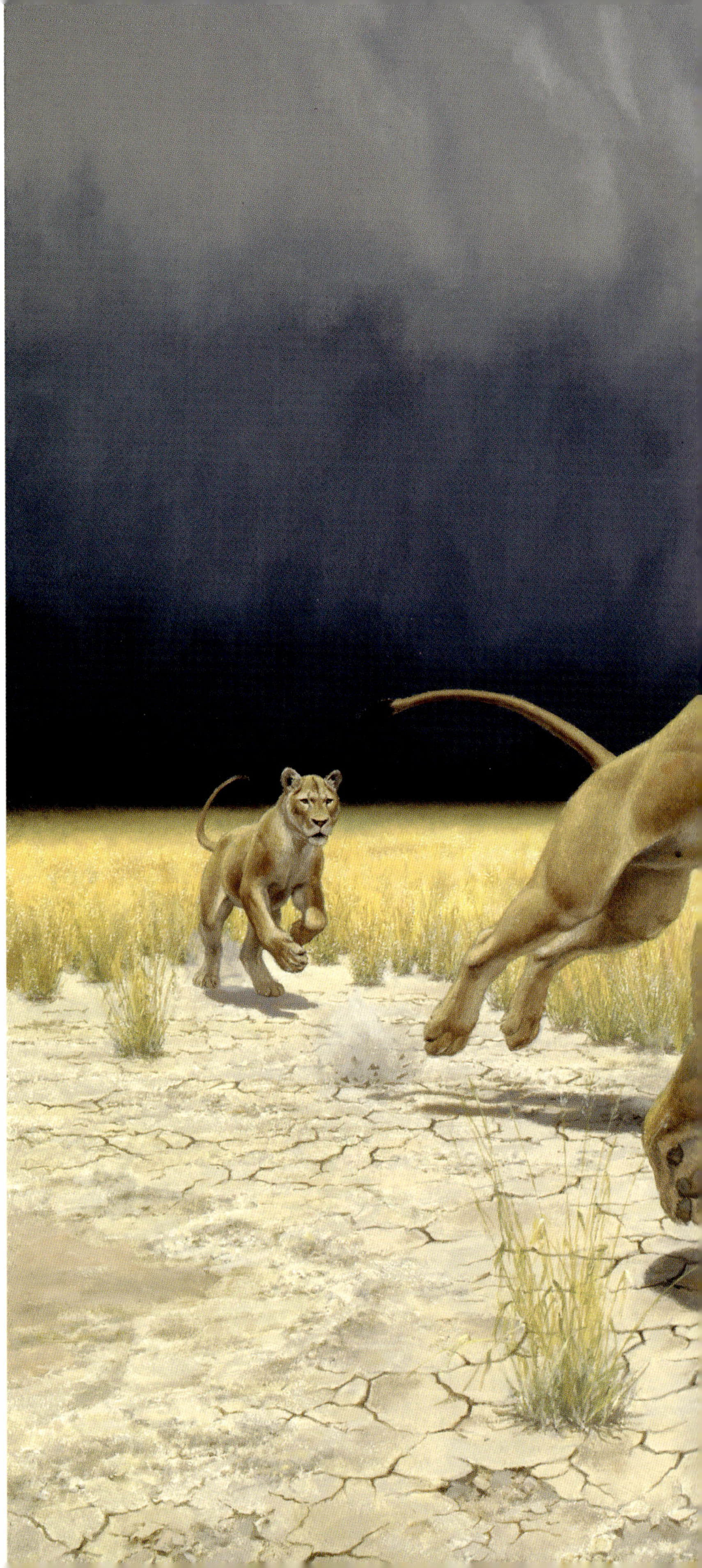

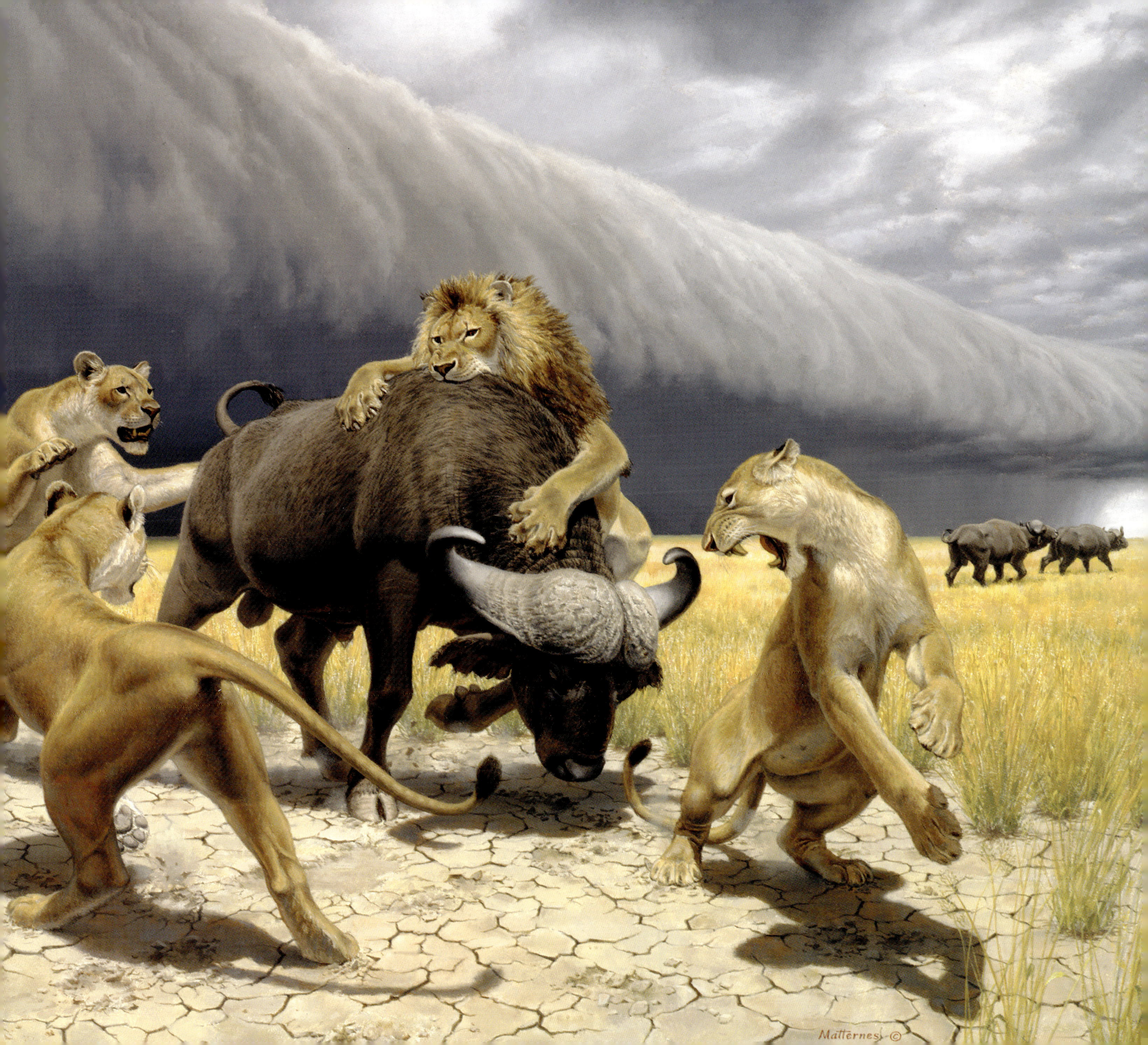
Matternes ©

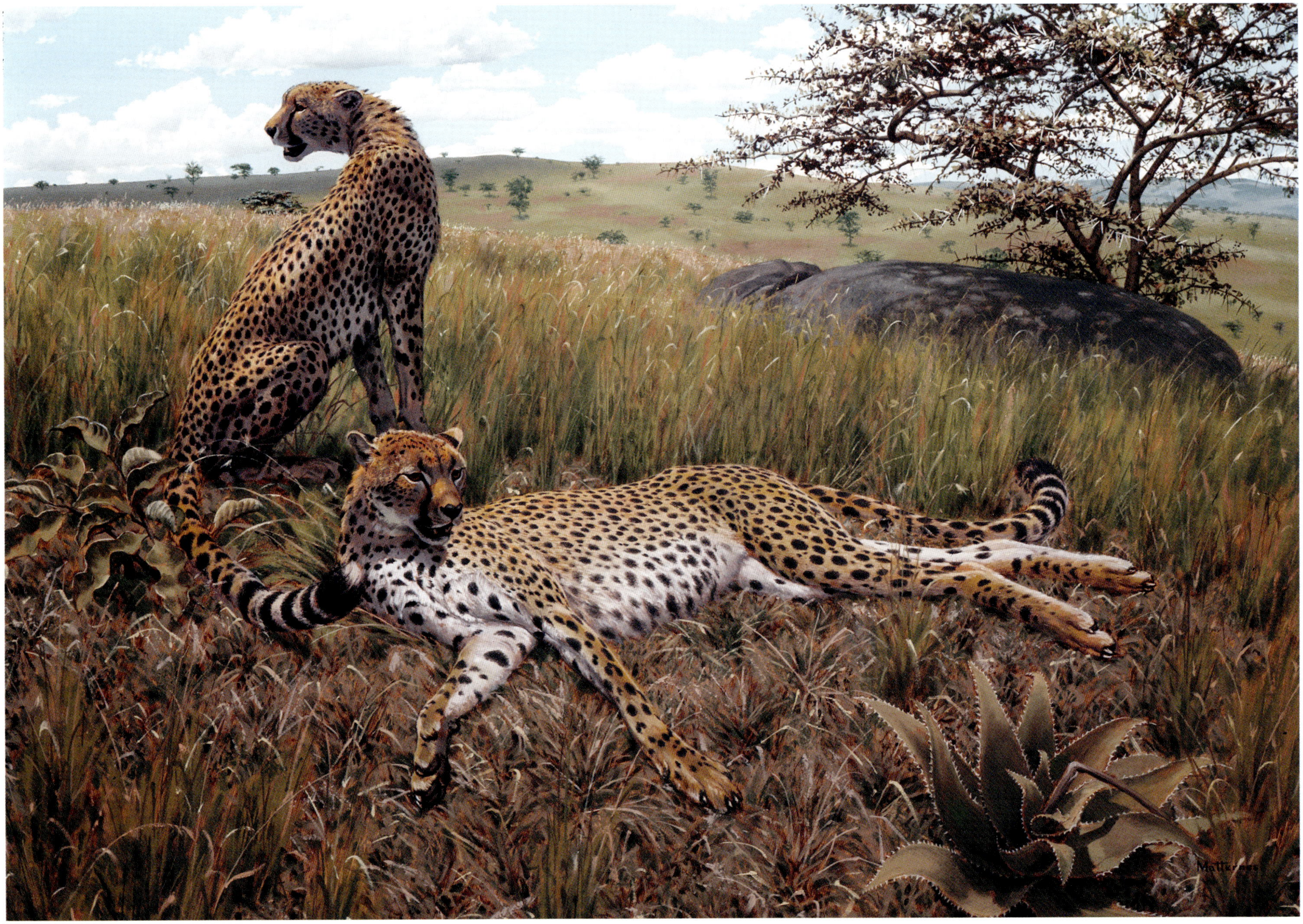

Cheetahs on the Athi Plains

When he returned from his first trip to Africa, Jay's wife Delphine asked him to paint a pair of cheetahs for their living room, a species she considered "elegant." On the Athi Plains near Nairobi, Jay had seen a pair of blasé cheetahs resting in the shade at midday, and retained a vivid memory on which he based this picture. Although they can run faster than any other mammal, Jay shows that they also know how to completely relax.

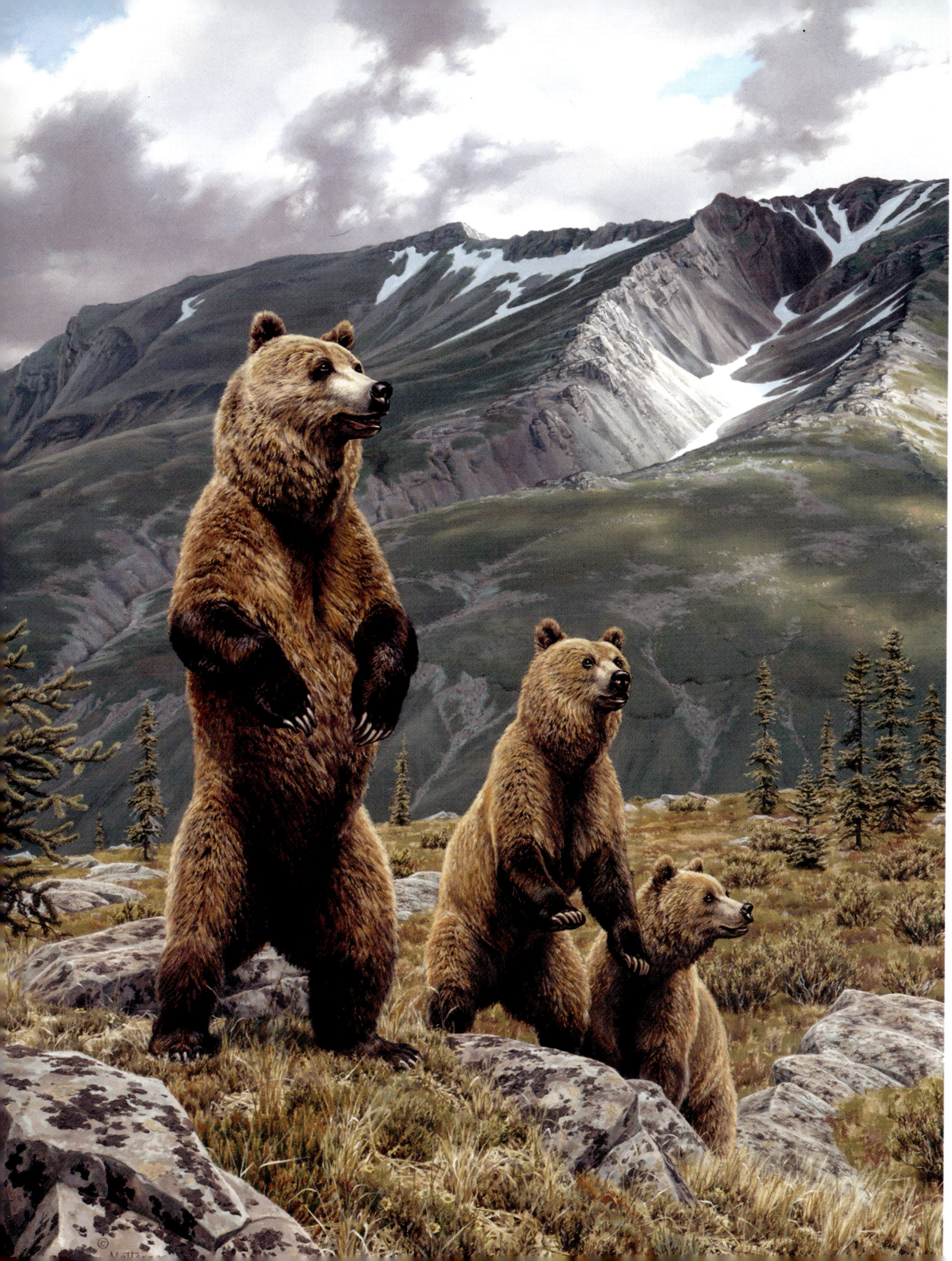

Golden Grizzlies

"During a brief visit to Denali National Park, Alaska, I spotted a large grizzly bear that was a lighter shade of brown than the others, and asked a ranger whether it was rare. He replied that the color phase was not uncommon, adding that Alaskans long ago dubbed such 'blonde' bears 'Jean Harlows' after the platinum-blonde movie queen of the 1930s.

"This grizzly sow was busy digging out the burrow of a ground squirrel, but the hard permafrost resisted her efforts to excavate, creating a mess that resembled a small bomb blast.

"When she finally succeeded in obtaining the small rodent, it seemed a disproportionate effort for such a meager reward—but grizzlies are very determined when it comes to meat. Like most wild animals, they avoid humans, but will attack if their young or their food are threatened. For the painting, I envisioned the sow grizzly with her two late-adolescent cubs attempting to discern something in the distance. The season is mid- to late summer, a time to fatten up for the coming winter's long hibernation."

ECOSYSTEMS INSIDE AND OUT

Day and Night in the Sonoran Desert

When the editors of National Geo's *American Deserts* (1972) asked Matternes to illustrate the wildlife of California's Sonoran Desert, he proposed a pair of matching day and night scenes inspired by the great J. Perry Wilson's paintings for *The World We Live In* (1955). Superb at conjuring moody nighttime wilderness, Wilson had also created spectacular nighttime dioramas at the American Museum of Natural History.

Matternes' daytime view depicts denizens of the desert foraging for food on a sunny morning in July. A chuckwalla lizard and Yuma antelope squirrel are headed for a succulent hedgehog cactus, while a nearby predatory bird, the iconic roadrunner, carries off its lunch of a freshly killed lizard. In the background are collared peccaries (a species of wild pig) rooting for tubers and insects. A desert tortoise reaches out for a prickly pear cactus, and a western red racer snake is about to enter a rodent's burrow. Nearby, a pack rat gathers cholla joints, and a horned lizard prepares to feast on harvester ants.

In an exquisitely rendered moonlit landscape, Matternes then depicted exactly the same area, only at night. Using meticulous tracings on vellum, he first copied the rocks, landscape, and desert plants from his daytime painting as the setting for his cast of nocturnal visitors. A great horned owl scans the ground from the air on silent wings as it searches for small mammals to ambush. A mature bobcat prowling under the owl singles out a pack rat for his focused attention. Meanwhile, a spotted skunk chomps on a giant desert centipede, while a western diamondback rattler prepares to engulf the rock pocket mouse he has just dispatched with a deadly strike. Nearby, a poisonous Gila monster lizard prowls, and a ring-tailed cat (relative of the raccoon) lurks near white flowers of the night-blooming *Cereus* cactus. One can also spot a jackrabbit, a coyote, and a big brown bat.

However, the magazine's editors were not completely satisfied, and kicked this picture back to Jay for revisions: they complained that he had not consulted the position of the stars as they would have appeared during July at the precise location he had chosen, and insisted that he reposition them to twinkle in a more accurate configuration!

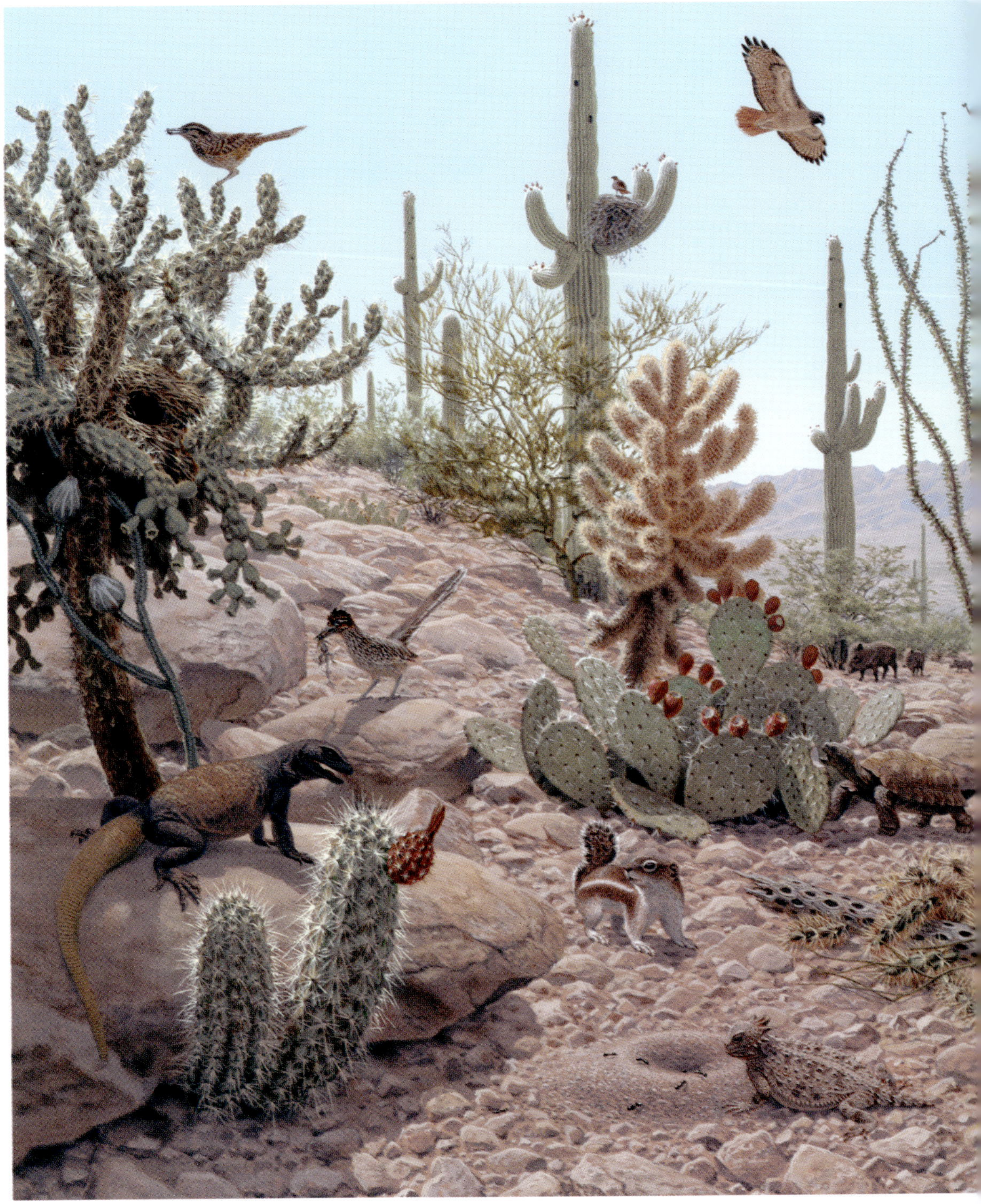

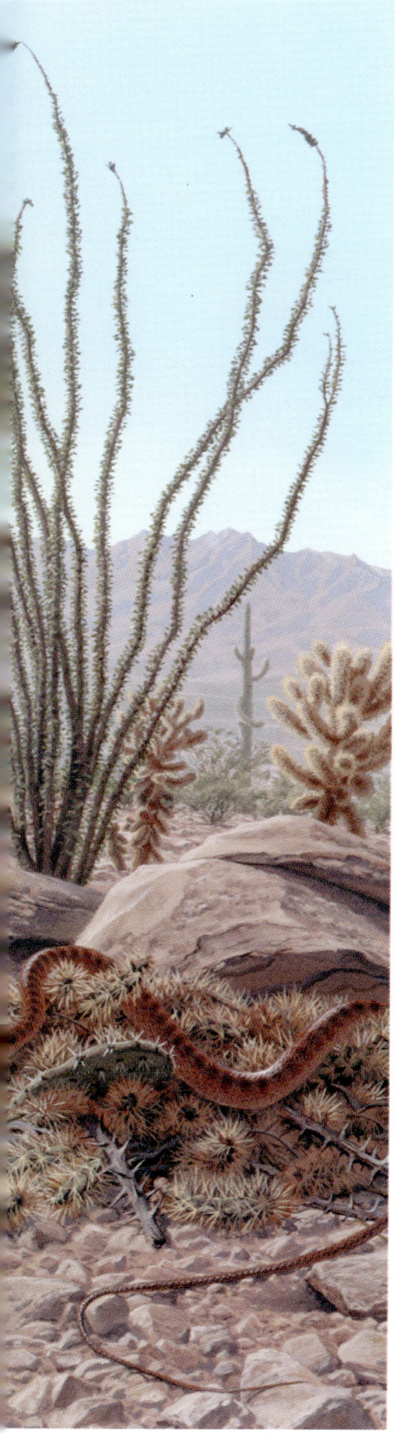

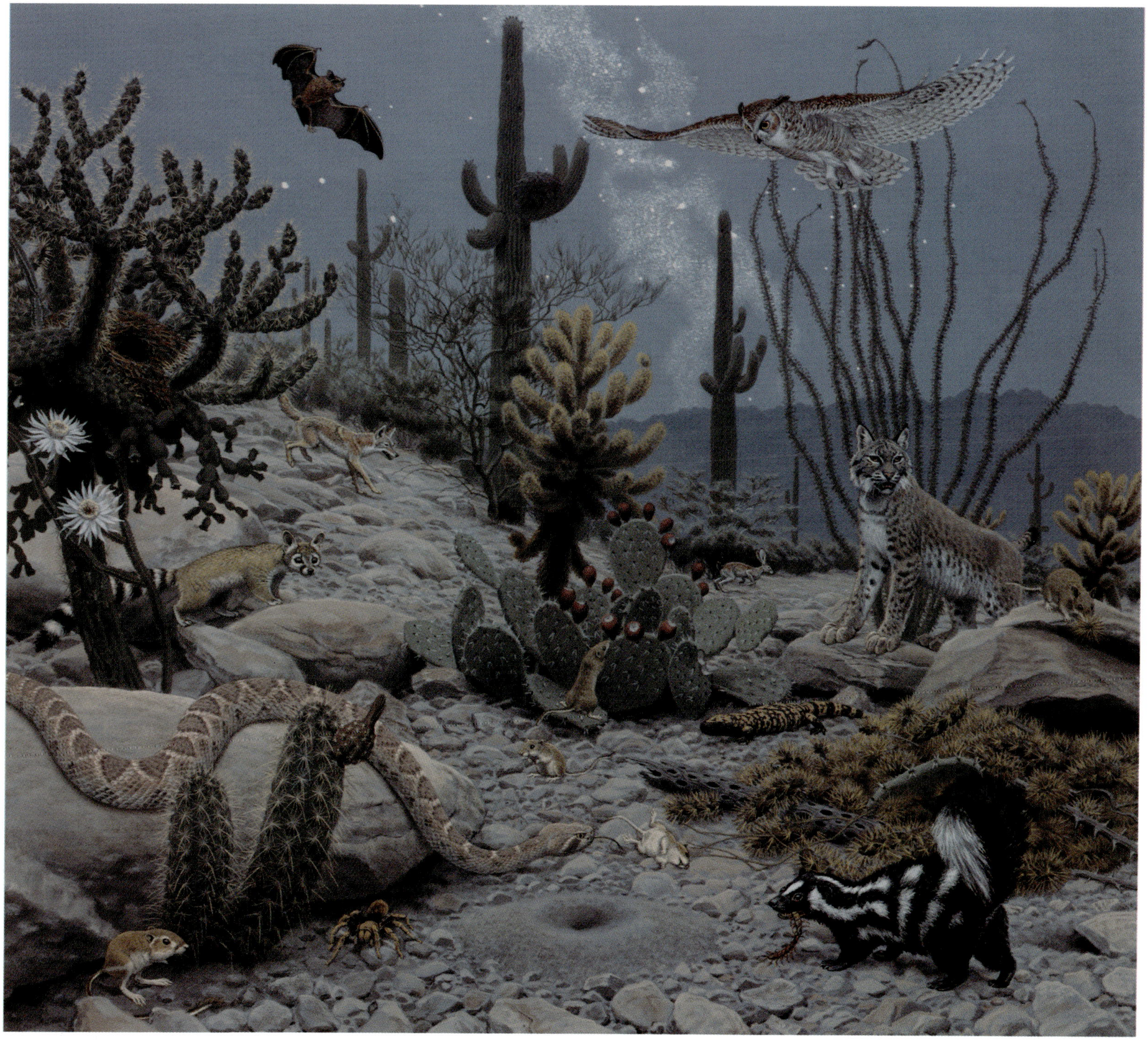

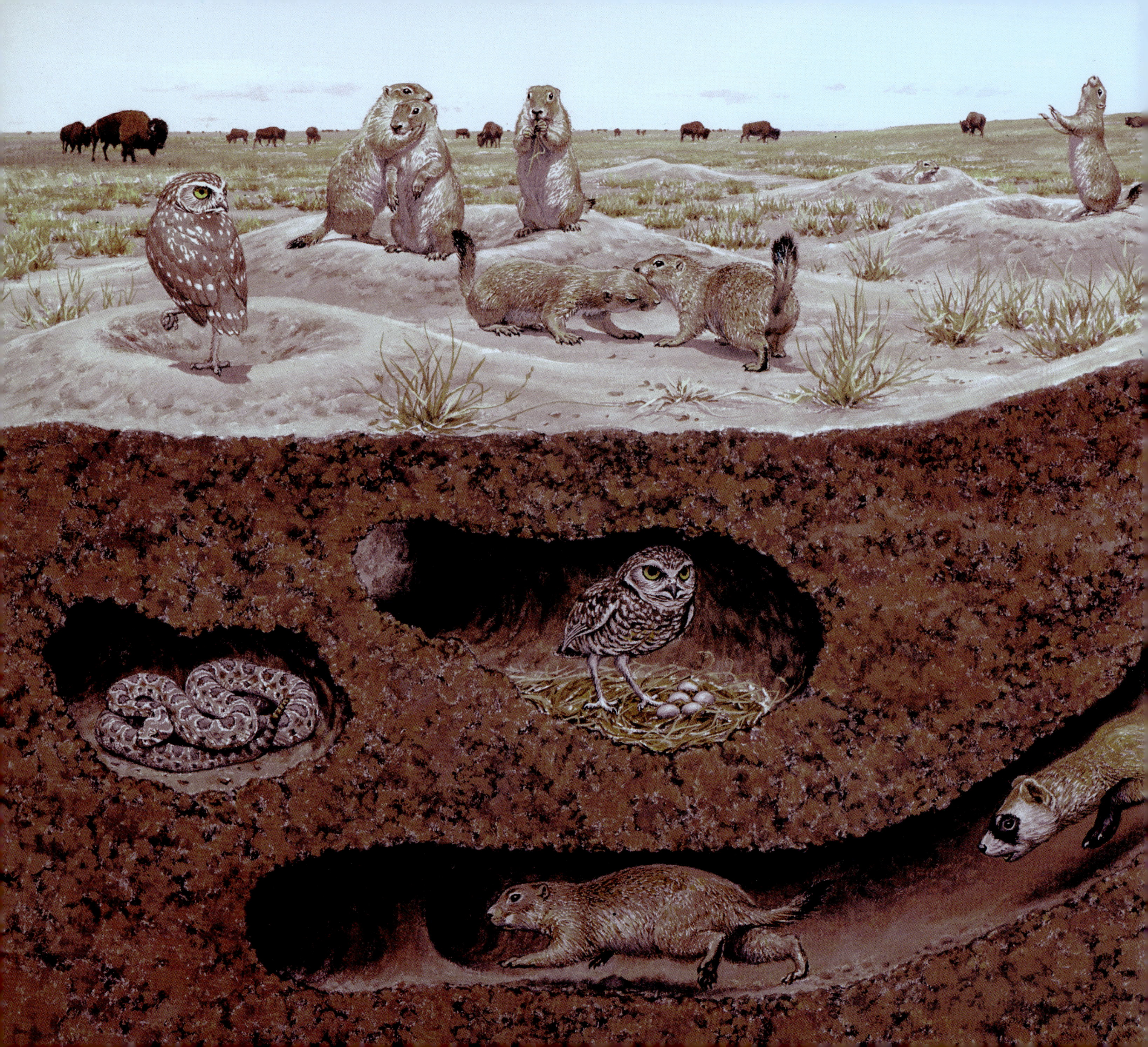

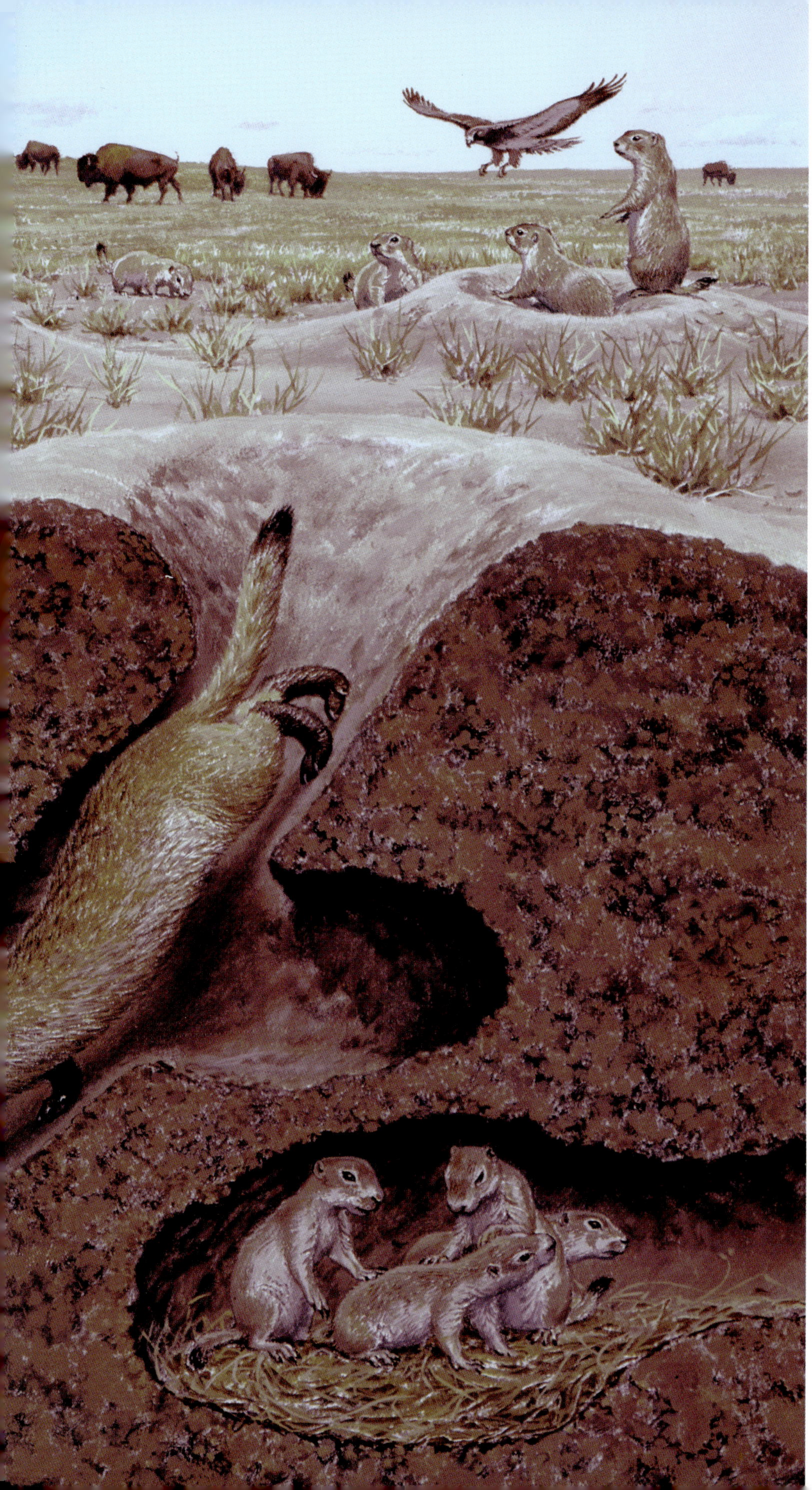

The American Prairie Beneath Where the Buffalo Roam

Prairies are flat, mostly treeless ecosystems in the Great Plains—vast areas of the United States and Canada east of the Rocky Mountains and west of the Mississippi River; they extend through Colorado, Kansas, Oklahoma, Montana, Nebraska, the Dakotas, Texas and Wyoming. While the most iconic animals of these grasslands were once great herds of bison and the ubiquitous prairie dog "towns," many kinds of animals still live under the Prairie in a dark, dangerous world.

A black-footed ferret hunts prairie dogs in their burrows, but is often outmaneuvered in their labyrinth of escape tunnels. Burrowing owls nest underground to avoid terrestrial predators, but prairie dogs and rattlesnakes often discover eggs and hatchlings underground.

Monsters of the Mississippi

An eight-foot alligator gar glides near the surface of the Mississippi River, while another chases a young blue catfish. Gar can grow to ten feet long, and the full-size blue catfish, right, may reach one hundred pounds. A long-snouted paddlefish sieves the water, mouth agape, for crustaceans and plankton. Near the bottom, shovelnose sturgeons rake mud near the bank for snails, crawfish, and insect larvae, which they suck up through tubelike mouths.

Jay relied on photographs and illustrations as references for these large riverine creatures, since—with the exception of the alligator snapping turtle—he had never seen any live specimens. His biggest challenge was to create the frontal twisted view of the huge alligator gar, working only from pictorial side views of dead specimens. At the time, most of these giant Mississippi denizens were completely unknown to the public.

Once plentiful in the southern United States, the prehistoric-looking alligator snapping turtle is now a rare and threatened species. Jay depicted this grotesque creature with the help of life sketches he'd made years before at Chicago's Lincoln Park Zoo. These heavily armored turtles, much larger than the common snappers, are often covered in algae and can be easily mistaken for rocks as they rest motionless on the river bottom. They can keep their powerful jaws wide open for hours, while wriggling a small wormlike appendage on their tongues. When a passing fish mistakes the lure for a worm and approaches, a lightning-quick snap of those formidable jaws—with a bite force of 1,000 pounds—spells instant oblivion for the inquisitive fish. When these powerful ambush predators ascend for a gulp of air, they may also grab any waterfowl they find swimming on the surface, like the doomed mallard in the painting.

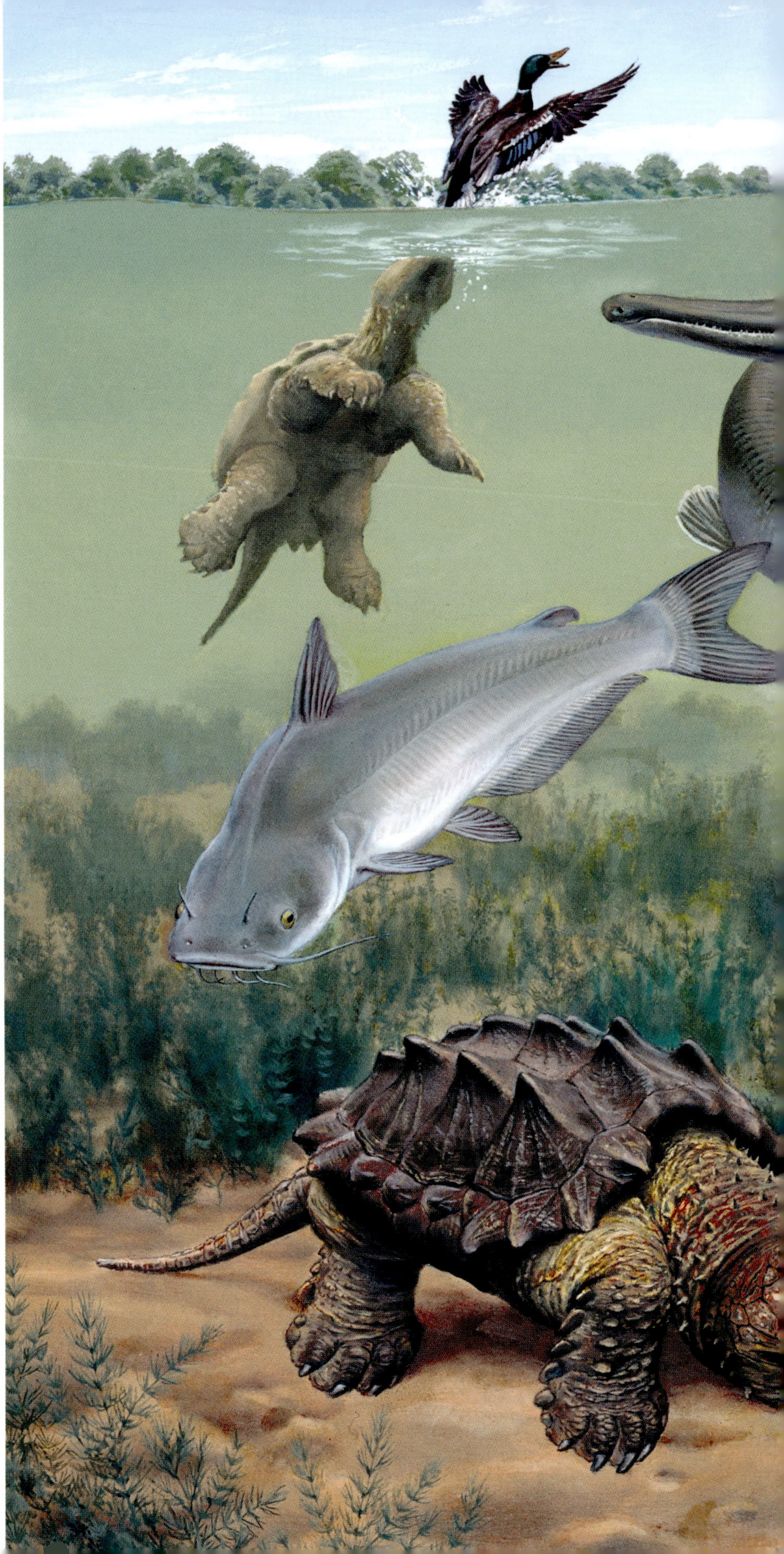

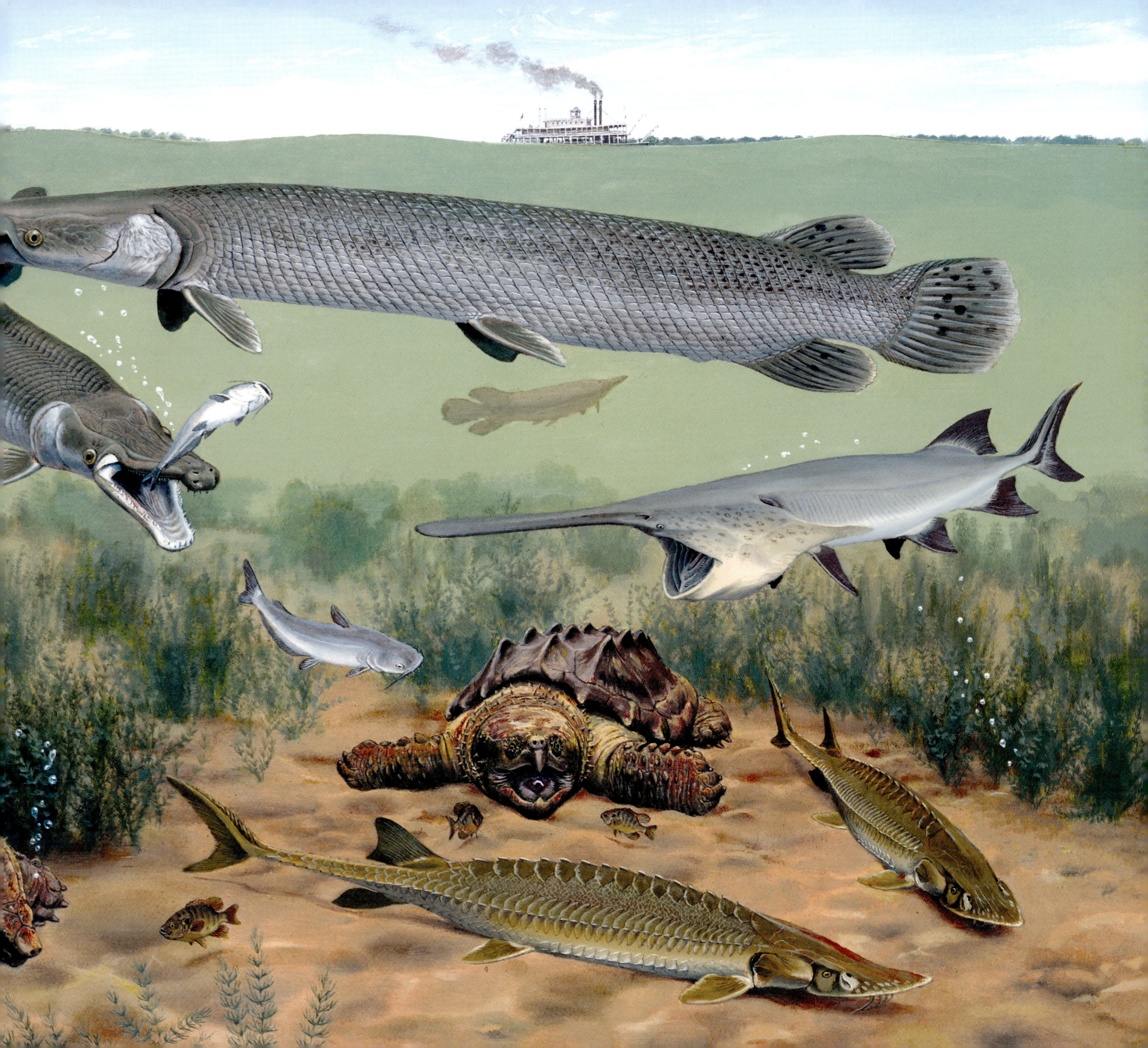

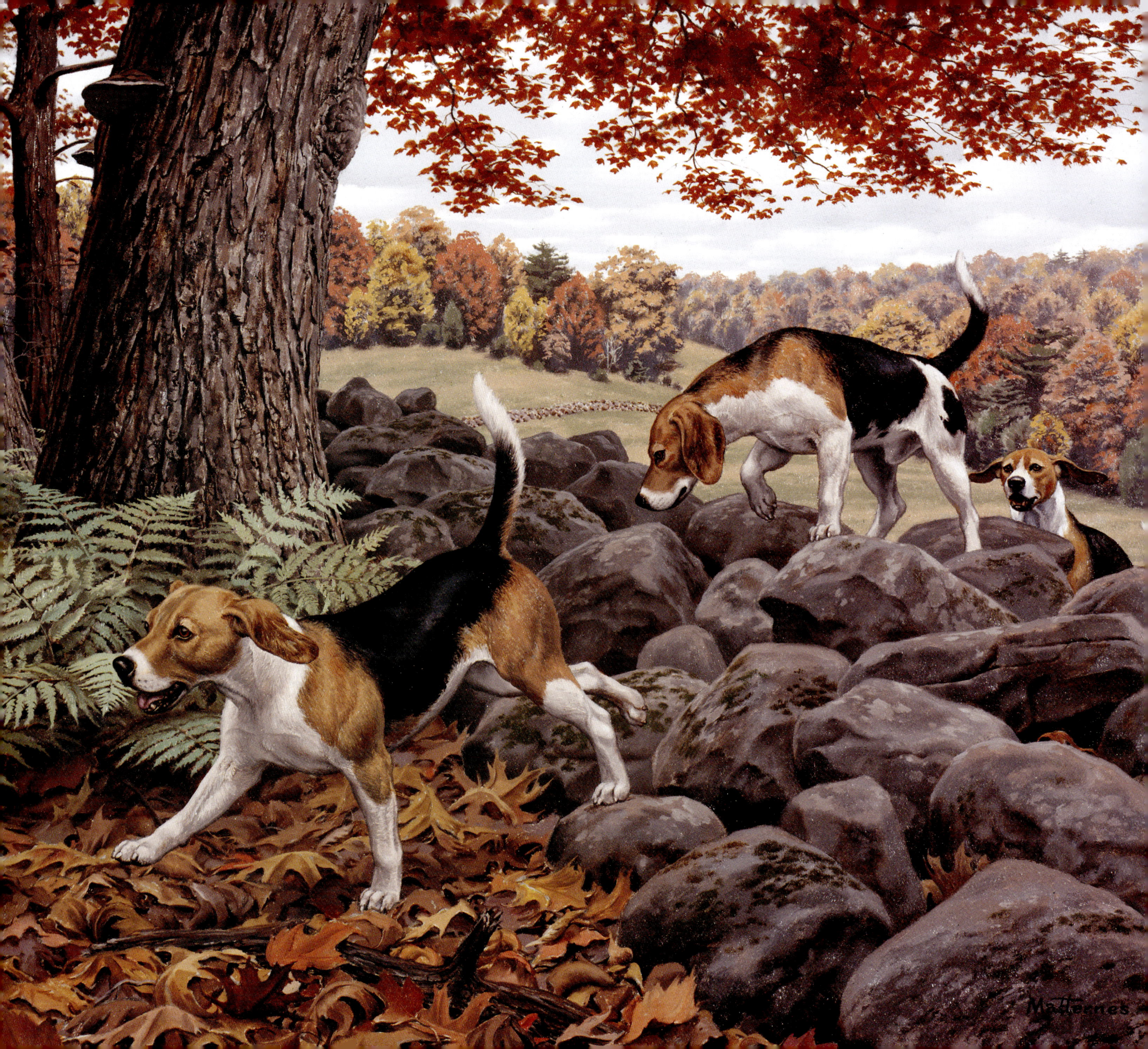
Matternes

HOMAGE TO HOUNDS

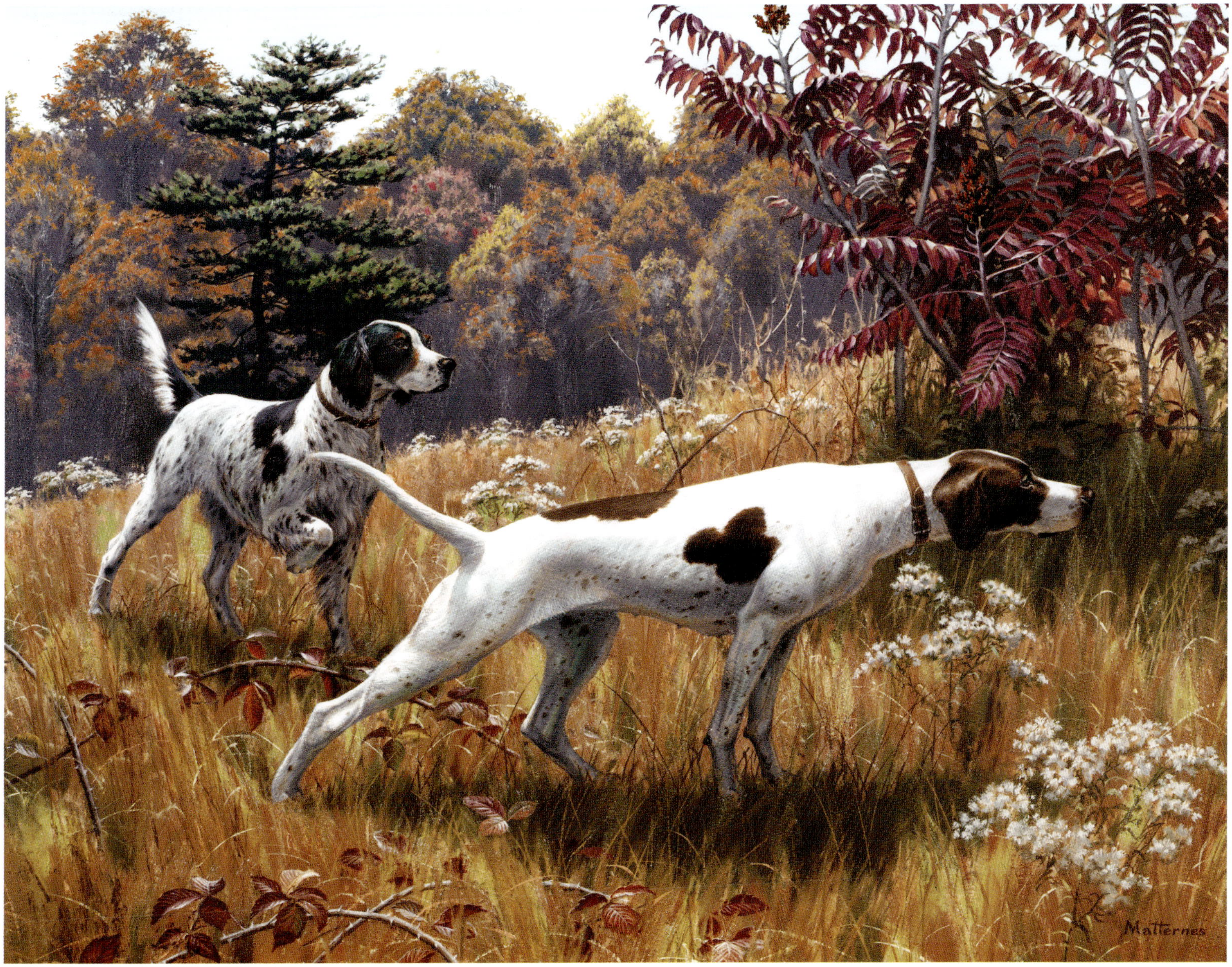

Field Partners

The Joy of the Chase was the first of what Matternes hoped would be an enjoyable and lucrative series of collectible prints on hunting dogs, sponsored by a dog food company. He set out to depict "the boundless energy and enthusiasm of the beagle on a hunt," set in the New England countryside during the peak of autumnal colors. The second print, *Field Partners*, shows a pointer and setter working together to hunt quail. The pointer has located a hidden covey of the birds, while his partner "honors" the point. Well-trained dogs hold the pose until the alarmed birds "flush" into the air, giving the hunter his opportunity. Unfortunately, when the promoter of the project suddenly died, the corporate sponsor abruptly canceled the remainder of the series.

The Joy of the Chase

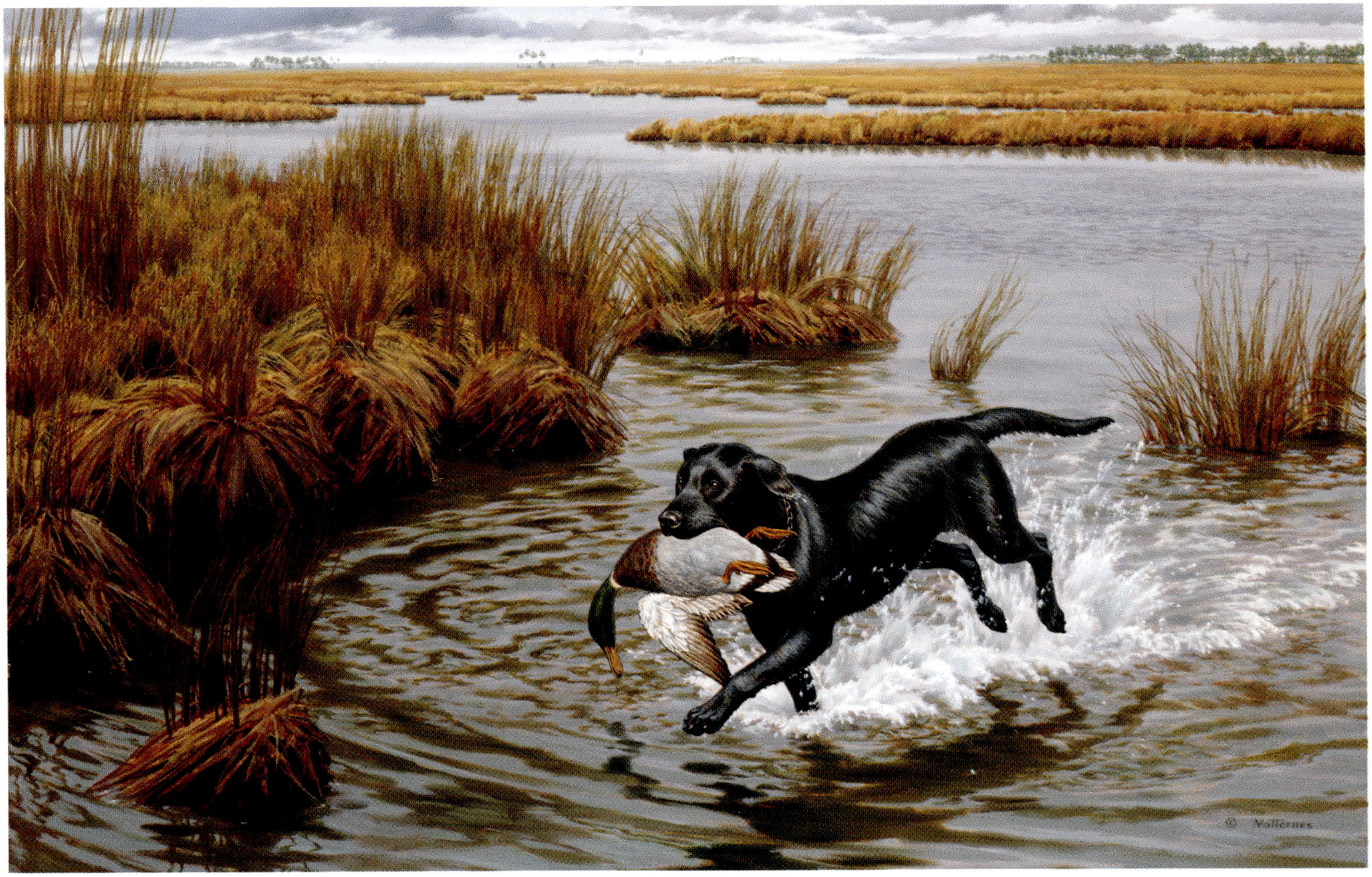

First Retrieve

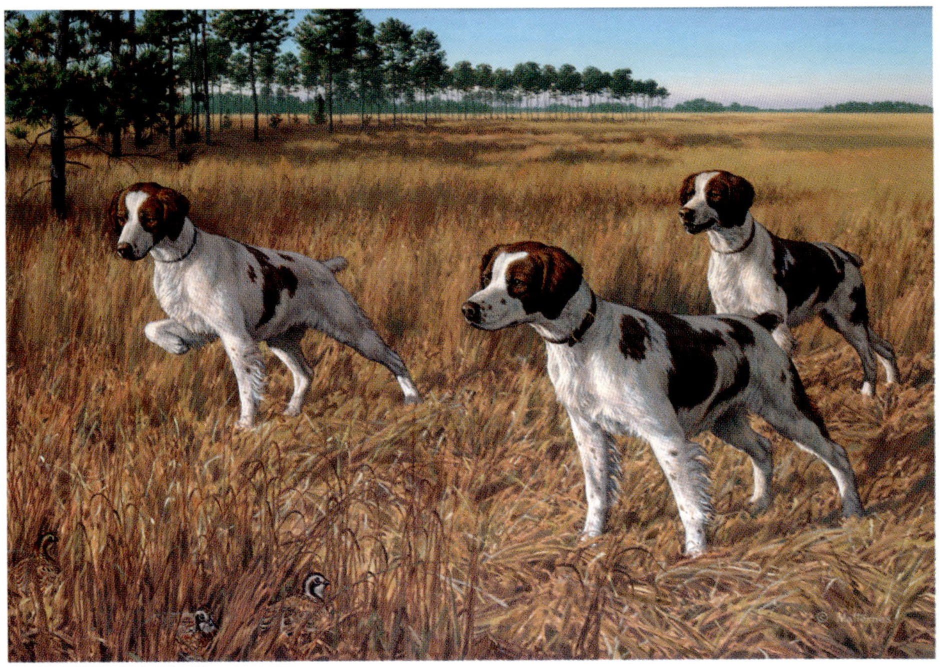

Britts on Point

Since he had already completed two hunting dog paintings, Jay's agent encouraged him to continue the theme, either as gallery paintings or prints, since they had been well received by hunters and dog fanciers. He began the new series with *First Retrieve*, which shows an eager young Labrador retriever on his first duck hunt bringing in a mallard. For *Britts on Point*, his final picture in the series, Matternes chose the Brittany spaniel, a larger breed very popular in the eastern United States. This delightful canine has a pink nose, a pied coat of white with butterscotch-colored markings, and "feathered" legs and belly, and is prized for its keen intelligence and gentle nature. It is the only spaniel that will "point" on locating the quarry, whereas other varieties plunge directly into foliage to force the birds out of hiding. Here, Jay depicted three Britts holding on point, having detected several hidden bobwhites just in front of them.

BIRD-WATCHING

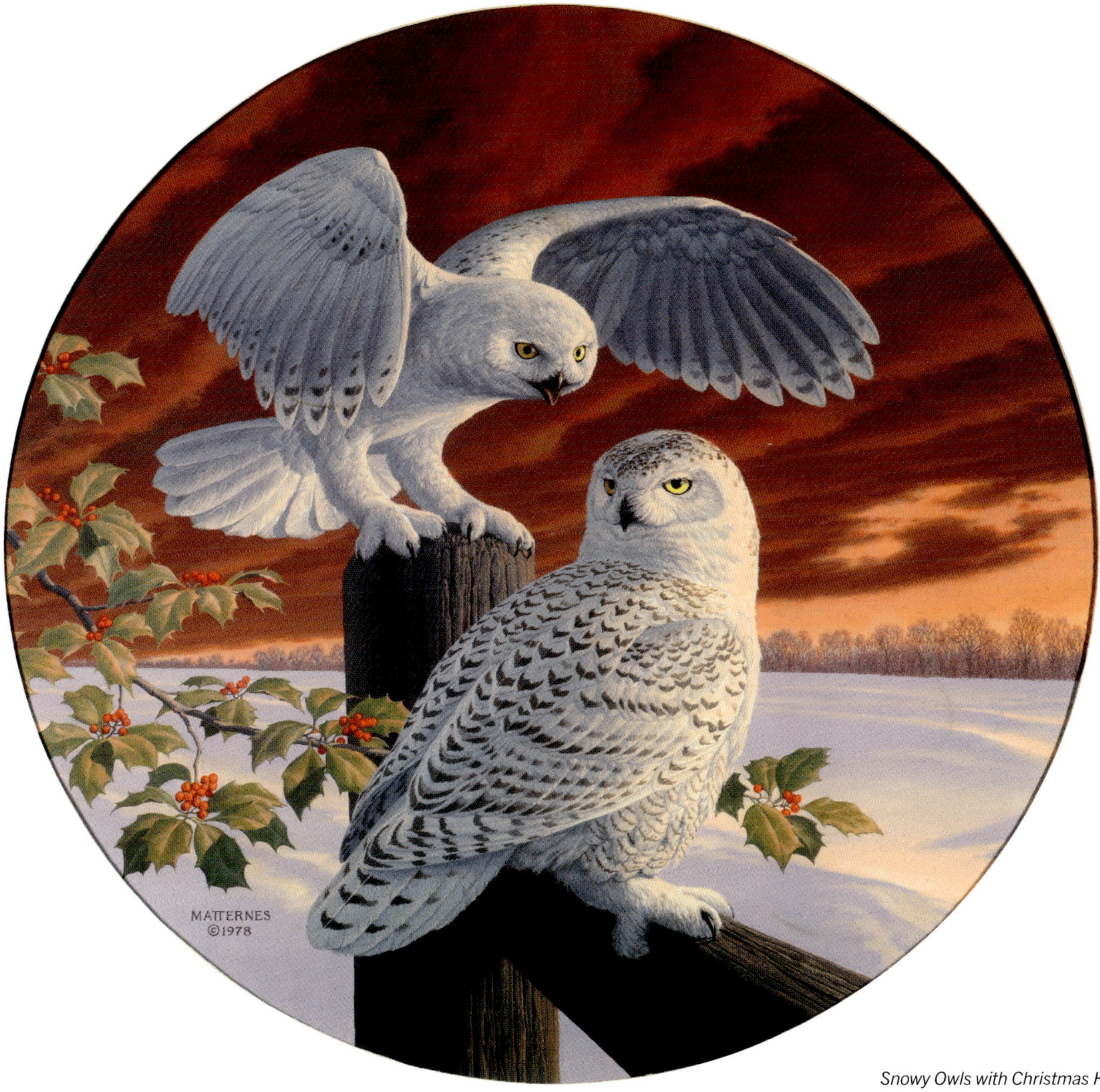

Snowy Owls with Christmas Holly

During a field trip to the Alaskan tundra for his Pleistocene mural, Matternes was intrigued by the snowy owl's adaptation to the harsh landscape. Snowy owls do not build nests, he learned, but deposit their eggs directly on the frozen ground, often among rocks. The chicks' grayish, mottled feathers blend in with the stones; when remaining perfectly still, they become invisible to predatory foxes and eagles. As they develop the ability to fly, the youngsters quickly acquire the dazzling white plumage that helps conceal them when they ambush small rodents in the snow. Females are larger, with barred feathers, while males become completely white.

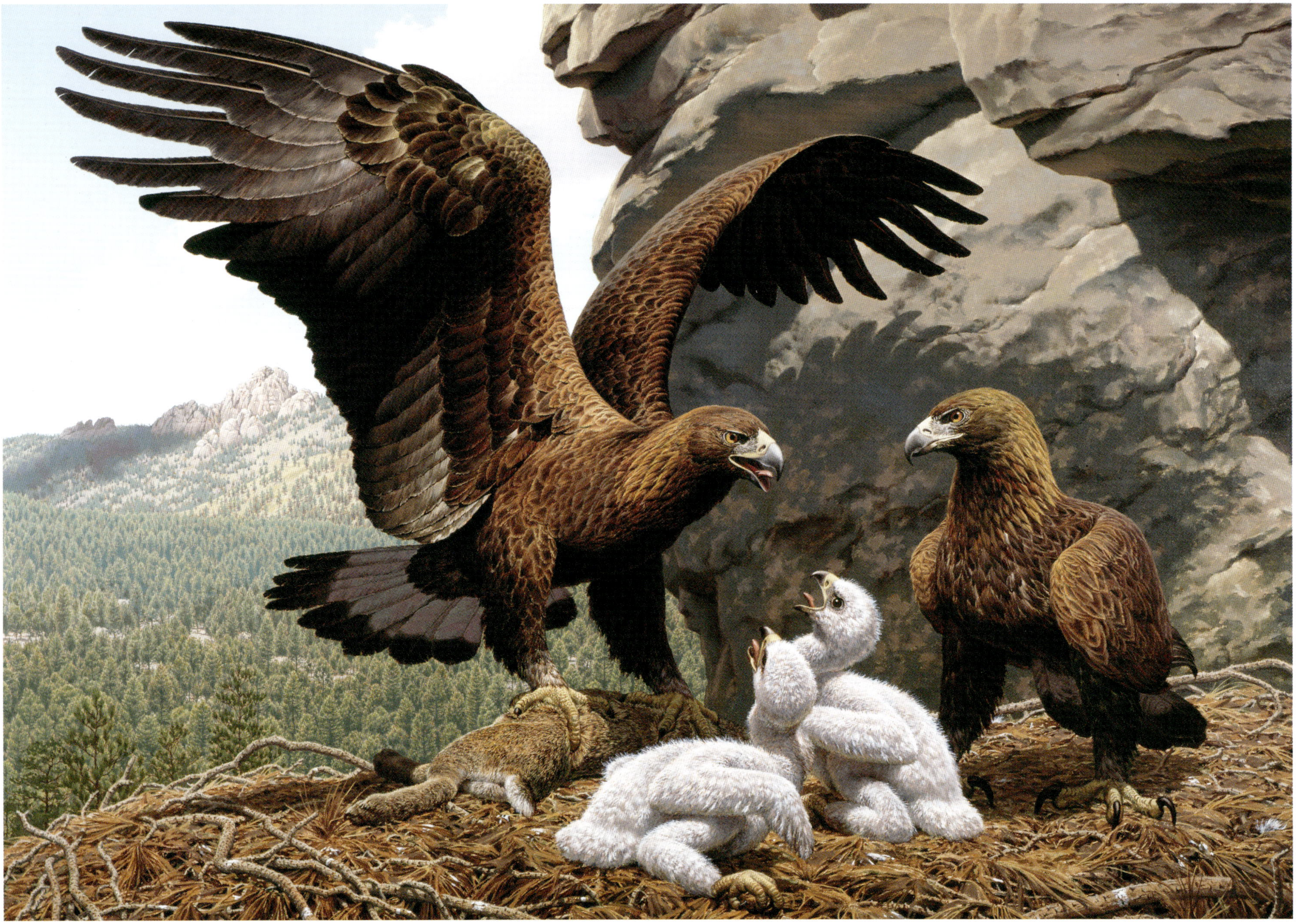

Golden Eagles in Mountain Eyrie

Because Matternes finds raptors "majestic and awe-inspiring," he was eager to paint this domestic scene of golden eagles feeding their brood. A collector purchased it, but his wife immediately insisted that they return the picture. Her problem? The ground squirrel's bloodied carcass grasped in the male's talons, which she refused to display in her home. Jay asked his wife Del's opinion, and was dismayed when she agreed that it would probably offend the sensibilities of the client's genteel dinner guests. A few years later, when Matternes painted another male eagle returning home, it carried an innocuous twig to the nest. Prior to beginning it, Matternes pinned a note to his easel: "NO MORE OFFENSIVE DEAD RODENTS!"

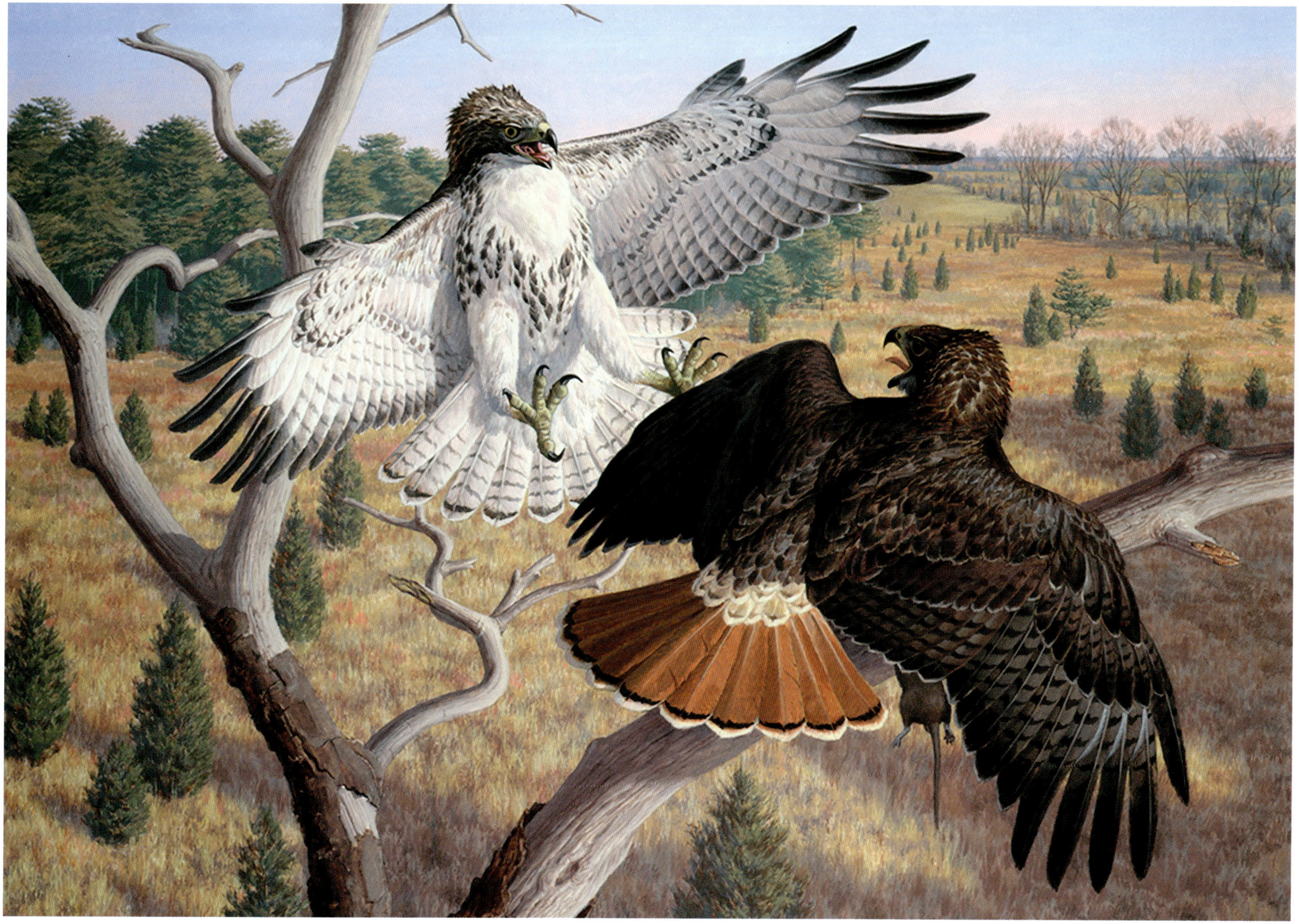

The Challenger

Jay's friendship with members of a local raptors society in Virginia led to his participation in their annual survey of migrating hawk populations. Volunteers used decoys and mist nets to trap the birds, collected data on their age and condition, then banded and released them. On one occasion, the group was permitted to work at an off-limits nature reserve adjacent to Dulles International Airport. Formerly farmland, the unused fields were still densely populated by rodents, offering a veritable smorgasbord for the predatory birds. Using binoculars, Jay made prolonged observations of several species who, surprisingly, didn't seem to mind the screaming engines of jet airliners using the nearby runways. A confrontation of two male red-tailed hawks over territorial boundaries inspired him to celebrate their dispute in *The Challenger*.

DUCK STAMPS

During the late 1970s, Jay entered three paintings in the Federal Duck Stamp Contest. Sponsored each year by the U. S. Fish and Wildlife Service, the winning entries are printed on pricey stamps that hunters and conservationists like to collect, with revenues helping to support wetland and wildlife conservation programs. Although there is no cash prize, the winners can profit from the legions of collectors buying quality prints of the winning designs directly from the artists each year.

A print bearing the artist's autograph increases its dollar value severalfold. Most prized are those with both signature and *remarque*—an original free-hand sketch, drawn on the white bottom border. Jay chose to paint species that were less frequently shown or that especially appealed to him. In later years, the competition authorities picked the species to be depicted.

Here are Jay's proposed duck stamps, painted in a colorfully bold, decorative style.

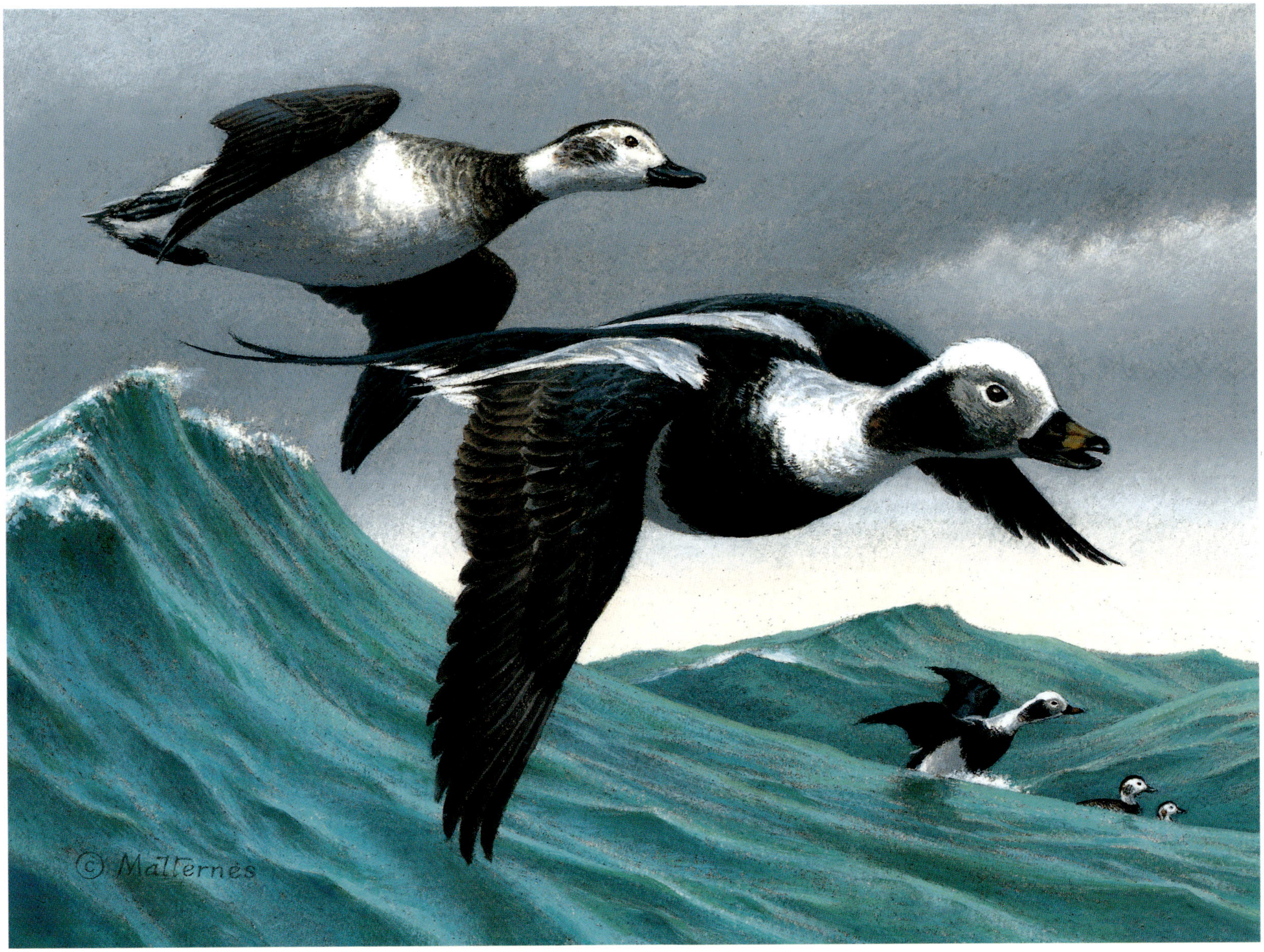

Oldsquaws—*Clangula hymalus*
The long-tailed duck, or old-squaw, is a North American sea duck, the only living member of its genus, *Clangula*. It breeds in ponds, streams, and other Arctic wetlands, and winters on open ocean or on large fresh-water lakes. Oldsquaws feed on aquatic insects, crustaceans, bivalves, fish, fish eggs, and plants. This attractive duck nests in shallow scrapes in the ground, which it lines with willow and birch leaves, and then with down. Nests are placed near the water's edge, often on islands or peninsulas, close to other long-tailed duck nests.

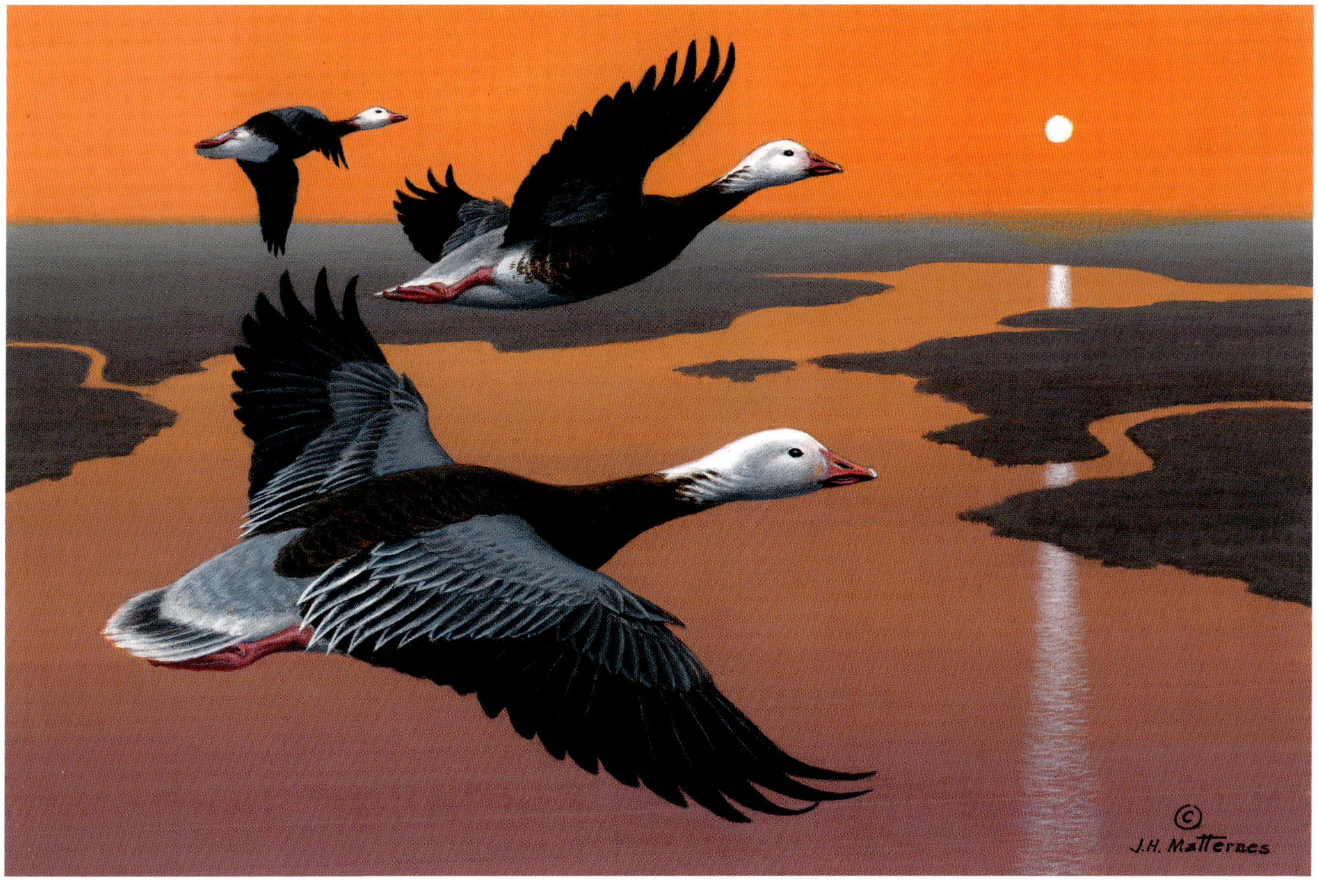

Snow Goose—Blue Phase—*Chen caenulescens*
A North American species of goose commonly referred to as "light geese," from their typically white plumage—although they also have dark-colored phases and mixed colorations. They spend the coldest seasons in southern coastal marshes, bays, wet grasslands, and fields, and are completely vegetarian. At winter's end, they migrate north to their breeding grounds in the Arctic tundra, during which they fly so high that they can barely be seen from the ground.

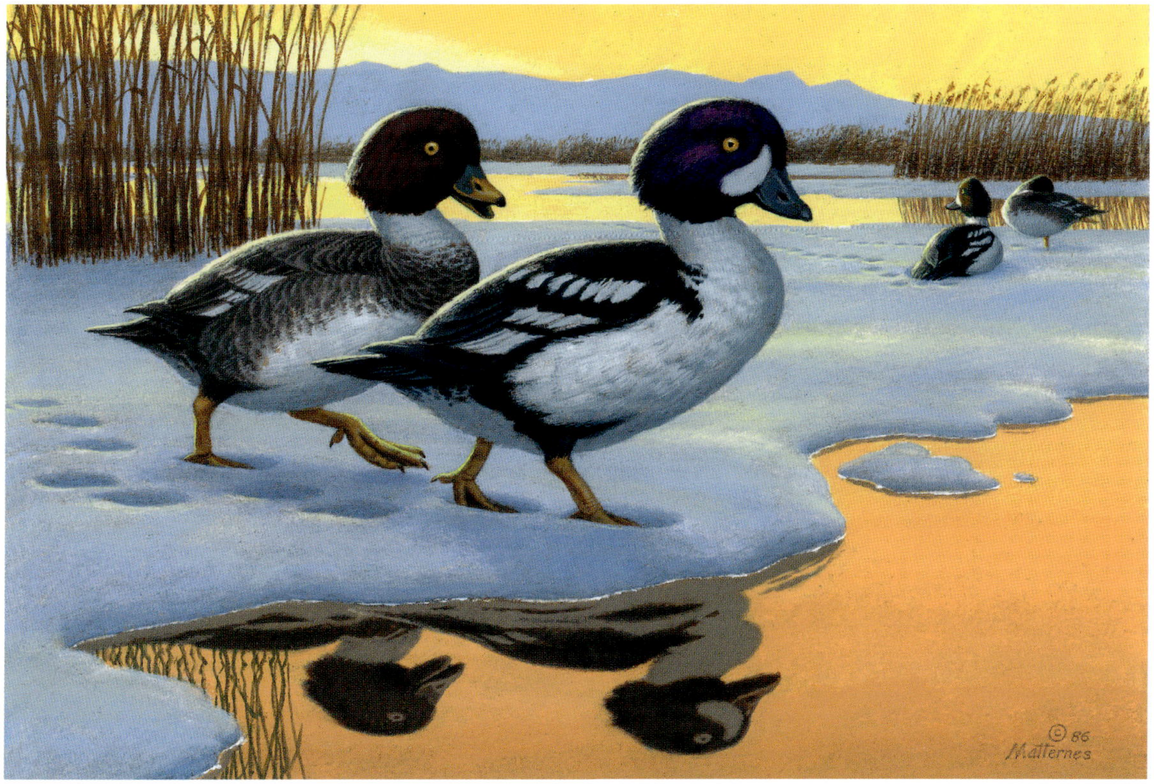

Barrow's Goldeneye—*Bucephala islandica*
A medium-sized sea duck from northwestern North America, the goldeneye feeds by diving for aquatic invertebrates, fish eggs, and small fish. Usually whole flocks dive together. It breeds along alkaline lakes in parklands and winters along rocky coasts.

MEDITATIONS ON ORANGUTANS

In 2023, Matternes found some sketches he had made during the 1950s of a female orang moving among metal poles at the Philadelphia Zoo, and reimagined her traveling through the rainforest canopy. Orangs shift their centers of gravity from tree to tree, enabled by their eight-foot arm spans. Each move is well thought out, without any daring acrobatics like those of their smaller gibbon neighbors; the fall of a 180-pound male from a 90-foot tree would be fatal. Nevertheless, they can make good progress traveling high in the treetops at a leisurely pace, without descending to the ground for weeks at a time.

In Jay's drawing, the ape grasps a branch with her left hand while also reaching out with her left foot. That maneuver would be impossible for humans, whose hip movements are constrained by ligaments that are absent in orangs, giving them almost unlimited rotation. Her youngster holds onto its mother's fur for dear life, as she cannot use her arms to support it. After the seasonal monsoons have drenched the forest, orangs feed on fruits in the mid-canopy, approximately 30 to 90 feet above the ground.

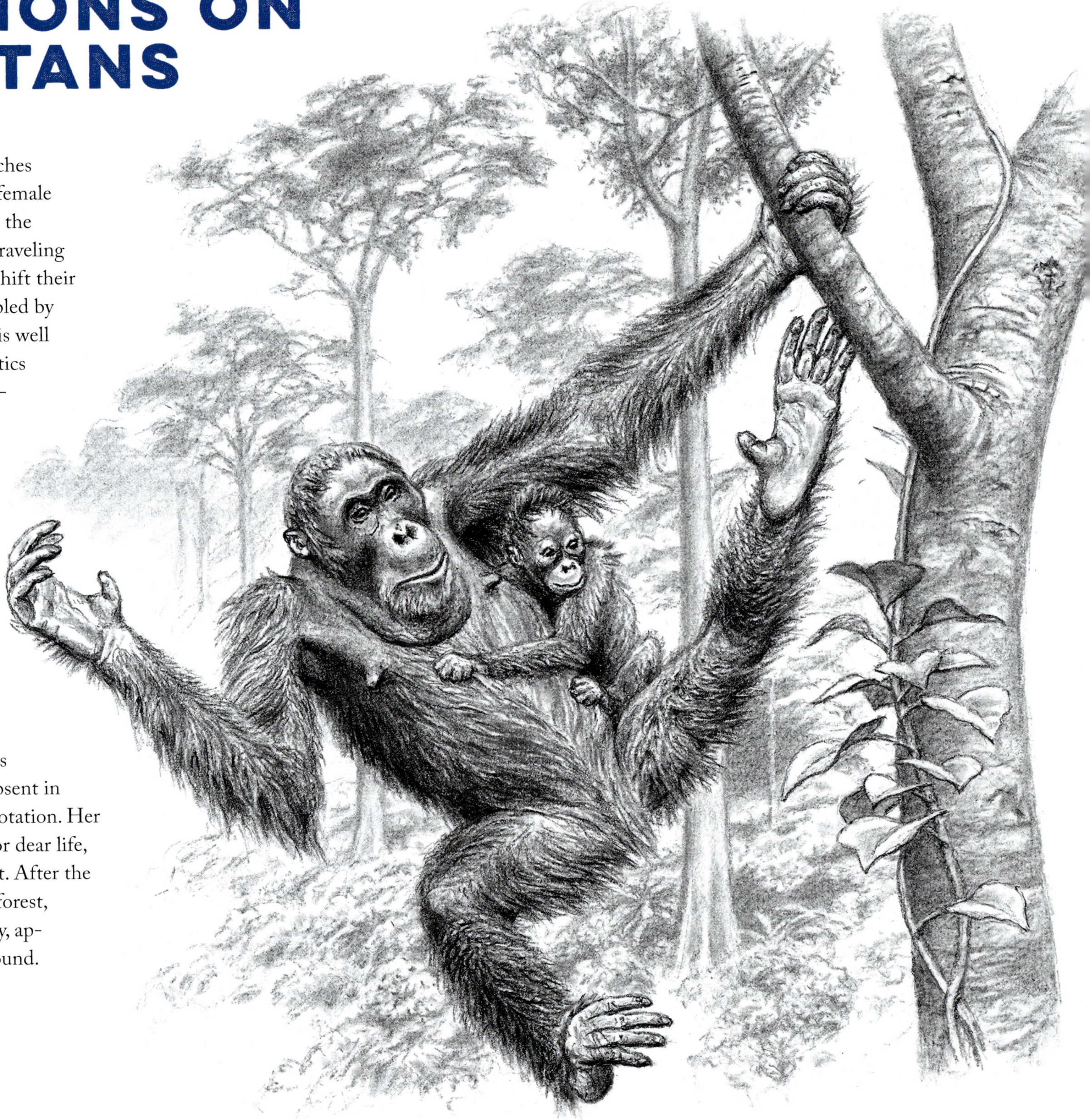

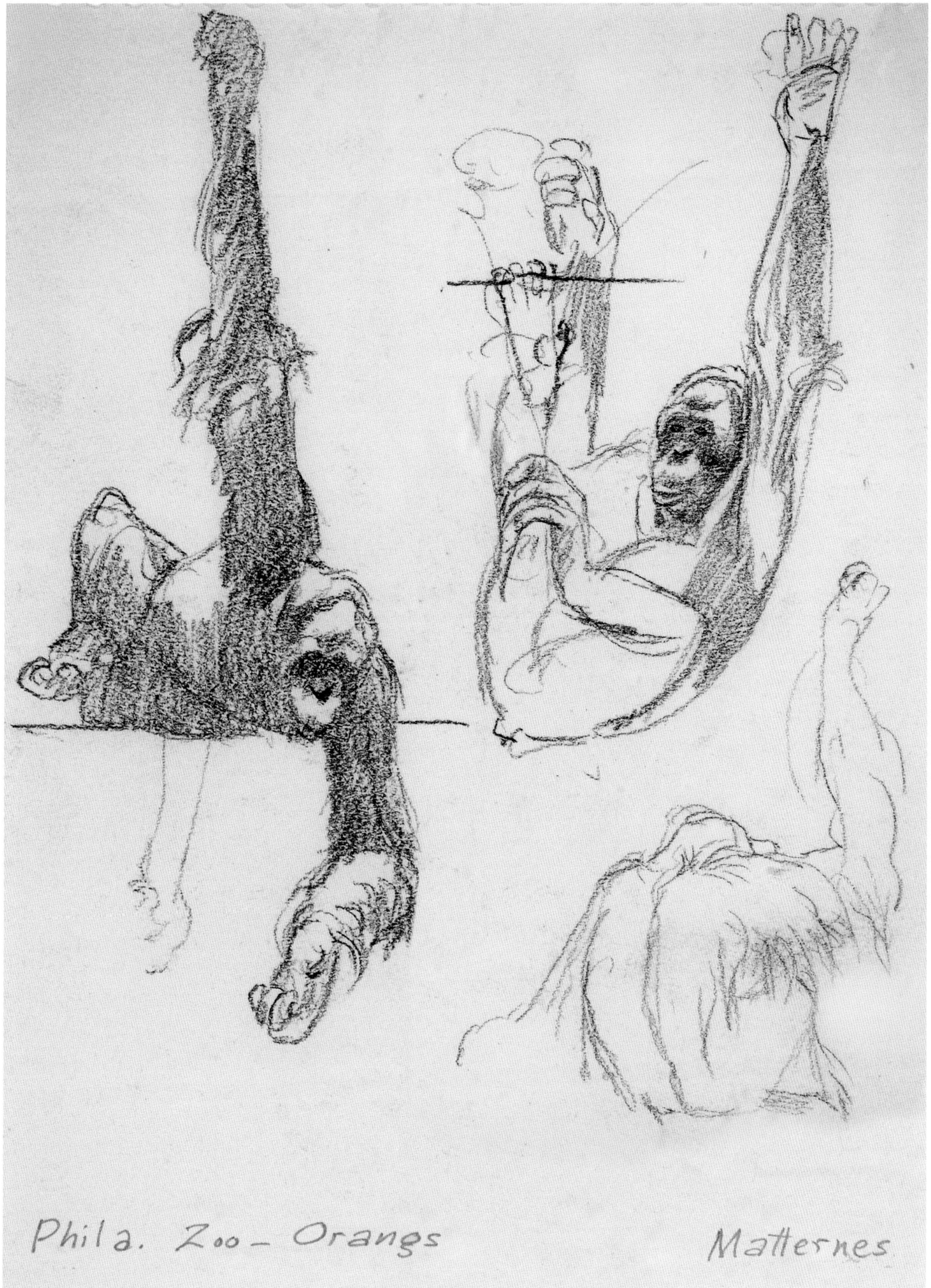

Orangutans have always been special favorites of Jay's. As a boy, he reminded his mother that Corregidor Island in the Philippines, his birthplace, is just a few hundred miles from the northeastern tip of Borneo. "If I had been born a little bit further to the south with a different mother and longer arms," he liked to say, "I might've entered the world as an orangutan." His mom was not amused by his feelings of kinship with the red apes, but he would seek out photos of them. "During World War II, when there was intense fighting in Borneo, one of my chief concerns was the safety of orangs and other island fauna. Over the years, I've done many drawings of these great apes and dissected a number of them that died in zoos."

Jay finds it fascinating that orangs have become specialized for a completely arboreal life: "How peculiar that a creature larger than a big man can live its entire life in the trees and gain its sustenance there. They seldom come down, although the Bornean species is sometimes found traveling on the ground. The Sumatran orangutan is more completely arboreal. A possible reason is that there are tigers in Sumatra, which makes the trees a much safer place."

The artist as comparative anatomist. Matternes dissects two orangs, a young female and mature male "Rango," the subject of the anatomical sketch opposite, made in 1972. Whenever one of these rare cadavers became available, Jay had to work quickly to extract as much information as possible; they had not been embalmed, so there was only a brief window before they began to rapidly deteriorate.

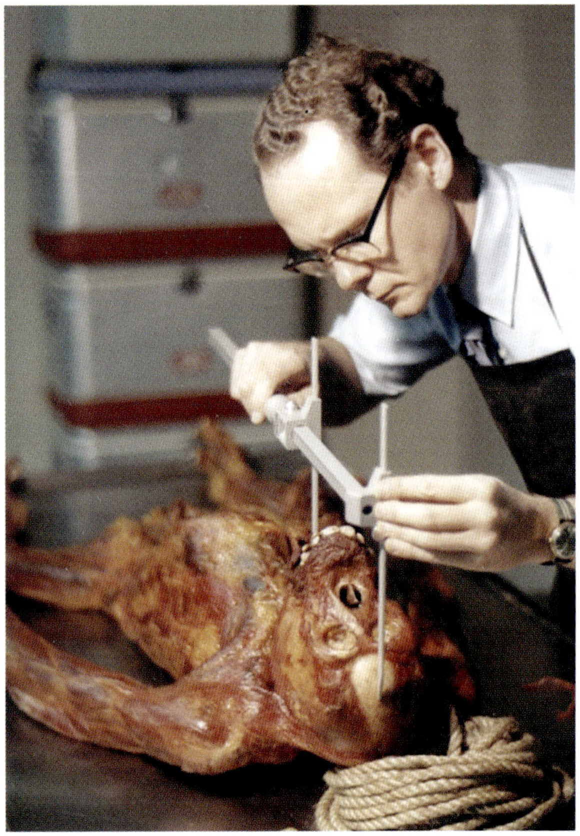

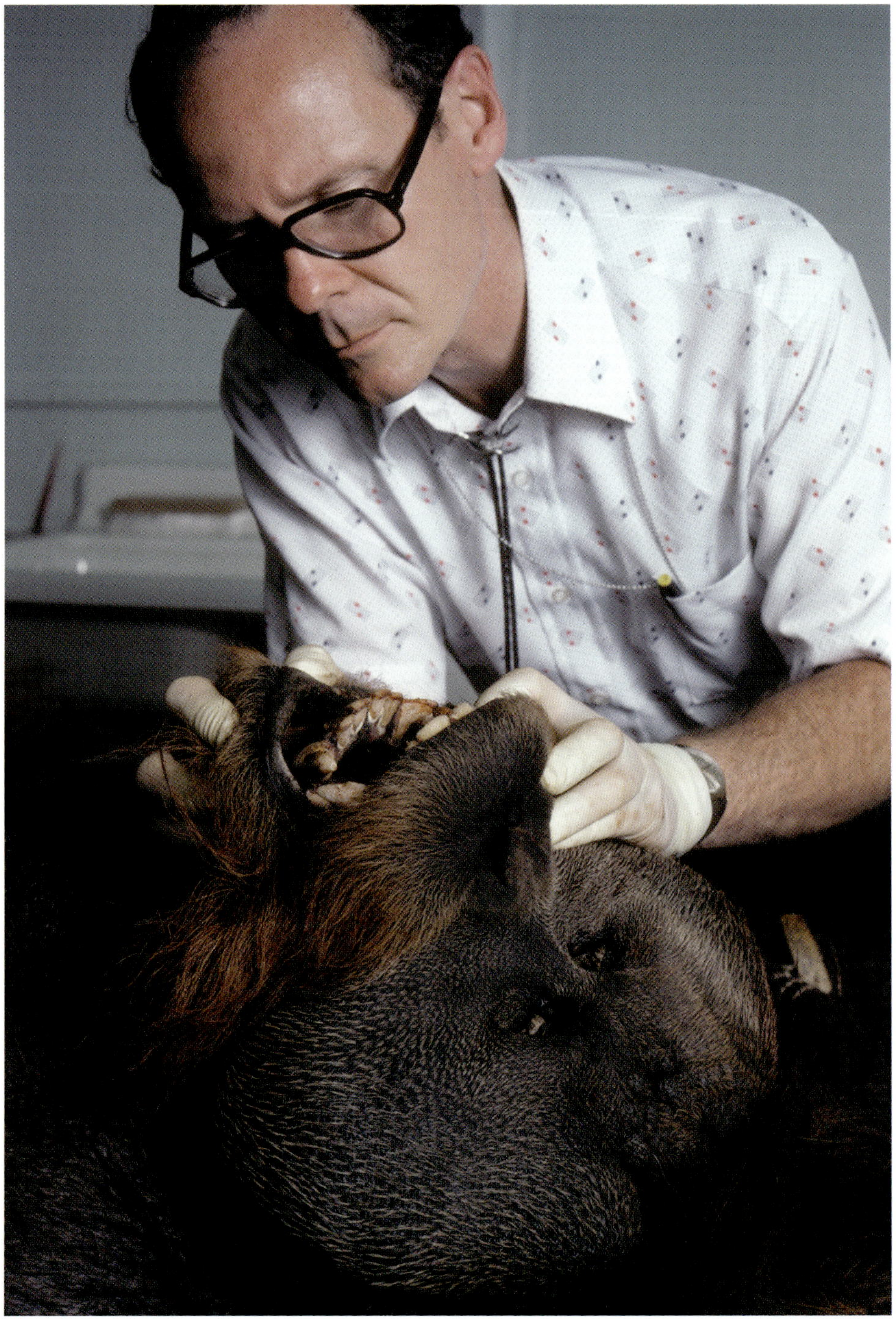

"In my dissections of orangutans, I noticed that many of them lack a femoral teres ligament, which helps to stabilize the hip joint in other apes and humans when they walk on the ground. I believe its absence shows that orangs inhabit trees not only because of habit and behavior, but also because they are anatomically adapted to a strictly arboreal existence. When I demonstrated to a comparative anatomist that orangs lack this ligament, he was initially incredulous, as it did not appear to be generally known at the time."

"Orangs impress me as being very thoughtful and rather pensive; they are rarely prone to outbursts of anger, like young chimpanzees. Unlike the more social apes, they remain solitary for most of their lives. When in captivity, escapes are quite frequent, because they are very clever at manipulating the nuts and bolts of their locked cages, and will sit still for long periods while they figure out the problem."

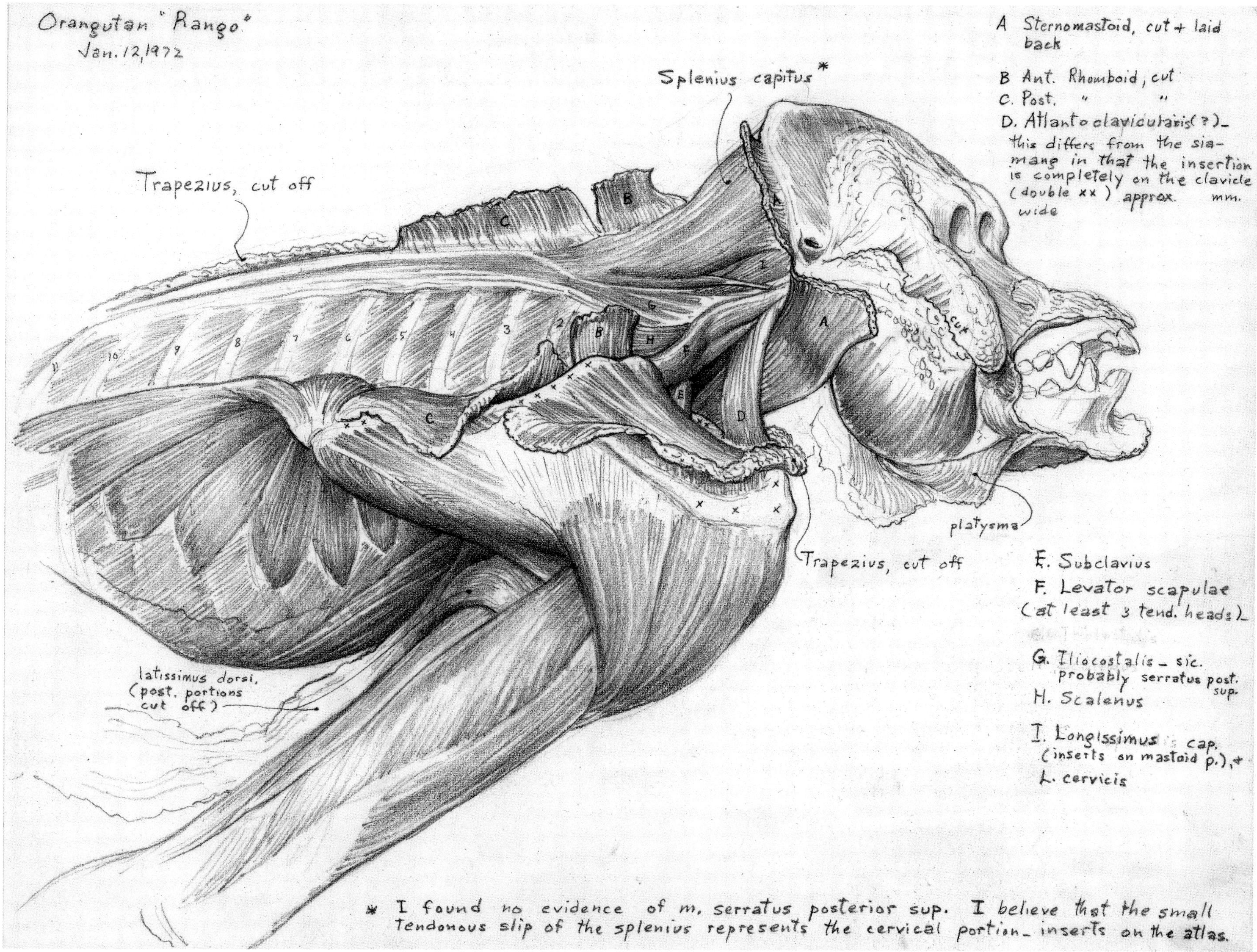
Orangutan "Rango"
Jan. 12, 1972
Splenius capitus*
Trapezius, cut off
A Sternomastoid, cut + laid back
B Ant. Rhomboid, cut
C. Post. " "
D. Atlanto clavicularis (?) – this differs from the siamang in that the insertion is completely on the clavicle (double xx) approx. mm. wide
platysma
Trapezius, cut off
E. Subclavius
F. Levator scapulae (at least 3 tend. heads)
G. Iliocostalis – sic. probably serratus post. sup.
H. Scalenus
I. Longissimus cap. (inserts on mastoid p.), + L. cervicis
latissimus dorsi, (post. portions cut off)
* I found no evidence of m. serratus posterior sup. I believe that the small tendonous slip of the splenius represents the cervical portion – inserts on the atlas.

Fig Feast

Southeast Asia's annual monsoon cycle includes periods of boom and bust, glut and starvation. After the rainy season, when fruit becomes abundant, orangutans, gibbons, monkeys, and hornbills gorge themselves on figs and other fruits. Orangs fatten themselves as a hedge against the lean times to follow. In the Indo-Malayan region, species and subspecies of hornbills are often confined to specific islands, but these large frugivores, Blyth's wreathed hornbills or Papuan hornbills (*Rhyticeros plicatus*), range widely throughout Wallacea and Melanesia.

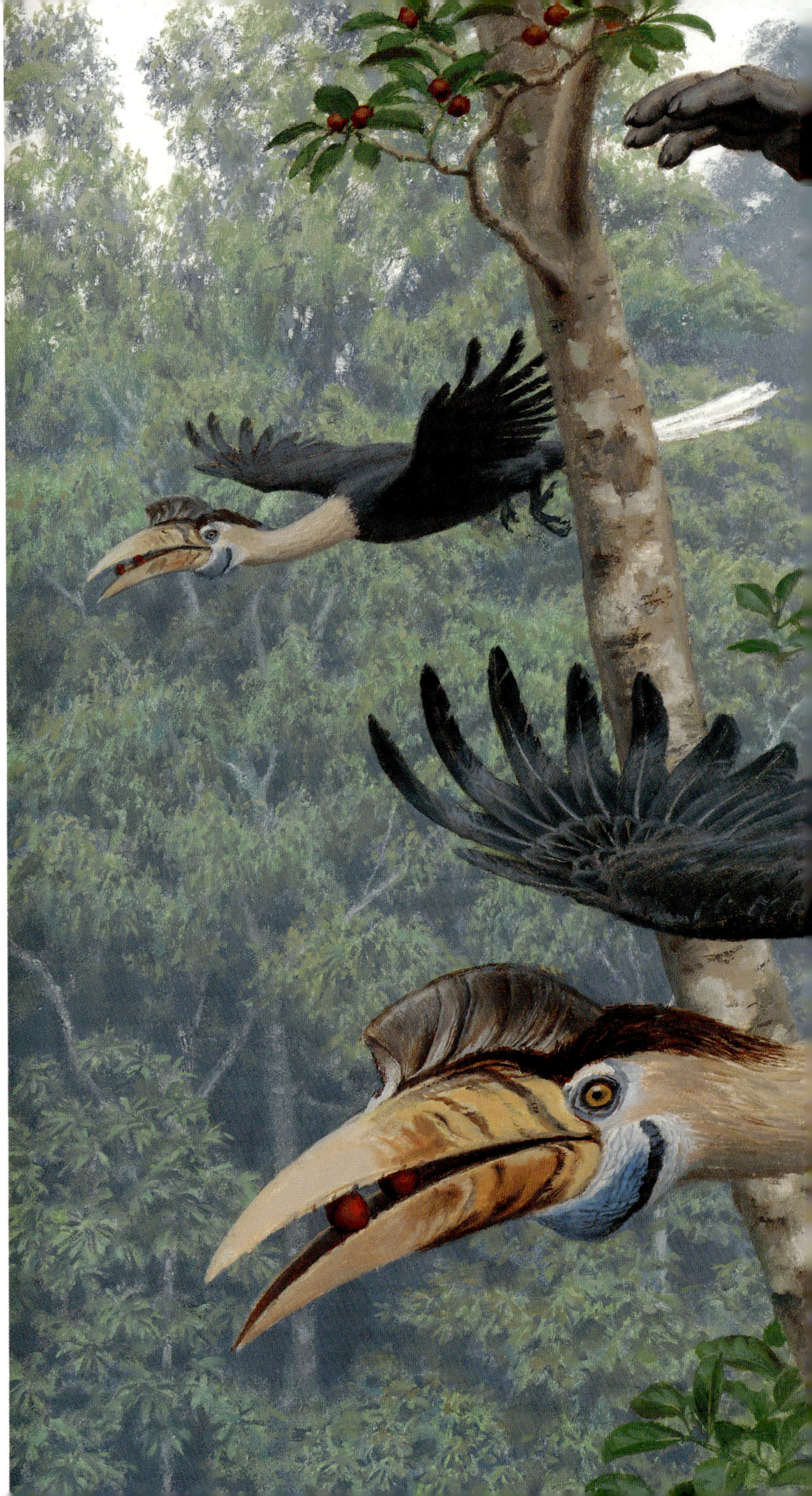

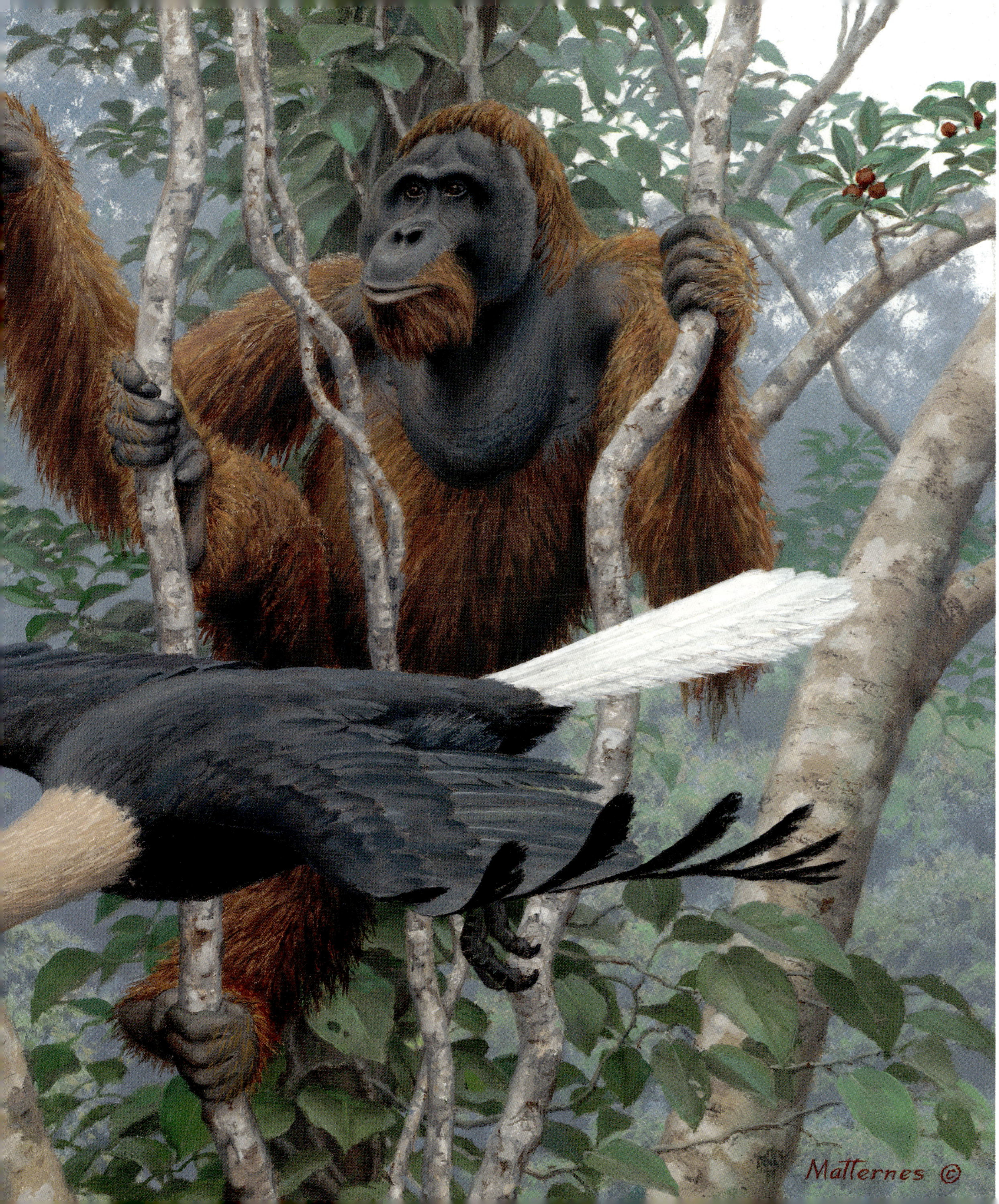
Matternes ©

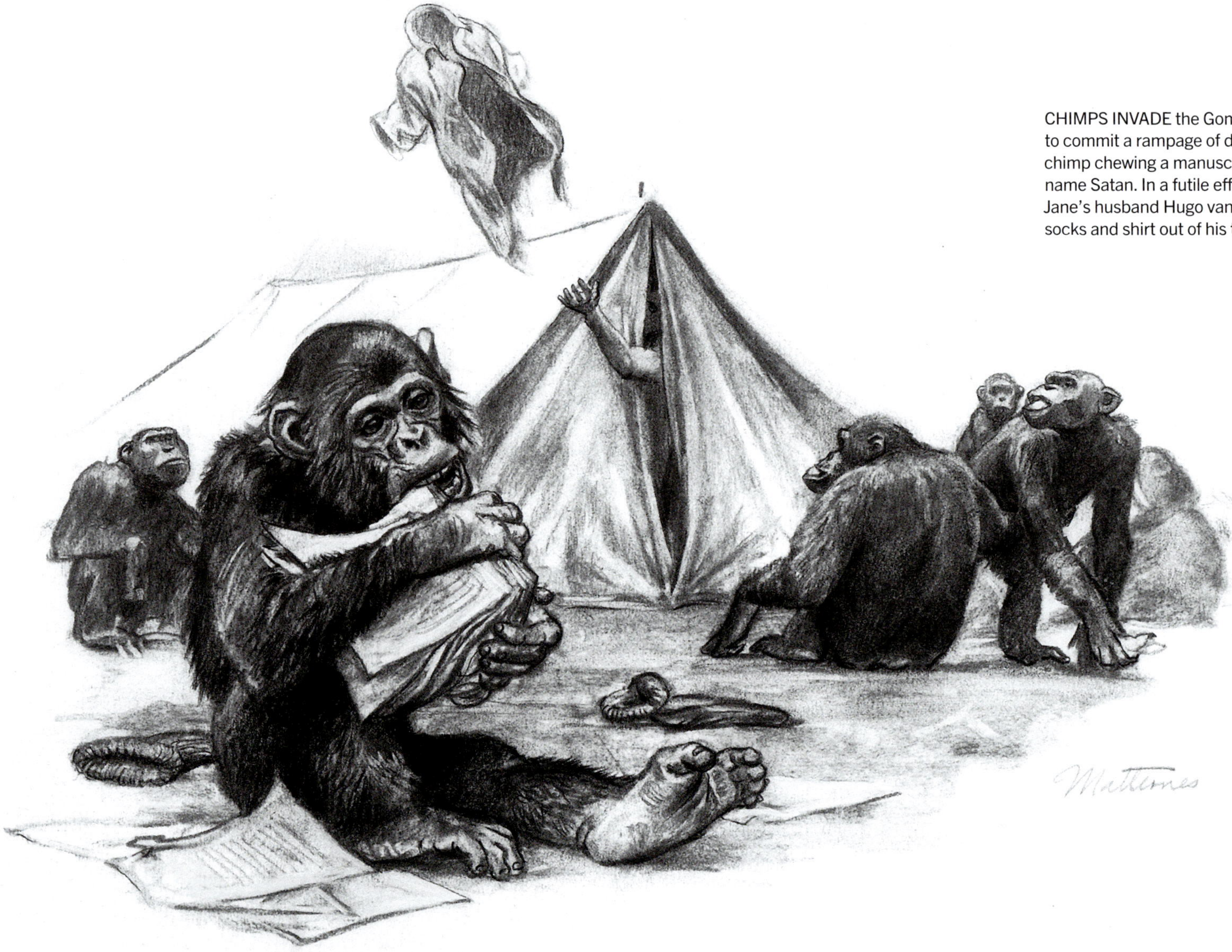

CHIMPS INVADE the Gombe Stream campsite to commit a rampage of destruction. The devilish chimp chewing a manuscript was then given the name Satan. In a futile effort to distract them, Jane's husband Hugo van Lawick has thrown his socks and shirt out of his tent.

CHIMPS, JANE GOODALL, AND A ROADSIDE ATTRACTION

Matternes was invited by the editors of National Geographic Books to create illustrations for Jane Goodall's groundbreaking volume *My Friends, the Wild Chimpanzees* (1967). Since it would also contain photos by lensman Hugo van Lawick, then Goodall's husband, Jay was asked to depict dramatic incidents that were *not* caught on camera. When a group spontaneously rampaged through the camp, scattering papers, books, and coffeepots, he sketched the chaotic scene from Goodall's description. During a heavy thunderstorm, about twenty adults ran about wildly in a "rain dance," that Jay illustrated from eyewitness descriptions.

He met with Goodall several times, but Jay's main contacts were with the editors of National Geo, who specified which behaviors to depict for her book. He was the second artist they hired. The first was a talented illustrator who glamorized Jane but had no expertise in drawing apes. After Jay was hired, he was puzzled when the editors initially rejected his drawings although they liked his chimps. His wife, Delphine, solved the mystery. She told Jay she thought that they had initially approved portrayals of Jane with

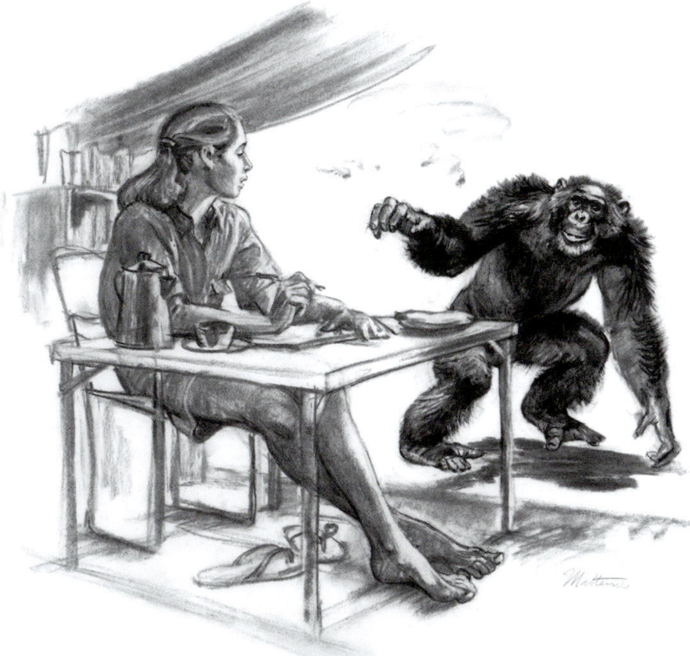

WHILE KEEPING EYE CONTACT with Jane Goodall, the chimpanzee David Greybeard audaciously enters her tent and swipes a banana from the tabletop. His daring theft, before they became friends, prompted Goodall to attract chimps by leaving bananas around the camp to lure them out of the forest.

a slender "Audrey Hepburn" torso and long movie-star legs—the current ideal of beauty. Jay had more accurately depicted her as somewhat athletic, with shorter and more muscular legs; she frequently had climbed trees while tracking her subjects.

Over time, Jay's renderings of the chimps' behavior changed from his first impressions: "I had started following Jane's initial descriptions of them as childish, unruly pranksters, but she later observed the dark side of chimpanzee behavior, including occasional kidnapping of babies, cannibalism, and the males' organized predation on small monkeys." (They shared the meat with kin, doling it out according to their own brand of etiquette.) Matternes was interpreting Jane's original vision of them as pacifistic, vegetarian, somewhat comical creatures, but then she observed a brutal male bludgeoning a young baboon to death. No scientist before Jane had reported that chimpanzees can be hunters and fierce carnivores, and she was deeply disappointed to observe their violent, bloodthirsty side. During a televised interview in 2020, she was asked whether she ever met a chimpanzee she didn't like. "Yes, I did," Goodall replied, "People think that they're my favorite animals, but they're not, because they're too much like people. Some chimpanzees I loved, and some I did not like at all. They're bullies and aggressive to each other and they're just not nice…"

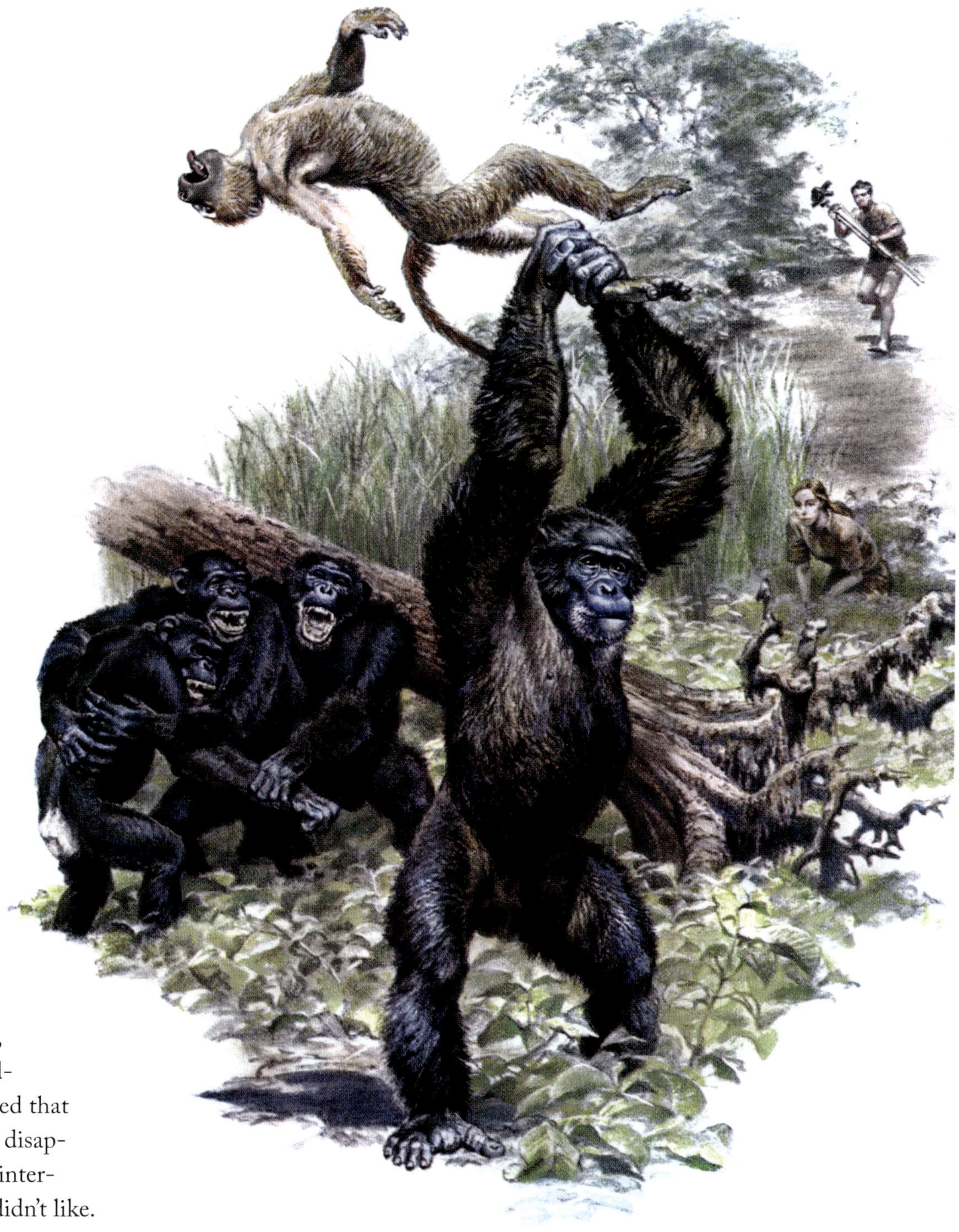

TIGHTLIPPED WITH FEROCITY, the dominant adult male Rodolf kills a young baboon for food, while his companions scream in a frenzy of excitement. Jane and photographer-husband Hugo (background) arrived too late to photograph the scene, so Matternes was assigned to reconstruct it for her book.

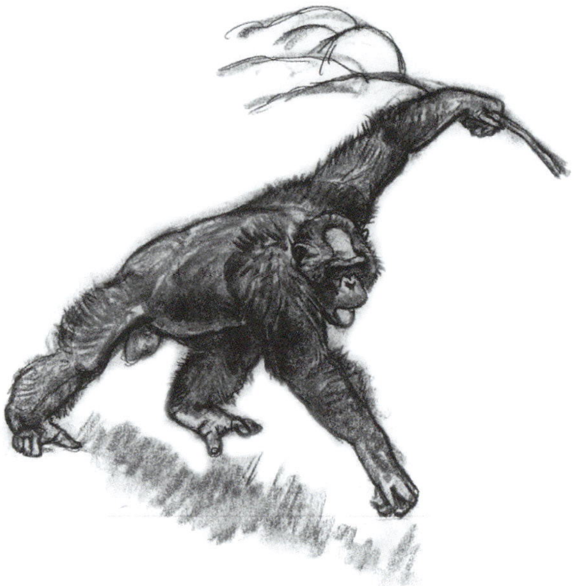

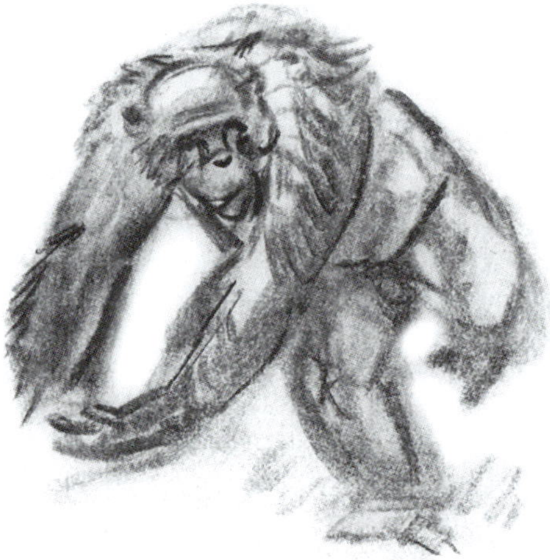

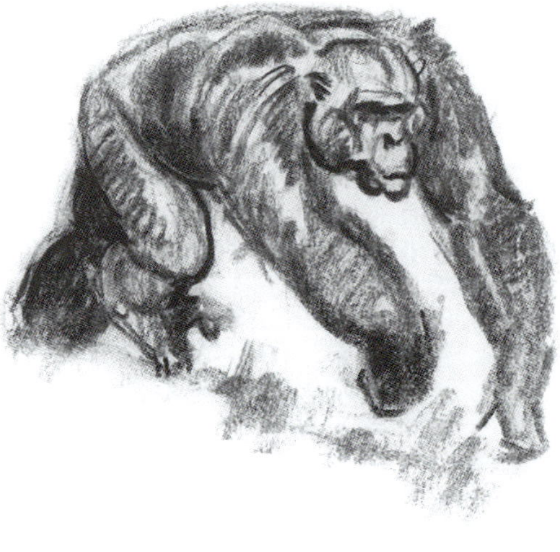

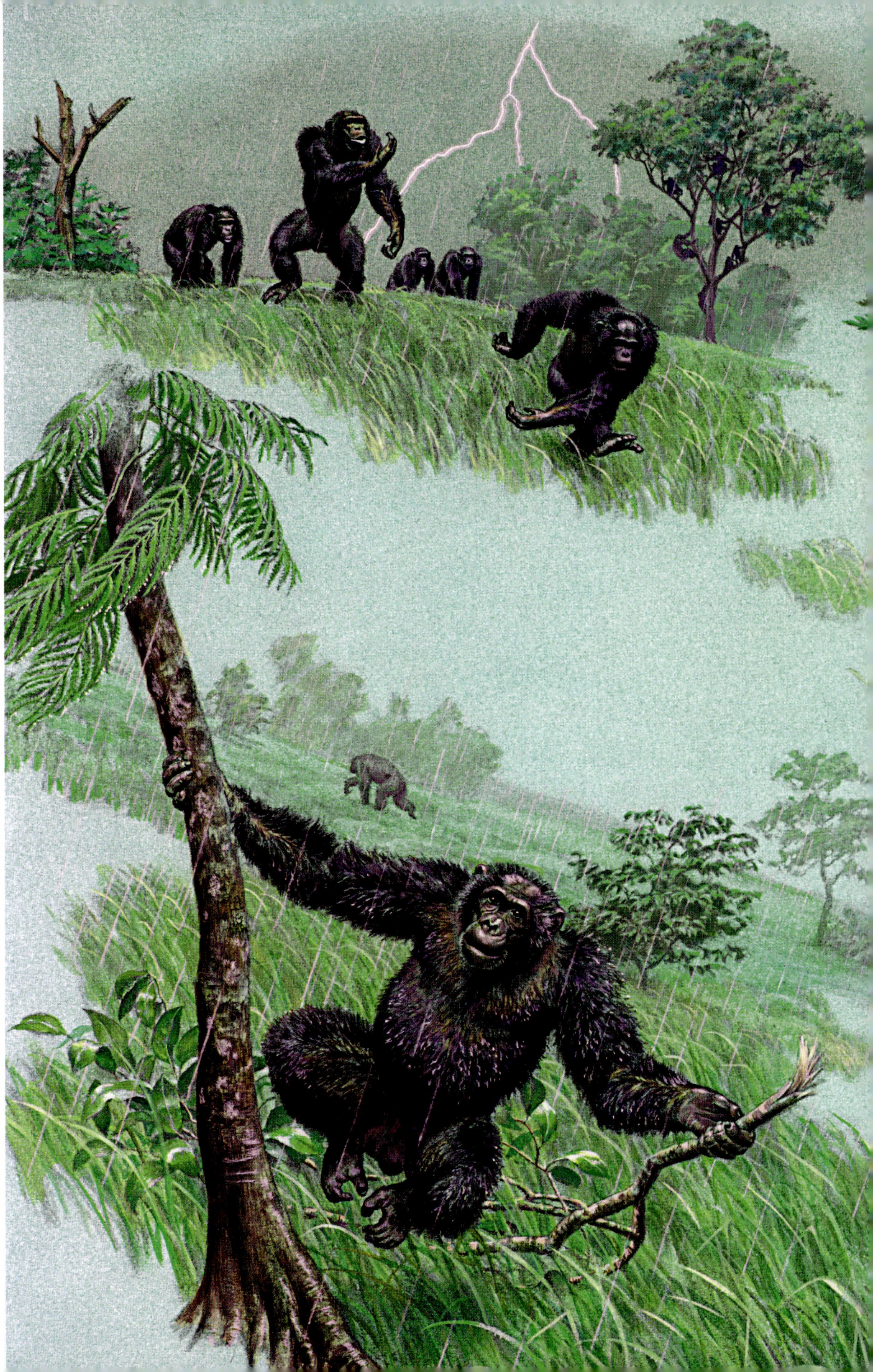

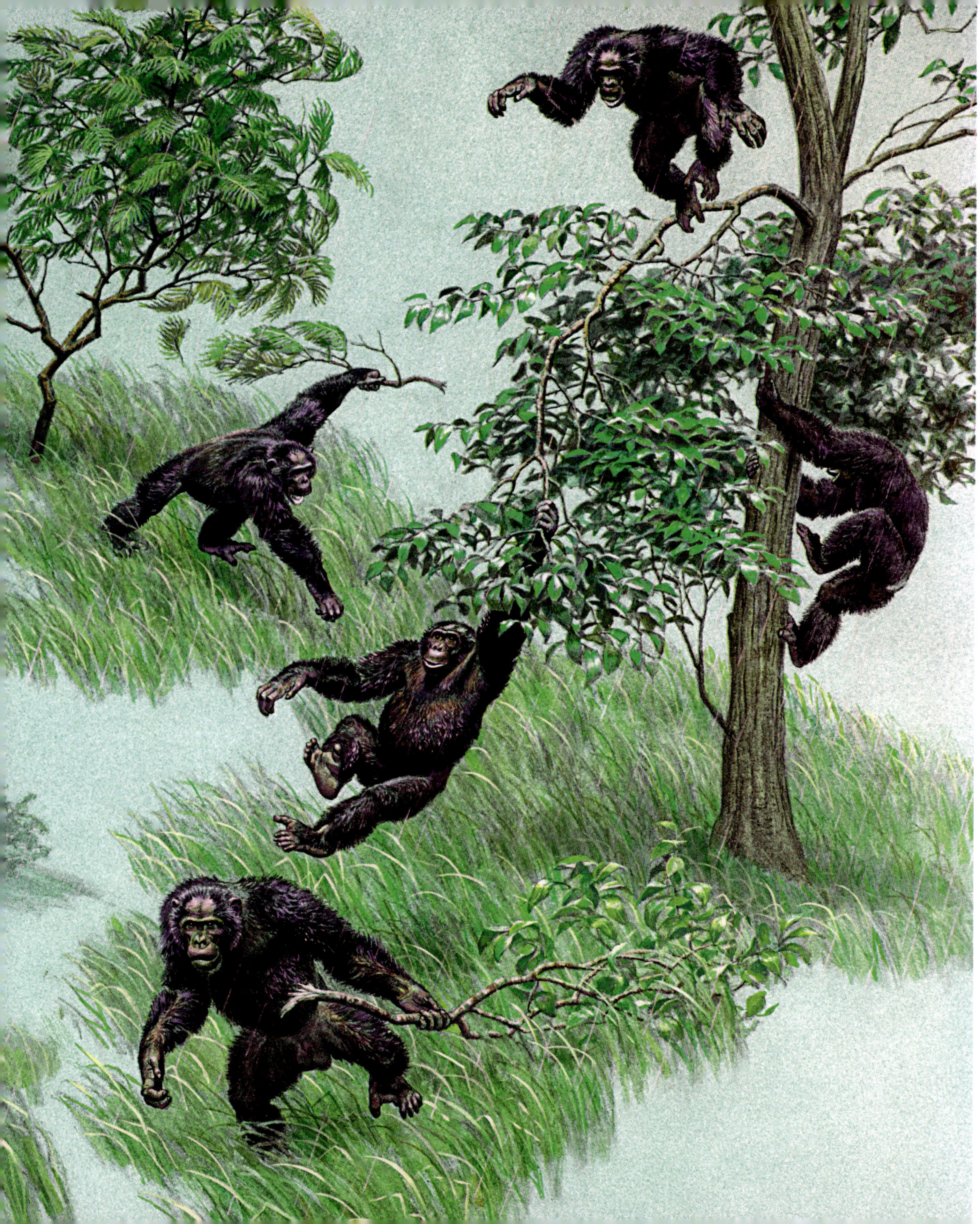

A FRENZIED RAIN DANCE
Hooting apes stage frenzied rain dances when pelted by a downpour. In this sequence, a chimp crouches, stands erect, grabs a branch, then slaps the ground as he charges downhill toward a tree and climbs its trunk. Hurling himself downward, he snaps off a bough and drags it behind him, swings around the tree, and finally returns uphill to turn around and charge downhill again. Mothers and their young watch from the ridge.

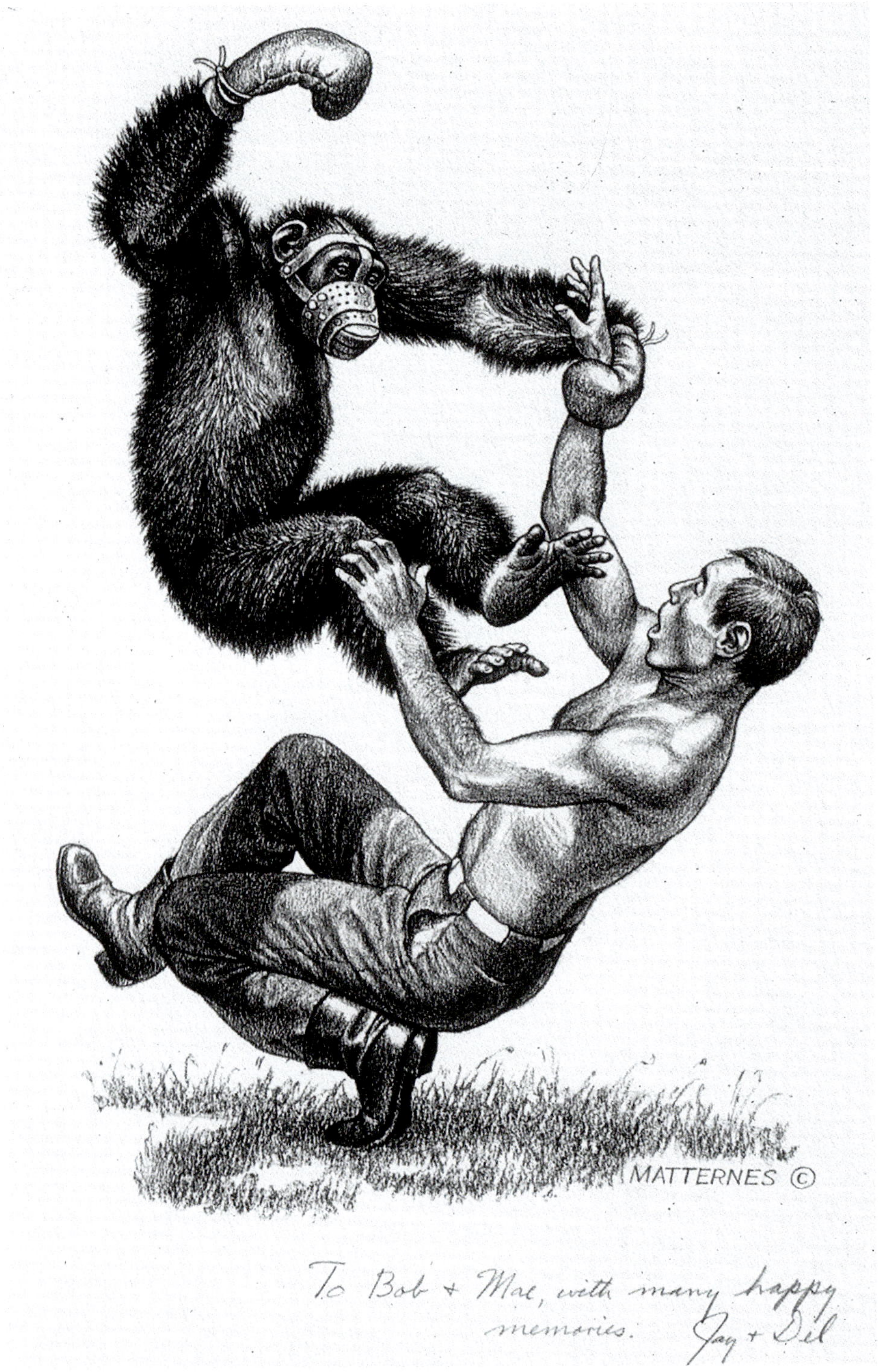

“The man was at the ape’s mercy . . . [and was] quite humbled.”

Beginning in the 1940s, years before Goodall’s field study, a married couple of country showfolks called Bob and Mae Noell ran a traveling chimp boxing and wrestling show out of Tarpon Springs, Florida. For years they maintained a small private menagerie—“Noell’s Ark”—that featured chimps and gorillas rescued from carnivals and circuses. Jay and Del visited their “roadside zoo” during a vacation trip, and formed a lasting friendship with the Noells. One day, Mae asked Jay to illustrate a cover for her book about their captive charges. He made a sketch based on a blurry photo of a well-muscled young man fighting a chimp, a nightly exhibition that showed the enormous disparity in strength between an athletic young man and the much smaller chimp (left). The picture uses a “cartoony” variation of Matternes’ “serious” style, demonstrating his versatility as an illustrator.

This particular chimp had mastered the effective surprise strategy of taking down a human opponent with a flying leap, and then pummeling the opponent with his feet. Once on the ground, the man was at the ape’s mercy, at which point Bob Noell would quickly end the roughhousing to prevent any possible injury to the hapless volunteer. Jay describes these encounters as “violent and fast, but all in good fun.” The athletic “challenger” of the evening usually took his instant defeat with good humor, but was quite humbled, having completely underestimated the speed and strength of his adversary. During all of these bouts, the chimp was heavily gloved and muzzled to protect his human sparring partner. Bob Noell had lost a couple of fingers during unprotected “arguments” with his charges, but emphasized that it was only chimps that had caused him harm over the years—never a gorilla. He pointed out the deep difference in temperament between these two species, which was the exact opposite of the public’s perception of them.

Bob and Mae Noell cared deeply for the welfare of their chimps, gorillas and orangs, as well as sundry other animals that they “retired” (or rescued) from exploitation, and offered them safe haven. In later years, the couple was constantly harassed by animal rights groups trying to close them down.

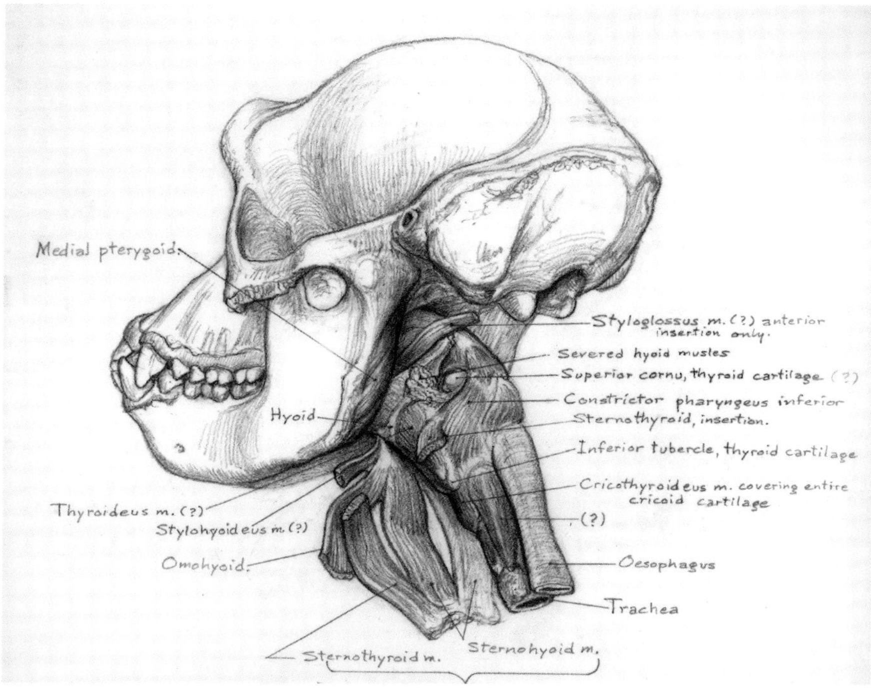

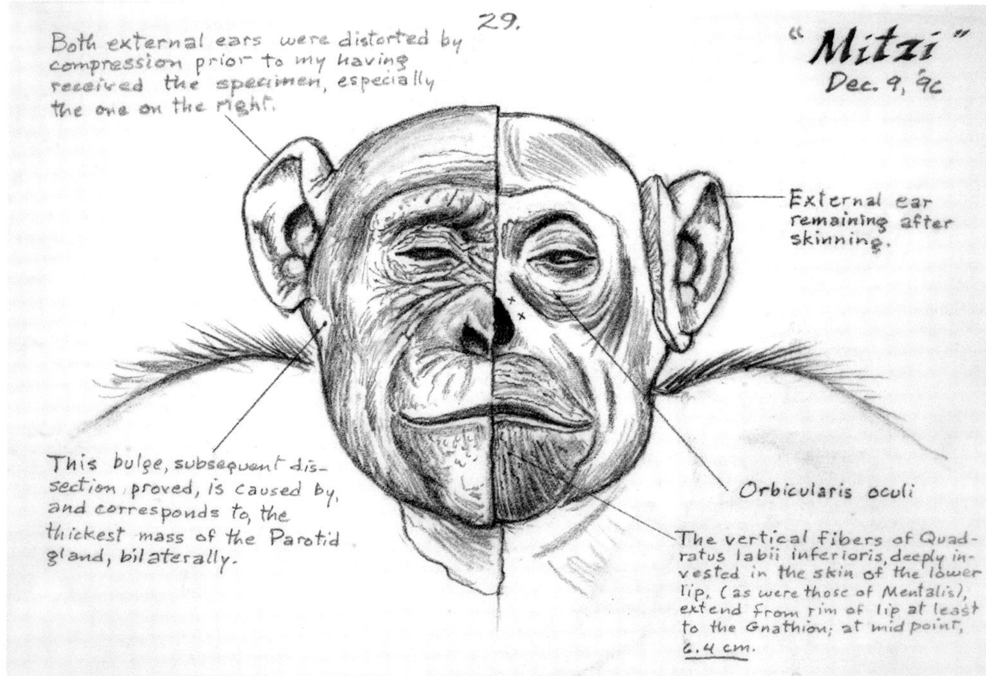

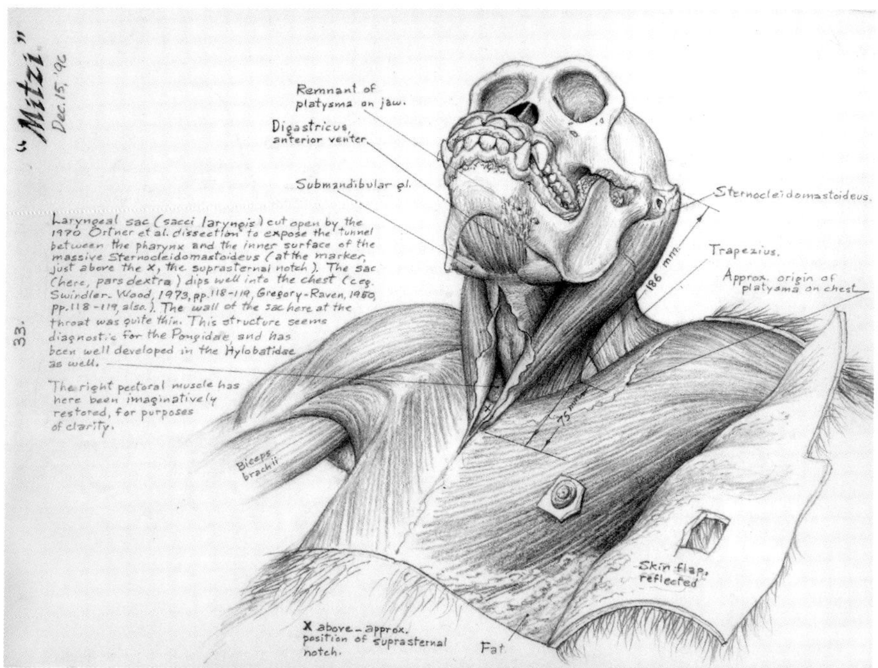

Times had changed, and public opinion was turning against live orca shows and performing circus elephants. "Pitting a human opponent against an ape for public entertainment would not be tolerated today," Jay opined, "and should be phased out."

However, what would have become of all the chimps and other apes in the Noells' menagerie if it were closed down? Mae had expressed to Jay her horror at the probable fate of their charges if they were confiscated. Her fears turned out to be well founded: over the next fifty years, hundreds of apes that were no longer useful to labs, entertainment, or medical experiments were ill-treated or callously killed. Jane Goodall has spent decades working to create ape sanctuaries to end the long, sad saga of callous exploitation.

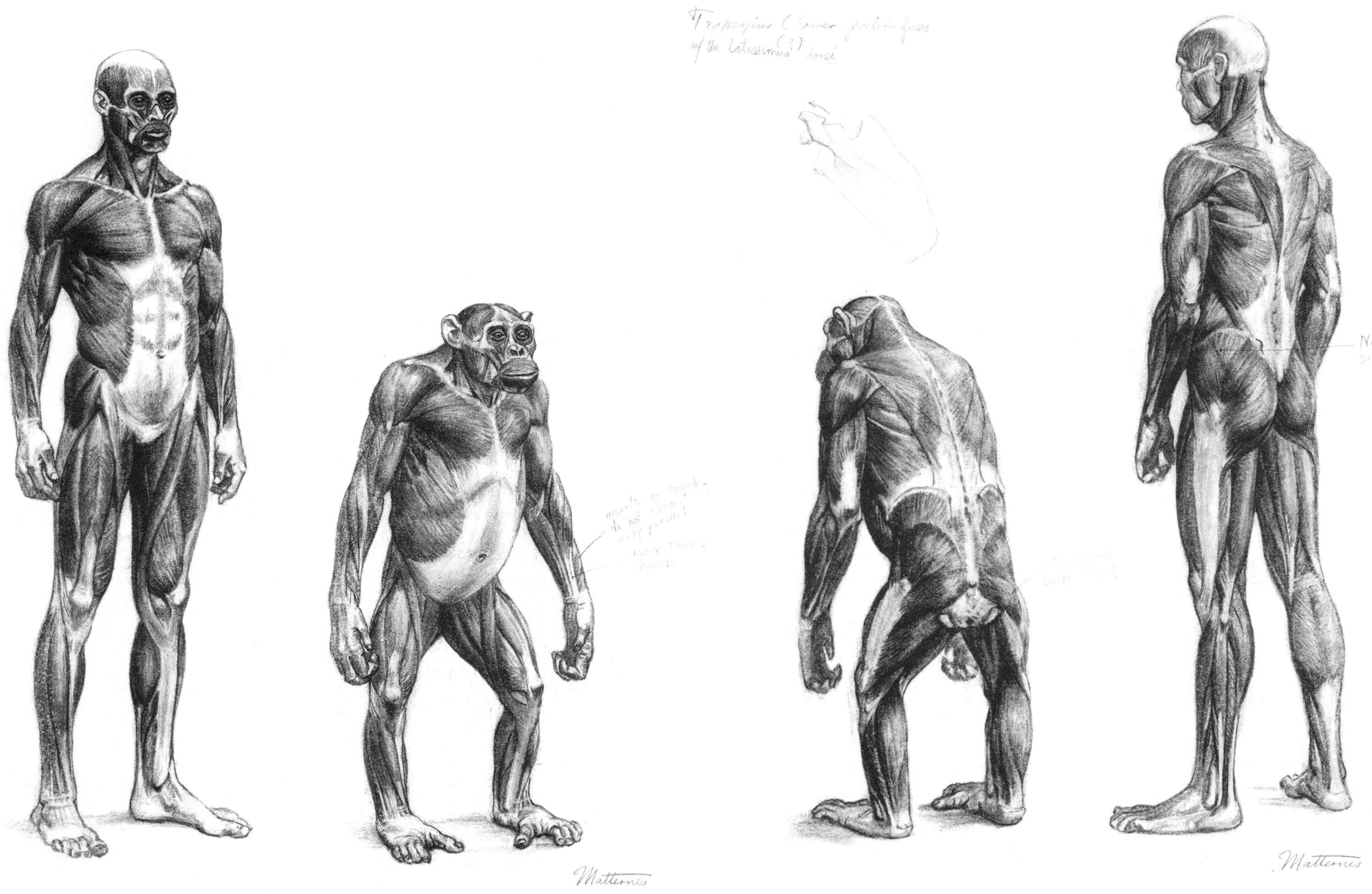

Matternes' delineation of chimp versus human anatomy illustrates the comparative size and placement of musculature. Unlike humans, the chimp cannot straighten its legs or lock its knees, but its longer arms and legs equip it for life among the trees.

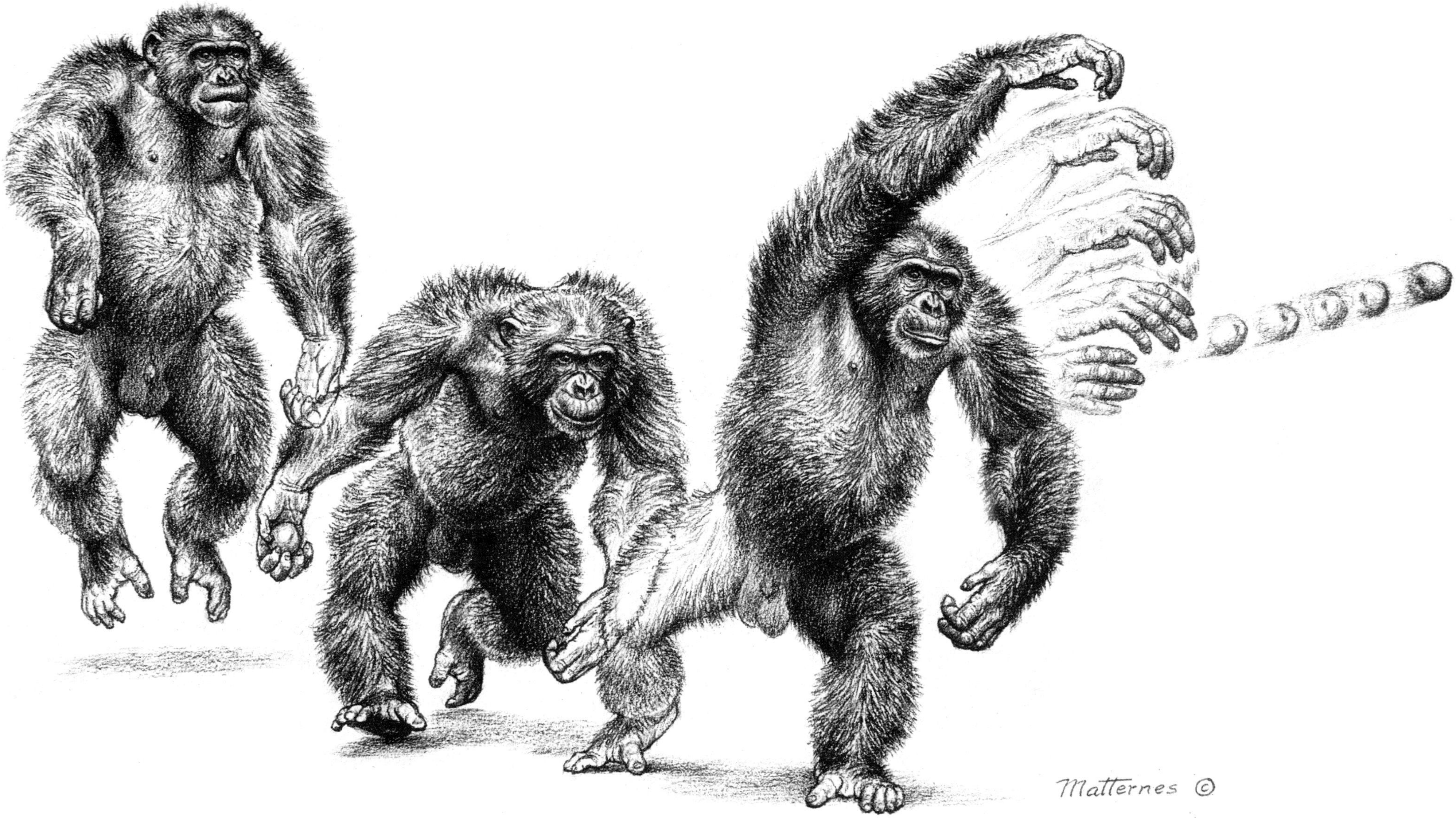

Mike, a strong male, performs a vigorous, aggressive display to intimidate others into acknowledging his dominant status. After a few exuberant jumps, he picks up an apple and hurls it, using a powerful, overhand windup. Chimps can throw things or spit with impressive accuracy.

GORILLA MY DREAMS

"Dian Fossey, the legendary investigator and protector of mountain gorillas, befriended me in 1968 when I first visited her Karisoke Research Station in the Rwandan forest. In September 1974, I returned there to continue sketching the free-living gorillas.

"Each day I accompanied one of Dian's students and an African tracker-guide to visit the first groups tolerant of human observers. We would approach on all fours, where they could see us well, making 'belch grunts,' which is proper gorilla protocol.

"On the first day, a curious adolescent called 'Pantsy' approached us and stood over me on all fours, smelling my hair, and lightly touched my ear with her pursed lips. I would not have thought it possible to be touched so gently by a wild gorilla.

"I always sketched as rapidly as possible. Later, by lantern light in my cabin, details were incorporated into the crude sketches while the subjects were still fresh in my mind.

"Those brief visits were the fulfillment of a lifelong yearning to observe and draw gorillas in the wild, and I shall always gratefully honor Dian's generous friendship. If the mountain gorilla survives into the next century, it will be because of her lifelong dedication to understanding and rescuing this magnificent species."

"I would not have thought it possible to be touched so gently by a wild gorilla."
—Jay Matternes

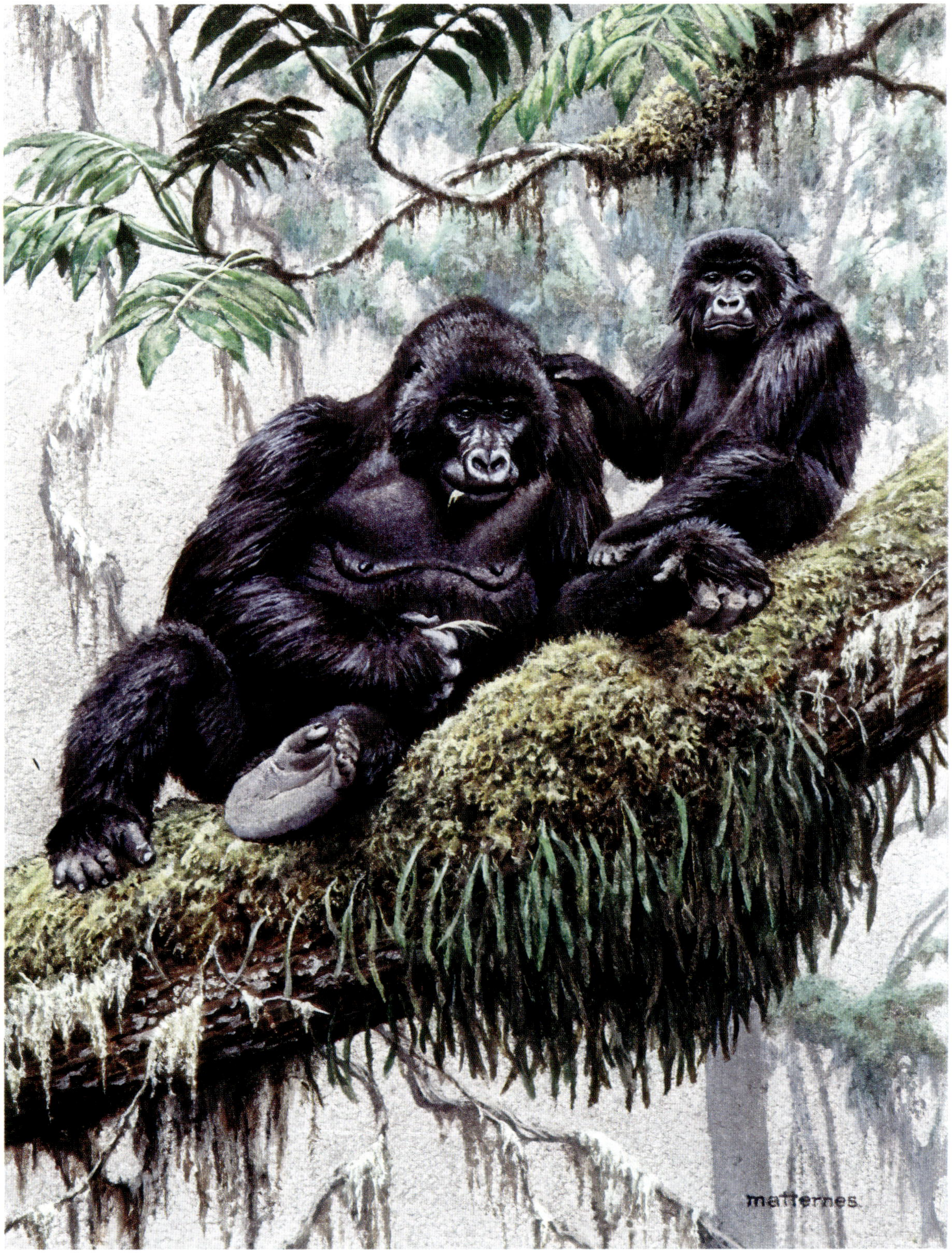

Mid-Morning Rest, watercolor on paper, 1969. Matternes painted this double portrait immediately after returning home from a brief visit to Dian Fossey's Camp Karisoke in Rwanda.

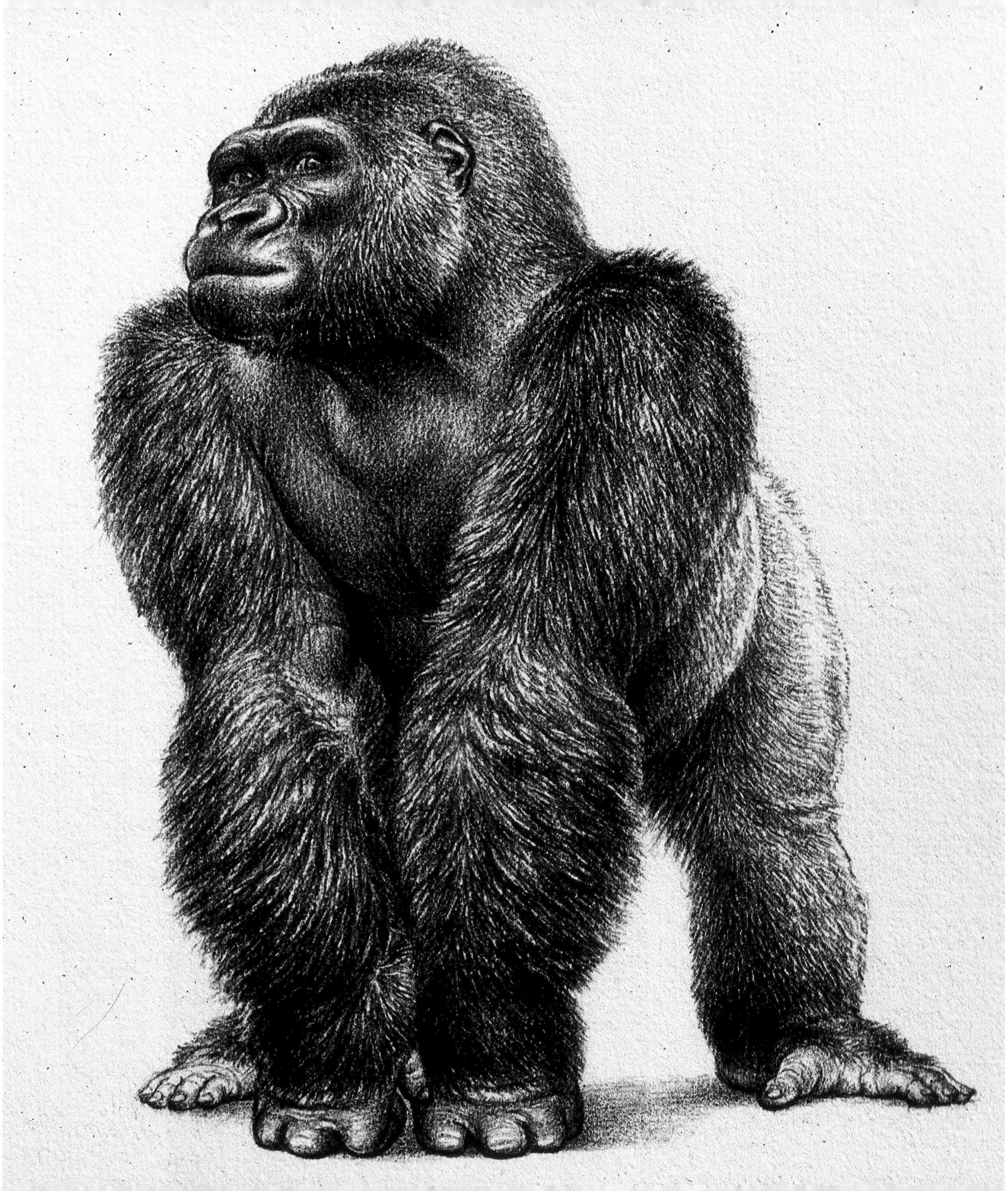

"Are You Looking at Me?" Otto, a captive lowland gorilla, in a petulant mood as he reacts to the artist studying him. With compressed lips and sideways glance, he displays annoyance and mild threat behavior. Sketched from life at Noell's Ark, a roadside menagerie in Tarpon Springs, Florida.

TWO BROOKLYN GORILLAS

This is Jay's loving portrait of Massa, the western lowland gorilla at the Philadelphia Zoo, who held the species record for confirmed longevity at the time of his death at age 54 in 1984. During the Matternes family's residence in West Philadelphia, young Jay spent every other Saturday drawing the city's zoo animals, with Massa receiving much of his attention. On alternate weekends he sketched mounted animals at the Philadelphia Academy of Natural Sciences.

Massa was born in a West African forest in 1930, captured as a small infant, and eventually bought by eccentric American socialite Gertrude Linz, who attempted to raise him "as a son" in her Brooklyn mansion. A short time later, she adopted a second juvenile gorilla called Buddy—who was to become world-famous as "Gargantua the Great" when she sold him to the Ringling Bros. and Barnum & Bailey Circus. After four years, both adolescent apes were becoming very dangerous and impossible to manage, so the Linzes sold Massa to the Philadelphia Zoo and Buddy/Gargantua to the circus.

Lurid posters proclaiming him "the world's most terrifying living creature" attracted gawking crowds credited with saving the Big Top from imminent bankruptcy. Gargantua's "performance" consisted of sitting in a rolling air-conditioned, glass-enclosed cage drawn around the arena several times by six white horses. Massa, his adopted sibling, spent his long life alone in a small cage at the zoo.

Matternes developed this 1967 painting from his zoo sketches of Massa in a quiet, contemplative mood, and showed him in an idealized West African rainforest with authentic vegetation. Jay's fascination with Massa was to foreshadow his work decades later at Dian Fossey's Karisoke Research Station in Rwanda, where he sketched her mountain gorillas in the wild. "This gorilla stirred my romantic imagination like no other in my developing years," Jay recalled later, "and I learned so much from my long hours of watching and drawing him, both directly from life, and from memory."

Massa died of a stroke in 1984, on the evening of his 54th birthday party, which Jay had attended. It featured vegetarian cake, a Dixieland band, and hundreds of well-wishers. Zoo officials permitted Matternes to dissect his cadaver and prepare the skeleton at the Smithsonian Institution, while making many notes and drawings. He got to know the 350-pound primate inside and out, and wrote years later, "Thinking back on what his solitary life must have been like in his few square yards of indoor or outdoor cages for decades, I can't help feelings of pity for the old boy."

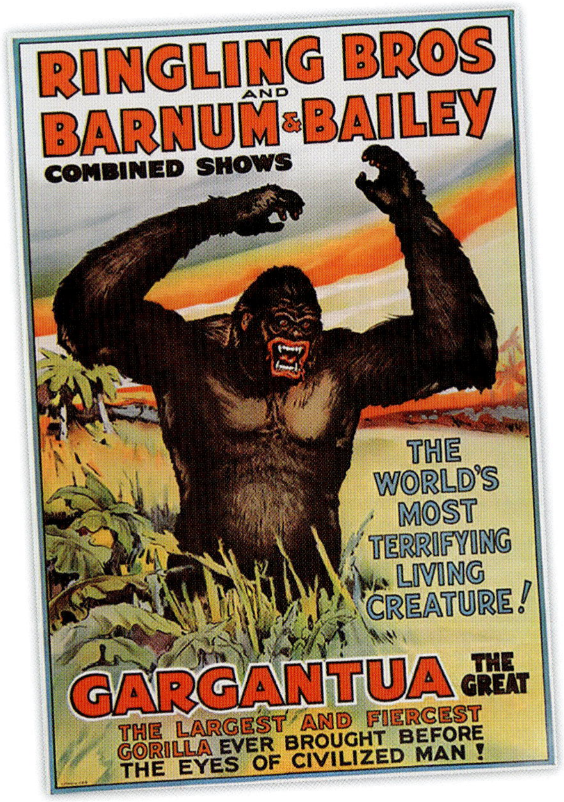

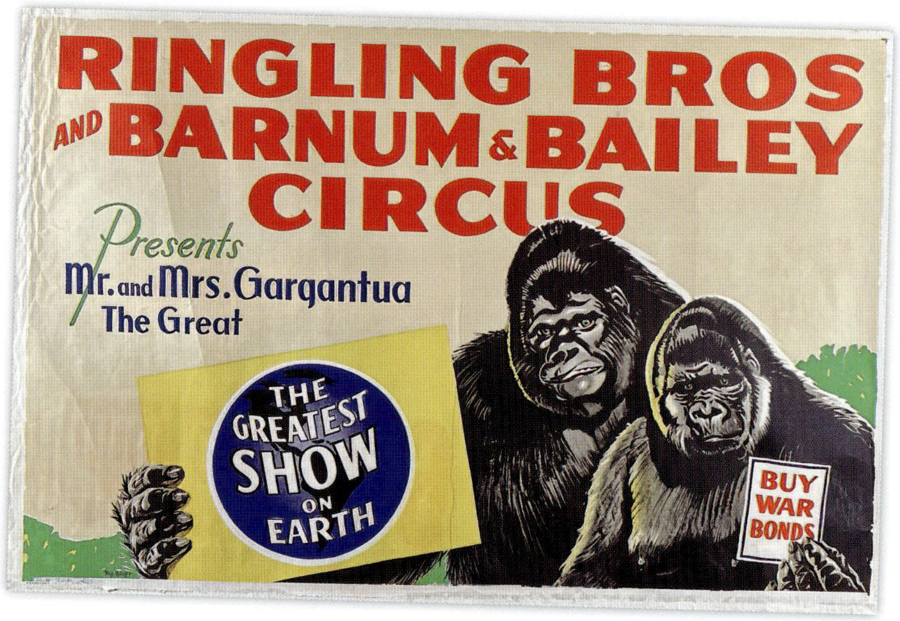

Above: Garish posters for the Ringling Brothers and Barnum & Bailey Circus (1940s) proclaimed "Gargantua, the World's Most Terrifying Living Creature," while he was actually a gentle "mama's boy" raised from infancy by an eccentric New York socialite. "Mrs. Gargantua" patriotically reminded the public to "Buy War Bonds." Opposite: *Massa*, oil on board. 1967. Based on sketches Jay made during the early 1950s at the Philadelphia Zoo.

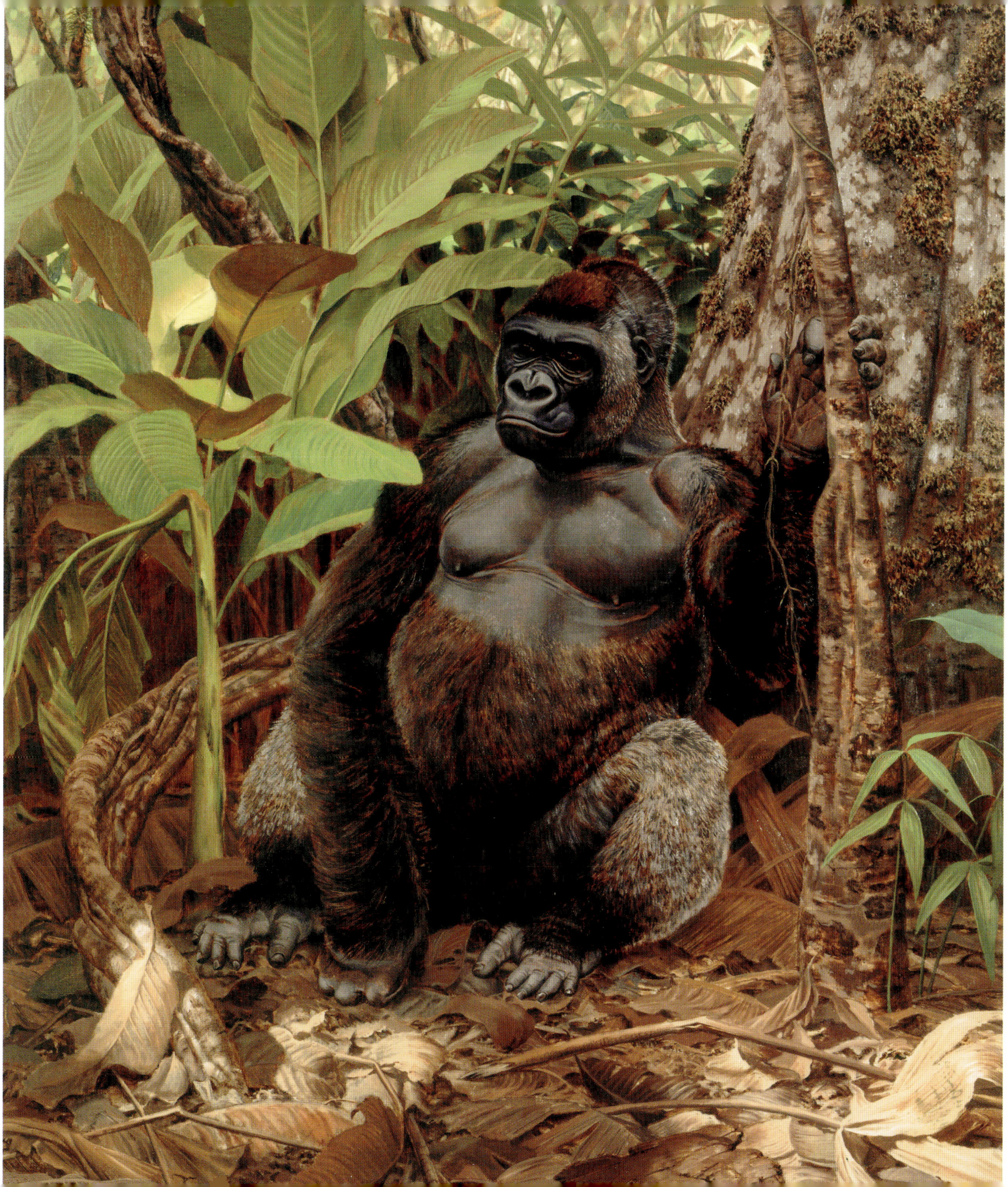

Mapping Ape Anatomy

When Massa died at the advanced age of 54, the Philadelphia Zoo granted Matternes permission to dissect the ape's body to glean additional knowledge of gorilla biology.

Although the dissection were made on the recumbent cadaver, Matternes illustrates the gorilla in an idealized upright stance, to show the pattern of fatty distribution in the body.

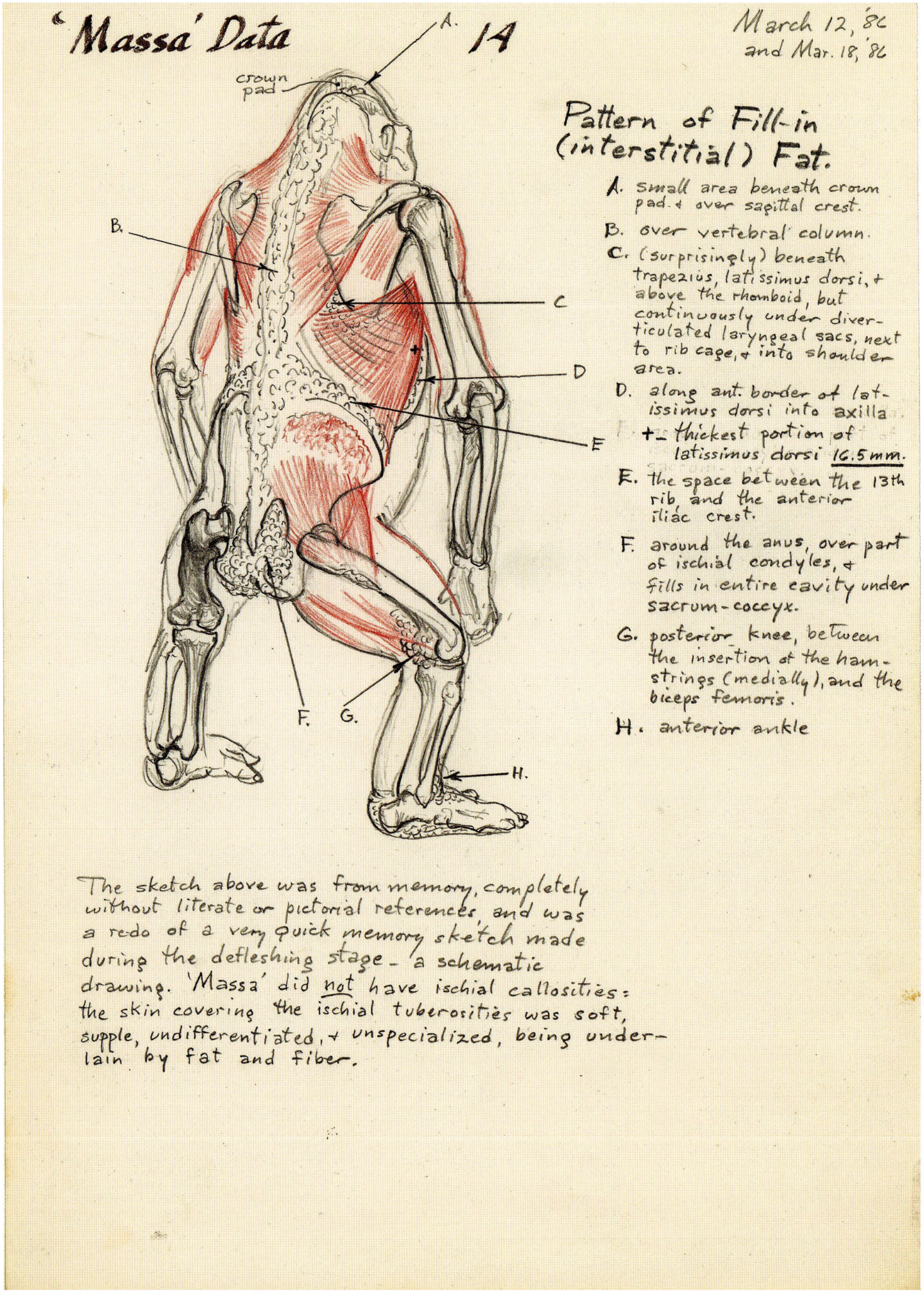

'Massa' Data 24.

June 20, '86

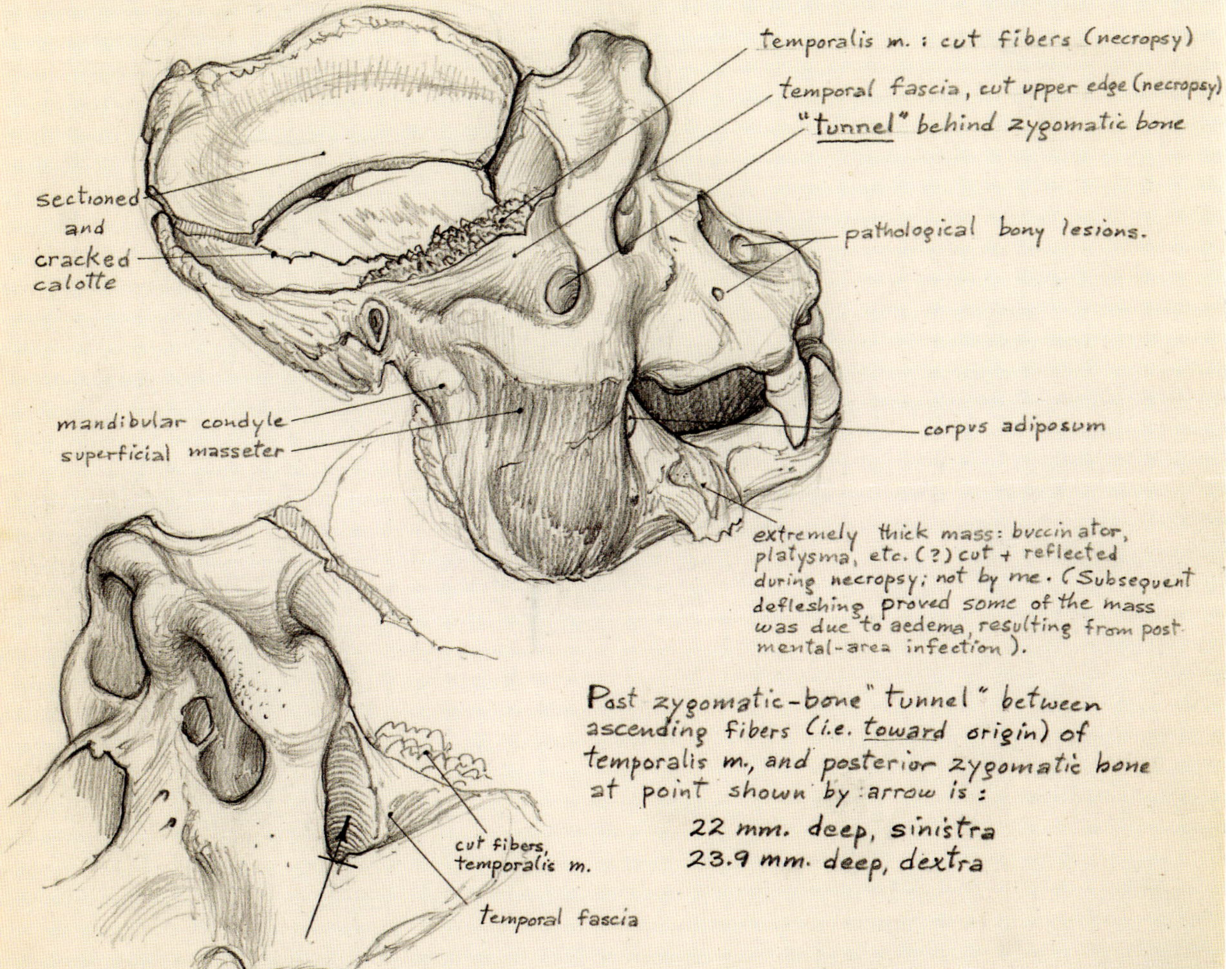

During my various reconstructions of fossil hominids (particularly sculptural ones, c.e.g. Zinj, etc.), and in the course of my human head dissections at Georgetown U. Med.-Dental School, the space within the temporal fossa between the temporalis muscle and the medial zygomatic arch was a matter of considerale interest to me. During his later, declining years, "Massa's" head appeared to be emaciated, indicating diminished muscular mass, so that the features of his skull stood out in high relief, particularly his extraordinary occipital crest, sagittal crest (of average height), and the upper margin of his widely flared zygomatic arches. Therefore, I was prepared for a wide <u>horizontal</u> space between the medial zyg. arches and the temporal muscle, but the occurance of this peculiar space behind the entire zygomatic bone, which forms an obliquely <u>vertical</u> tunnel (sketches above) as wide as 24 mm. by mm.* (latter, antero-posterior dimension) was a complete surprise. In all my primate dissections, I have seen nothing like this. Generally, in

* this measurement not taken, before defleshing.

/over

ADVENTURES WITH THE GORILLAS OF KARISOKE

In 1968, the editors at *National Geographic* hired Jay Matternes to illustrate a book on East African mammals to be written by famed anthropologist Louis Leakey. They provided him with photo equipment and sent him off to Kenya for the trip of a lifetime. His mission: to study the Leakeys' fossils of early mammals and hominins in the Kenya National Museum, and photograph trees, botanicals, and landscapes as references for his paintings. However, the trip took a fateful twist when he followed an impulse to visit the recently established Karisoke Research Station in the Rwandan forest, where a dedicated, intrepid American, Dian Fossey, had begun her groundbreaking study of the rare and at that time extremely reclusive mountain gorillas.

When he set out for Africa, Jay did not expect to meet her or see wild gorillas, or even to visit Rwanda at all. In the end, his fateful spur-of-the-moment decision to head for mountain gorilla country led to one of his most rewarding experiences, and an enduring friendship between the artist and the field biologist. One happy outcome was the first (and arguably best) sketches of wild gorillas in their forest habitat ever made. The remarkable story of their collaboration is told here in their own words from a 20-year-long correspondence and Jay's original sketchbooks, which have languished in a closet at his home in Fairfax, Virginia, for almost fifty years.

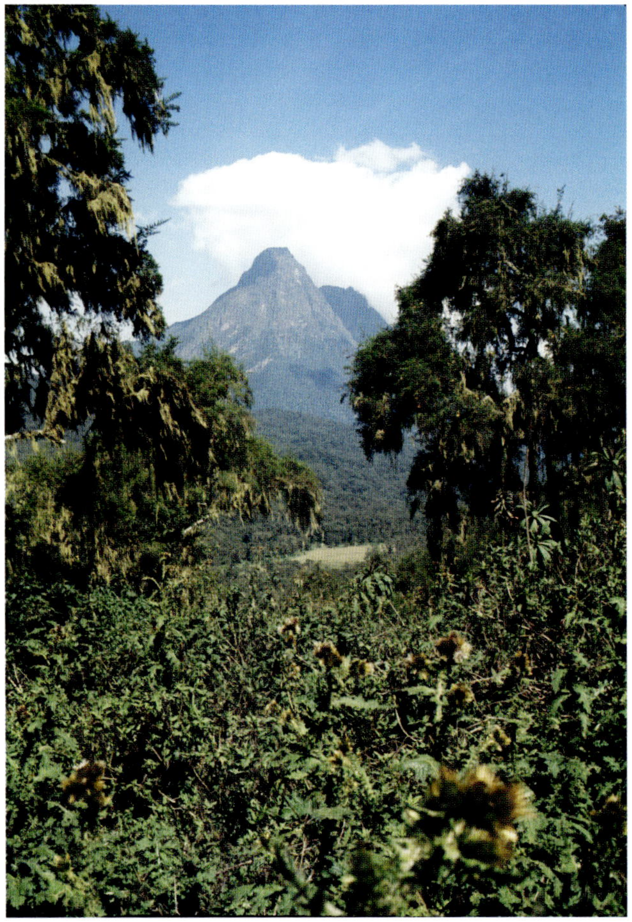

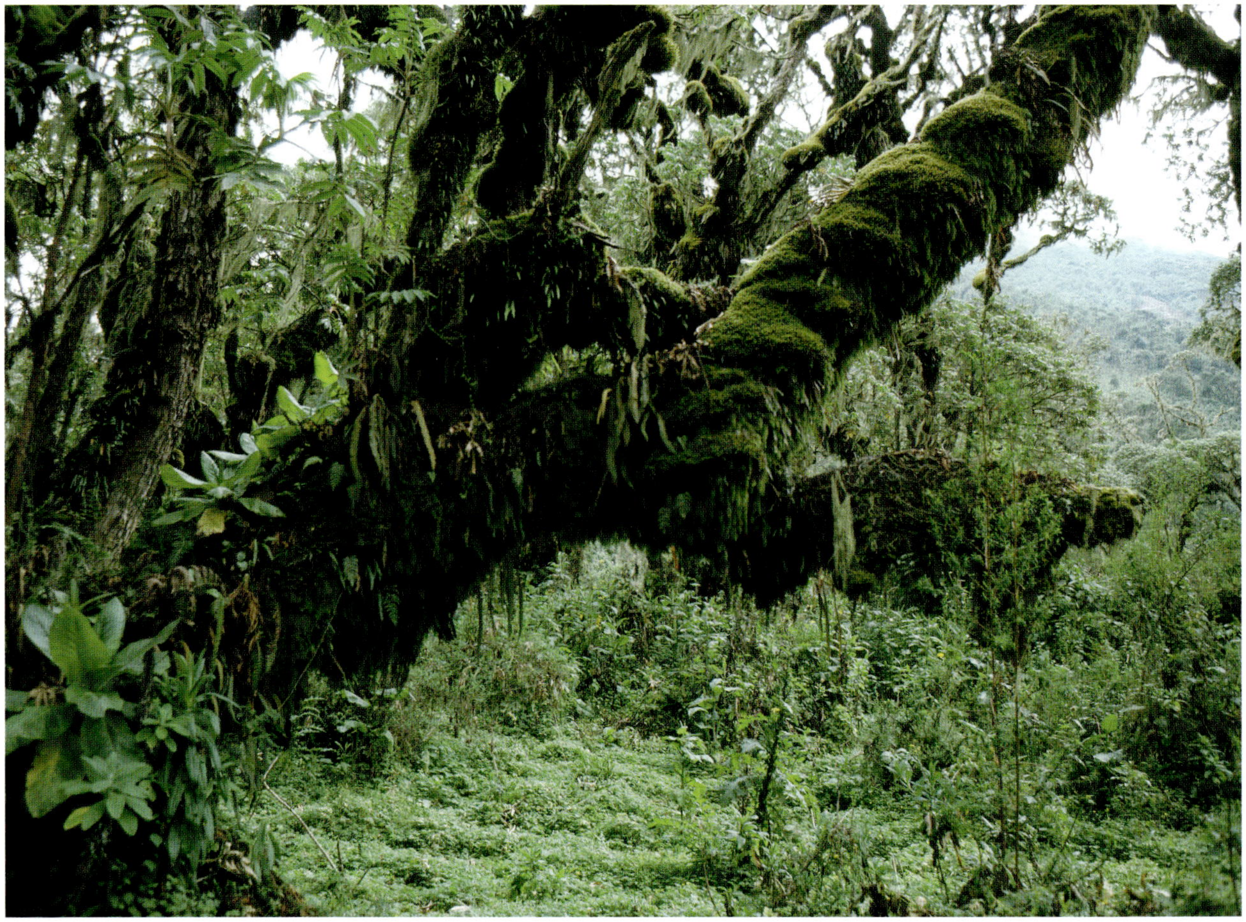

Jay's first sight of Mount Mikeno, which contains the iconic forest habitat re-created in the American Museum of Natural History's famous gorilla diorama. Later he climbed Mount Visoke with his Rwandan guide Nmeye, making their way through the *Hagenia* trees, which grow massive oblique branches.

Matternes Leaves Nairobi on Safari Arranged by Richard Leakey

12:30 Sunday night, the 12th May, 1968

My dearest Mother and Dad,

This will be my last letter from Nairobi. Richard Leakey has arranged a tour for me of the African bush [although he did not accompany Jay on this expedition]—a real Safari w/Land Rover and guide…He says this will cover every type of terrain that this country has to offer.

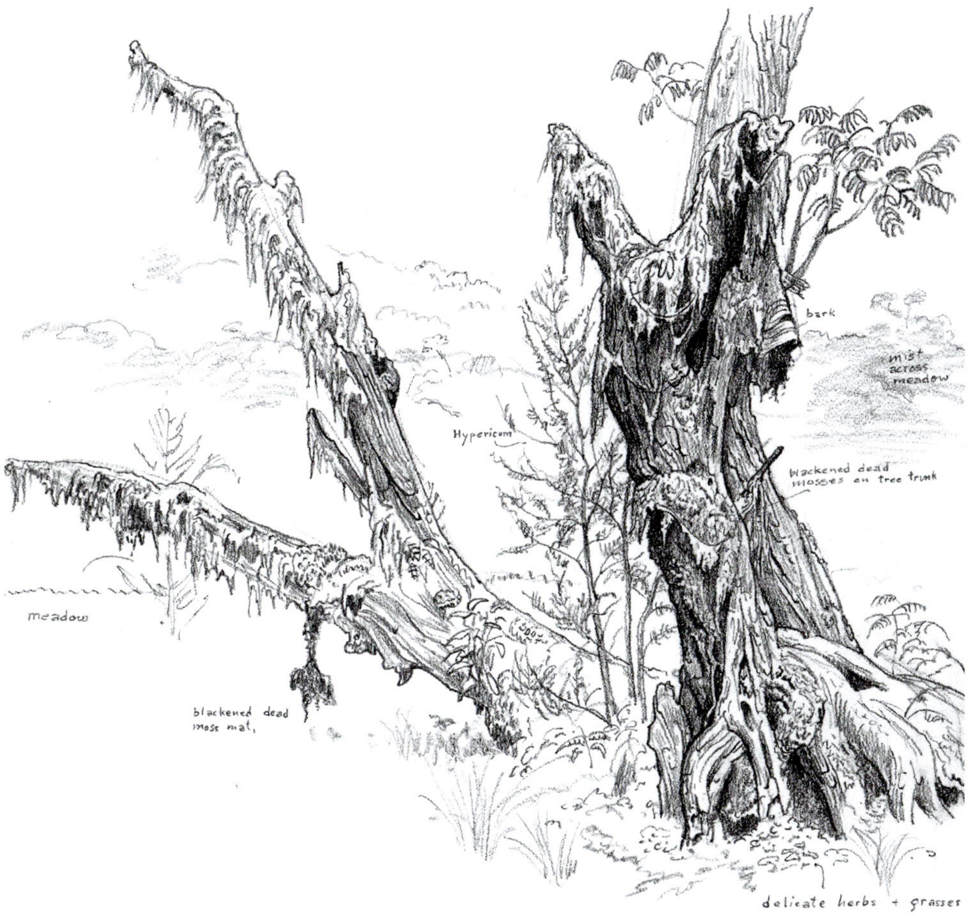

Giant *Senecio* trees, with their upward-curving leaves, and other exotic vegetation appeared to Jay "like something unworldly from a fairy tale and never to be forgotten."

First we shall go southeast to Tsavo Park and then to Mombasa on the coast. Next, we swing east into Tanzania, visiting Ngorongoro Crater, Mount Kilimanjaro, and the Serengeti Plains—then back into Kenya and the Mara Reserve, and finally, I end up in Uganda and tour the "Mountains of the Moon"—the Ruwenzori ranges, heavily forested areas, some of which contain chimpanzees, but no gorillas—I do not mean to see these apes, for this is extremely unlikely, but I want to photo-document botanical details.

This will be a full 27-day tour, and we may be camping out on some of those nights. I have been working very hard every day

> Richard Leakey arranged a month-long safari that "will cover every type of terrain that this country has to offer."

(including today) at the Kenya National Museum with Leakey's fossils, and making arrangements, etc.…had a wonderful tour of the game in Nairobi National Park last Sunday, however. You would love the climate here—never hot, never cold, just optimal…Although I have been busy here, I have also been quite lonely—I do miss Del so much. God bless you. I'll drop a letter to you from time to time while on Safari—Don't worry!

Your son, Jay

Jay's Fateful Change of Course to Include Karisoke

During his grand safari, Jay met the renowned wildlife photographers/filmmakers Alan and Joan Root at Mzima Springs, in Tsavo Park, where they were shooting a wildlife film for the BBC. Root suggested that Jay should try to visit Dian Fossey's camp, as he knew she was presently without a photographer

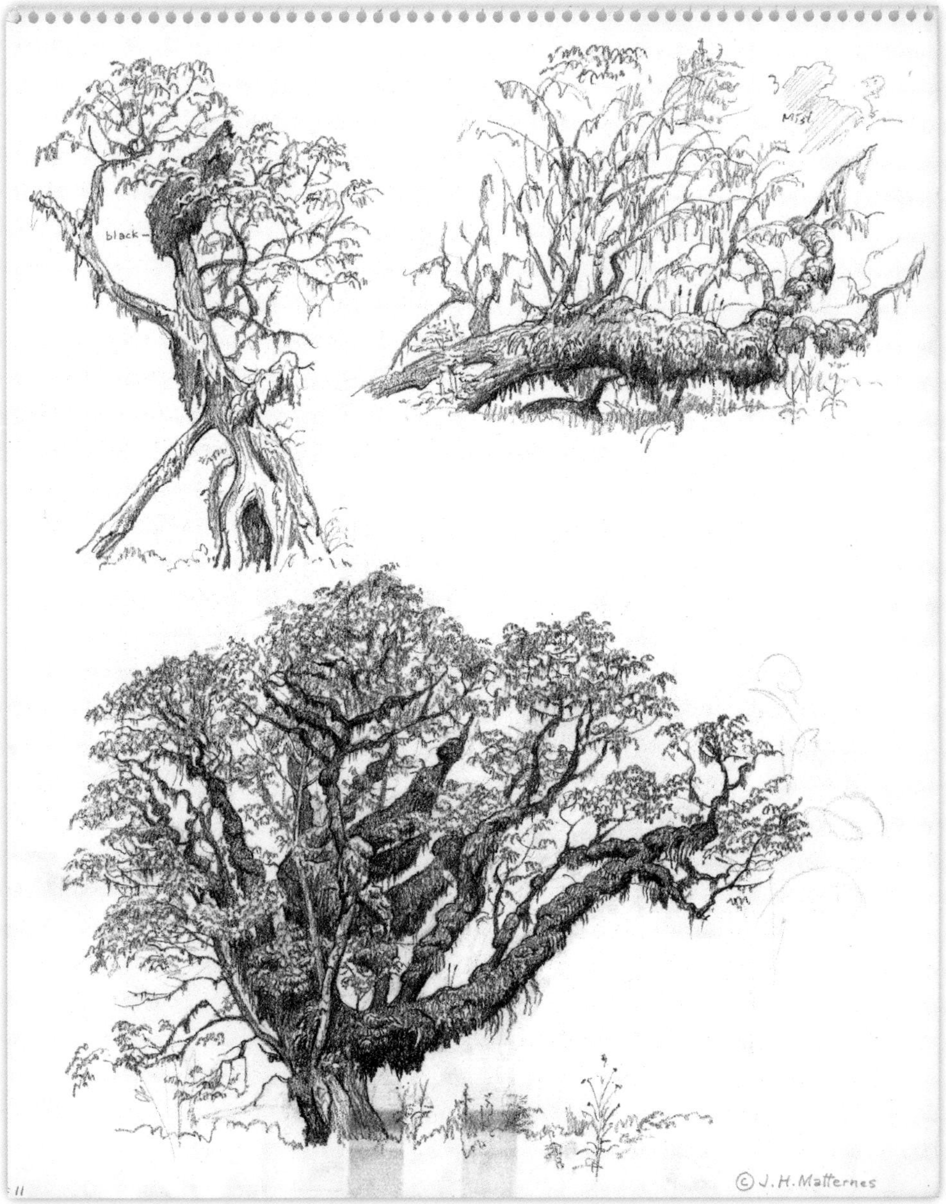

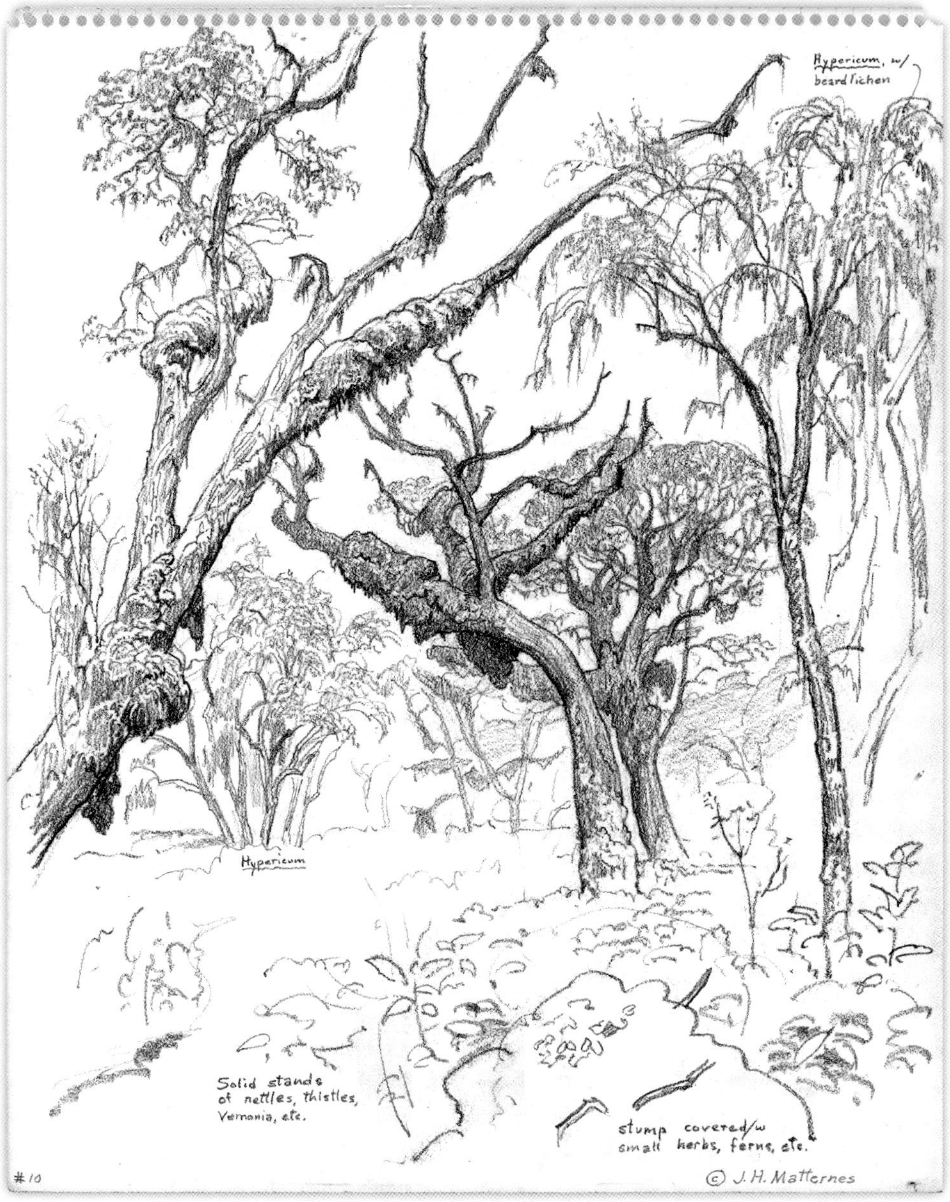

Hagenia trees and and *Hypericum* on Mount Visoke. The oblique branches of *Hagenia* are festooned with an overgrowth of mosses, algae, pendant ferns, and graybeard lichens. The forest floor contains many species of undergrowth, including ferns, blackberries, *Gallium* vines, *Lobelia*, and stinging nettles.

to document her gorillas. Jay then asked his driver to cross the border into Rwanda, which was risky since he didn't have a visa, and soon discovered that finding Fossey's base camp was not so easy. Its location was a well-guarded secret; no one seemed to know where it was or how to get there. Eventually, through a succession of introductions and interrogations by a few British and European ex-colonials who had Dian's confidence (including a planter's widow and the owner of the Traveler's Rest Hotel), he finally arrived at the mountain trail to Camp Karisoke.

However, just as he was ascending the mountain (before mobile phones enabled instant communication), Fossey had already come down for a trip to town, and they completely missed each other. So Jay and Dian never met on his first visit in 1968, and he saw no gorillas, but was captivated by the area and knew that he must come back. After Jay returned to Virginia, the two began a correspondence. It turned out that a few months later, Fossey visited Washington, DC, to report to her sponsors at National Geographic, which was near Jay and Del's home, and the three of them finally met and spent an

Jay's only snapshot of Dian with Nmeye, taking a cigarette break after a visit with Digit, one of her favorite gorillas. A heavy tobacco smoker, Fossey suffered from emphysema during her final decades.

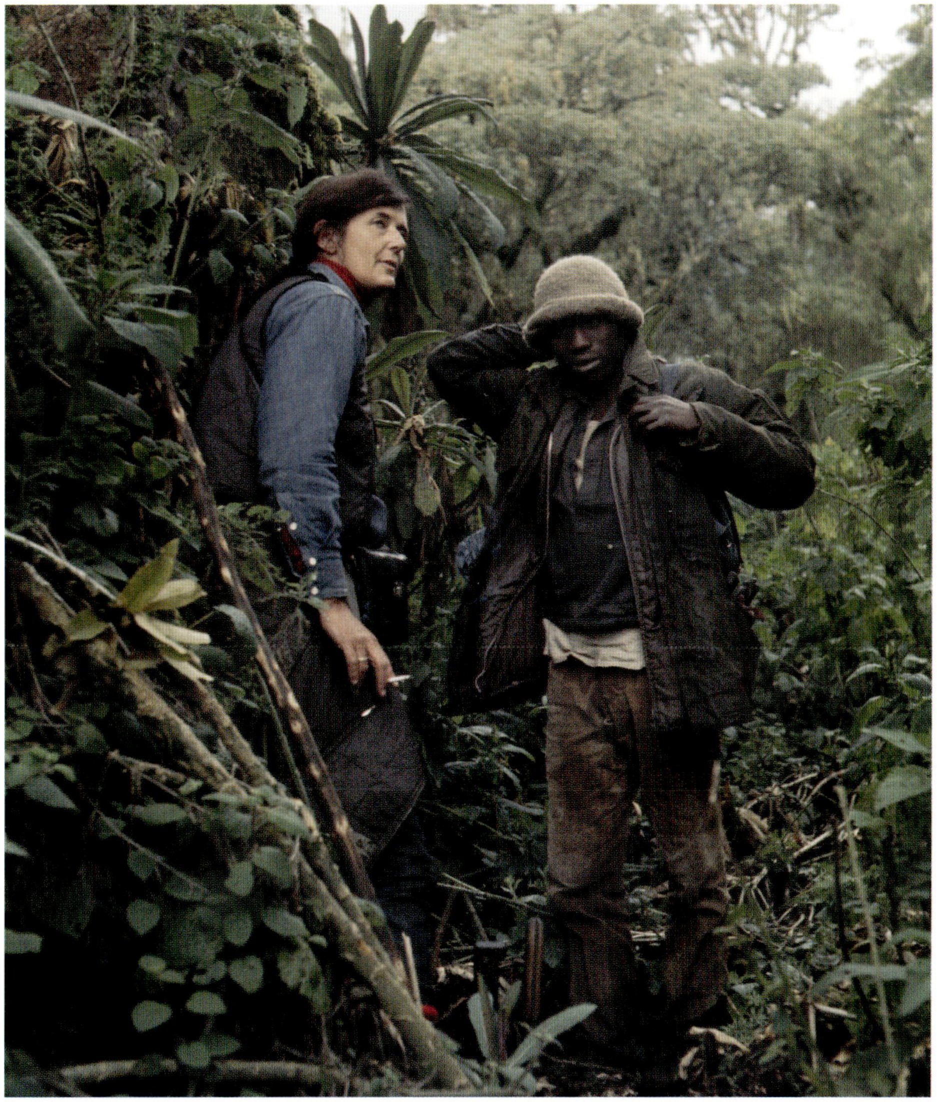

enjoyable evening getting acquainted while dining out together. That night Dian invited Jay to return to her camp in Rwanda for an extended stay, which didn't come about until six years later. Del, who was also invited, elected to stay home to care for their beloved poodles.

Dian's letters to Jay were often field reports on the status and behavior of the gorillas in her study group. A selection from these follows:

Dian's Personal War with the Poachers

In January of 1969, Jay received news that Dian was now actually warring with game poachers among local tribesmen. Fearing for her safety, he asked if they were now using firearms or other sophisticated weapons, and received this not very reassuring response.

From Dian Fossey, January 24, 1969:

> No, the poachers are still in the spear and bow-and-arrow class, and this brings up my latest progress report...On Jan. 7th the Bambuti poachers kidnaped my dog "to get even", but not knowing if a Bambuti or Watutsi (the cattle people) took her, I kidnapped 8 Watutsi cattle; built a string and pole corral in 2 hours that was strong enough to hold them; then issued the proclamation that a cow a day would be killed until my dog was returned. This caused me some concern, as, with my usual lack of forethought, I'd sealed the entrance shut with tin sheets and nails and had visions of the bodies piling up as the days went on. Utilizing proper psychological warfare techniques I made one man of mine yell this "proclamation" in the general direction of the Watutsi camp once an hour around the clock all night long and sat up all night to guard the cattle—it was easy to stay awake as 24 elephant [*sic*] came to camp that night! The next morning at 5 o'clock a Watutsi came to camp—he was the same one whose milk gourds I'd stolen some 10 days previously by frightening him out of his wits with one of the Halloween masks and offered to lead my men to the hidden Bambuti camp where the dog was being held if, of course, I wouldn't kill his cattle. We never would have found this camp for it was high on Karisimbi and nearly on the other side. For their help, I've signed a temporary peace pact with the Watutsi, but open war is declared on the Bambuti!

Field Notes on Gorilla Behavior, 1971

From Dian Fossey, November 23, 1971:

> The work continues to do so well with the acceptance of the animals of Group 4 superseding my wildest expectations. A young blackback of that group, Digit, allows me to scratch him and he in turn leans on us and plucks at our clothing. I've shown him a mirror which fascinates him and he preens like a teen-ager primping for a prom—twisting his head from side to side, sometimes with semi-pursed lips. Twice he has reached behind it and "felt" for the handsome devil returning his stare! Digit is the only adult left in this group now...He no longer has any peers which is one reason I think he is suddenly turning to us (he used to be somewhat reclusive) and it is also possible that we are affecting his dominance status within the group, but this will take some time to verify.

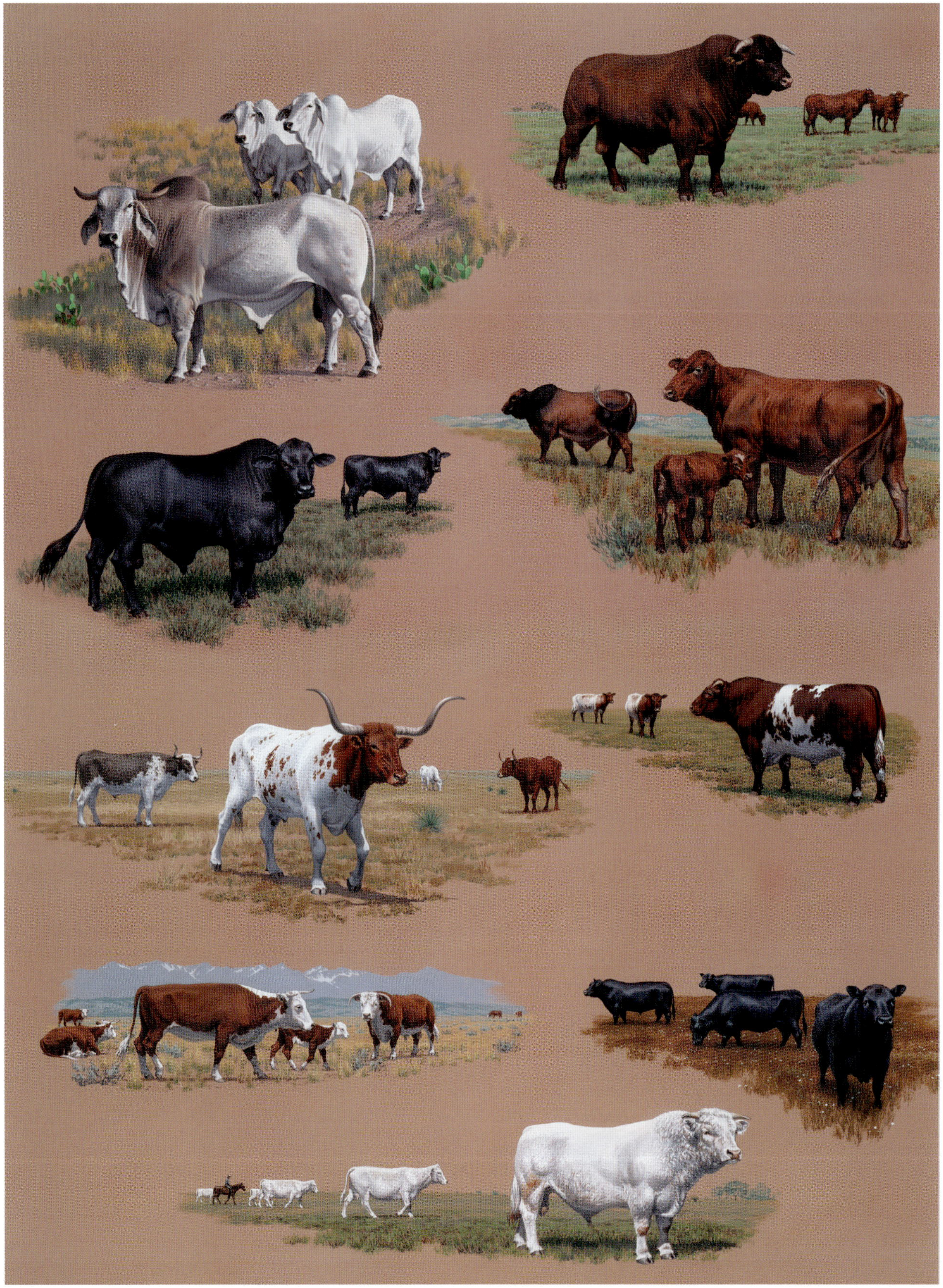

Matternes' paintings of cattle breeds in National Geo's *The American Cowboys* (1972) fascinated Fossey's African crew, who found them the most interesting pictures in the book. First in North America were the hardy Longhorns, top, brought by early Spanish explorers. Later settlers imported more tractable breeds from the British Isles, including the Shorthorn, White-faced Hereford, Black Aberdeen, Santa Gertrudis, and the humped Brahman or Zebu from India.

Since we can now literally travel, "feed", and day-nest amongst the group members, dung eating has been more frequently seen. (Bob has cine of the silver-back Uncle Bert eating away and appearing to "relish" it!) The orphan juvenile, Simba, born in April of 1968, is still sleeping and traveling with [him] and although the other juveniles can play next to us and touch us, Simba is carefully guarded by Uncle Bert, who gets most upset when he is in proximity to us. Both Bob and myself feel that the day is coming when U.B. is going to want to assert his dominance over us by a whack or two—I don't think he will mean much harm by it—it would only be to put us in our places so to speak...

Field Notes on Gorilla Behavior, 1972

From Dian Fossey, March 22, 1972:
Digit at times appears jealous when in play and proximity with us, and objects to other youngsters trying to join in. I don't think he's trying to protect them from our presence, though this does occur sometimes when the group is intent upon traveling. It is so comical to watch Uncle Bert and some of the females as they sit back observing our play with Digit; they seem to wear an expression of absolute disdain. I must say, though that Digit STINKS from all ends...Wow! When he flings his arm out on a warm day and splits one out at the same time—Ugh!—It's almost too much to bear. His breath is equally as repulsive, especially when he has his nose buried in your hair. When are you going to pay a visit and get your share of Digit's joy juice!!!????

Your cattle art is "much admired. . . . You know how these people are about cows!"

—Dian Fossey

From Dian Fossey, May 30, 1972:

It still continues to pour rain day in, day out. I really think it's a good thing that Bob [Campbell, the National Geographic photographer] left when he did. It's impossible to try to get any work done in this weather, and it's extremely depressing. . . . By the way, I'm sure in the market for a husband. If you know of anyone slightly over 40, slightly over 6 feet tall, who loves the outdoors and animals and can hold a camera, please send him on collect. I'm quite serious, believe it or not. . . To make matters worse, during our last three days together, Bob and I had four guests—bigwigs—thus six of us went out to see Group 4, and it was terrible. The group was really terrified of so many people and acted like completely unhabituated animals. I was so disappointed. Bob and I were able to get over to the Saddle Area between Muhavura [volcanic peak] and Gahinga and had the good fortune to find a wild group who really charged us. We held our ground during the roaring and screaming charge, which puzzled the lead silverback, who ended up giving rather frustrated ground thumps and foliage whacks. We then advanced a bit and sat down to make contented gorilla noises, which brought the whole group—seven that we saw—forward as though hypnotized. They were so very curious and ended up sitting around like babies, feeding, even though somewhat obscured in the bamboo. It was very exciting.

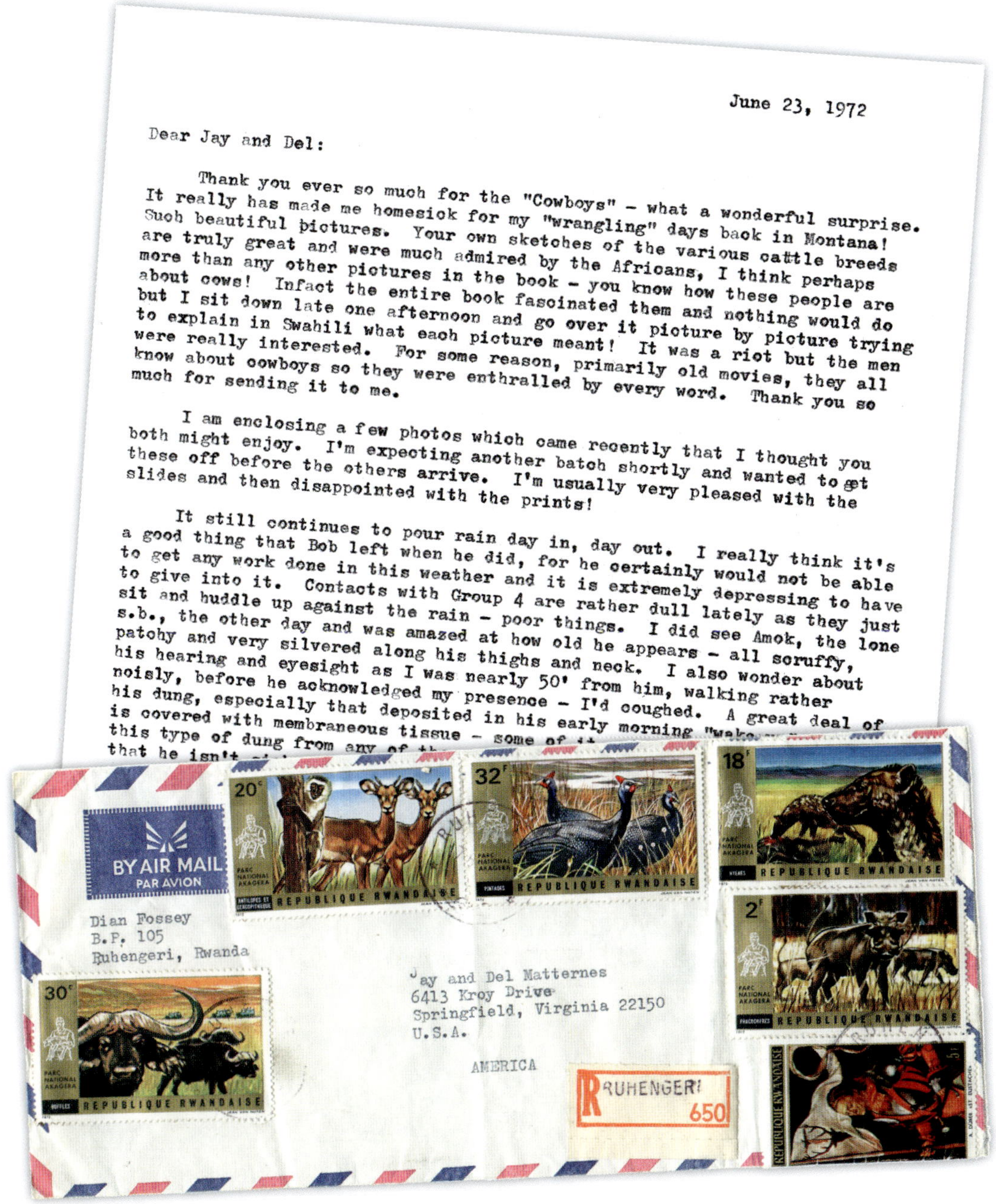

June 23, 1972

Dear Jay and Del:

Thank you ever so much for the "Cowboys" - what a wonderful surprise. It really has made me homesick for my "wrangling" days back in Montana! Such beautiful pictures. Your own sketches of the various cattle breeds are truly great and were much admired by the Africans, I think perhaps more than any other pictures in the book - you know how these people are about cows! Infact the entire book fascinated them and nothing would do but I sit down late one afternoon and go over it picture by picture trying to explain in Swahili what each picture meant! It was a riot but the men were really interested. For some reason, primarily old movies, they all know about cowboys so they were enthralled by every word. Thank you so much for sending it to me.

I am enclosing a few photos which came recently that I thought you both might enjoy. I'm expecting another batch shortly and wanted to get these off before the others arrive. I'm usually very pleased with the slides and then disappointed with the prints!

It still continues to pour rain day in, day out. I really think it's a good thing that Bob left when he did, for he certainly would not be able to get any work done in this weather and it is extremely depressing to have to give into it. Contacts with Group 4 are rather dull lately as they just sit and huddle up against the rain - poor things. I did see Amok, the lone s.b., the other day and was amazed at how old he appears - all scruffy, patchy and very silvered along his thighs and neck. I also wonder about his hearing and eyesight as I was nearly 50' from him, walking rather noisly, before he acknowledged my presence - I'd coughed. A great deal of his dung, especially that deposited in his early morning "wake... is covered with membraneous tissue - some of ... this type of dung from any of th... that he isn't ...

Rwanda's colorful wildlife stamps adorned Dian Fossey's chatty letters to Jay, keeping him updated on the drama of shifting gorilla relationships.

From Dian Fossey, June 23, 1972:

Thank you ever so much for [sending the National Geographic book] *The Cowboys*. What a wonderful surprise. It really has made me homesick for my "wrangling" days in Montana. Such beautiful pictures. Your own sketches of the various cattle breeds were much admired by the Africans, I think perhaps more than any other pictures in the book. You know how these people are about cows! In fact, the entire book fascinated them, and nothing would do but I had to sit down with them late one afternoon and go over it picture by picture, trying to explain in Swahili what each picture meant. It was a riot. But the men were really interested. For some reason, primarily old movies, they all know about cowboys, so they were enthralled by every word.

A juvenile called Tiger performs a chest-beating display for Matternes.

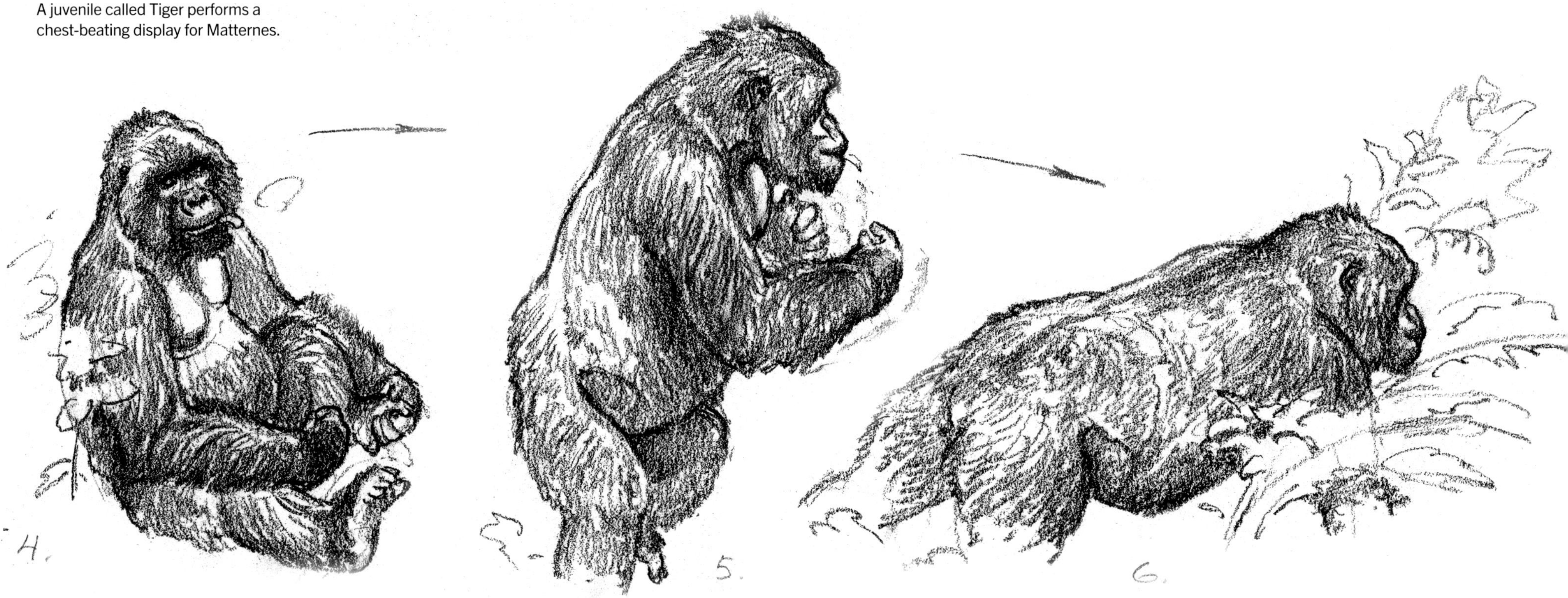

Jay's Artistic Encounters with Wild Gorillas at Karisoke

"These sketches were made during a week and a half's stay in mid-September, 1974, at Camp Karisoke, Rwanda, as Dian's guest. I had just returned from a trip to Lake Turkana, Kenya, having participated in the making of a television documentary (*Bones of Contention*, 1976) on Richard Leakey's fossil researches being produced by the UK-based Survival Anglia Ltd. [See *Hippo Feast at Ileret*, pp. 142–45.]

"By previous arrangement, Dian and I met in Nairobi, where she had just returned from Britain to have a broken fibula re-set, and she was not fully mobile. We flew by *petit avion* to Ruhengeri, Rwanda, and made the arduous trek with porters up to Dian's 10,000+ ft. base camp. I was given my own cabin and privy, and each day I accompanied one of Dian's visiting students and an African tracker/guide to make contact with one of two gorilla groups that had become tolerant of human visitors. Dian had previously coached me on proper contact behavior (e.g., avoiding standing upright, or direct eye contact, sudden movements, loud noises, etc.) in order to reassure the gorillas of our friendly intentions.

"We would gingerly settle down at the best viewing point we could find at the fringes of the group, and remain there, peering over the intervening vegetation, as the gorillas would raise their heads periodically to check on us. It was very important never to surprise them, which might cause them to move off, or possibly provoke aggression.

"Often the youngsters would improve our view by flattening some of the undergrowth in their boisterous play with each other, revealing other reclining or feeding group members that were previously unseen. I sometimes had the impression they were showing off for our benefit, since they would often glance in our direction.

"On the first day of my contact with Group 5, an adolescent female called by the researchers 'Pantsy' (all individual study subjects had a moniker) approached me, noting that I was a new face, and stood over me on all fours, smelling my hair, and even lightly touched my ear with her lips; after having satisfied her curiosity, she moved back to the group. I would not have thought it possible to be touched so gently by a wild gorilla. She was so close I could almost sense her body heat.

"The sketches were made on an 11" by 13½" spiral-bound artist's sketch pad which I carried in a backpack. I sketched as rapidly as possible, since my subjects were likely to move about, and there was so much going on within the group, I was afraid that if I looked at the paper for more than a few seconds, I would miss some significant interaction, positional behavior, or behavioral sequence. After each day's contact, by lantern light in my cabin, I would refine or

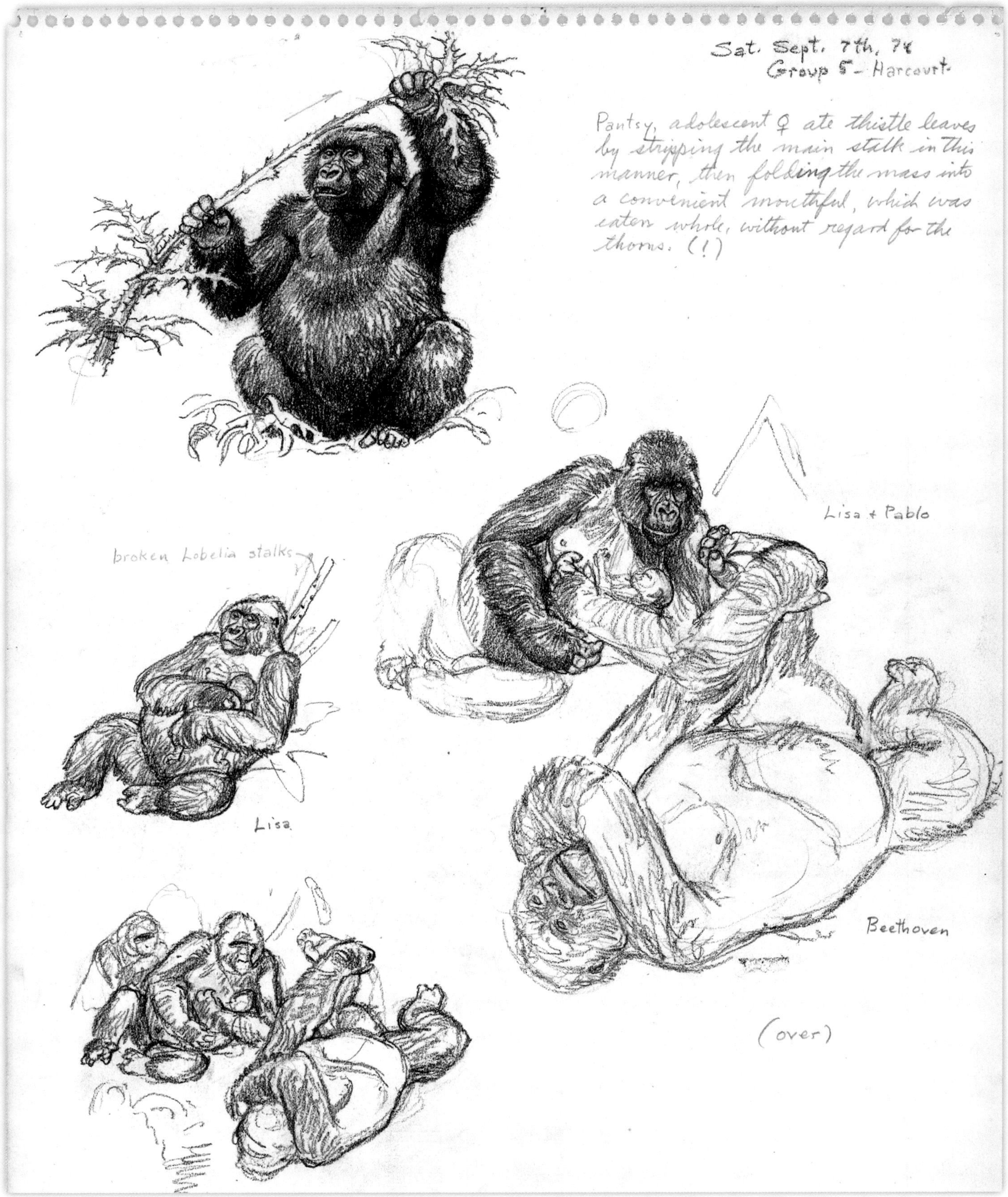

Free-living mountain gorillas sketched during Jay's 1974 visit to Karisoke, six years after his initial visit there.

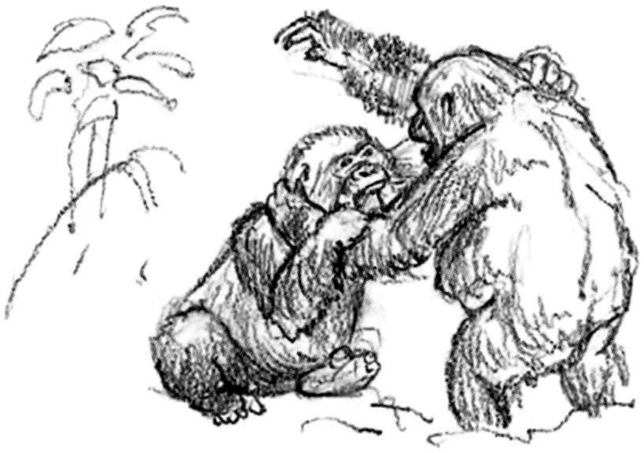

Two rambunctious juveniles play-fight, attracting attention of a silverback male, who decides not to intervene and walks away to feed on vegetation.

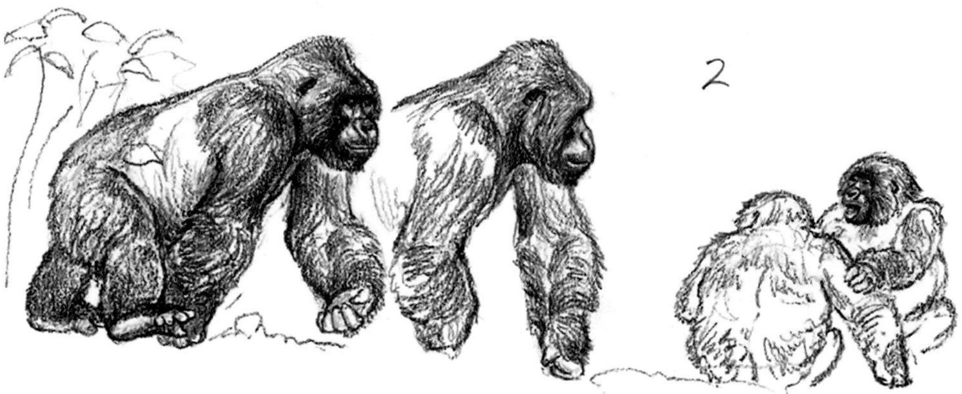

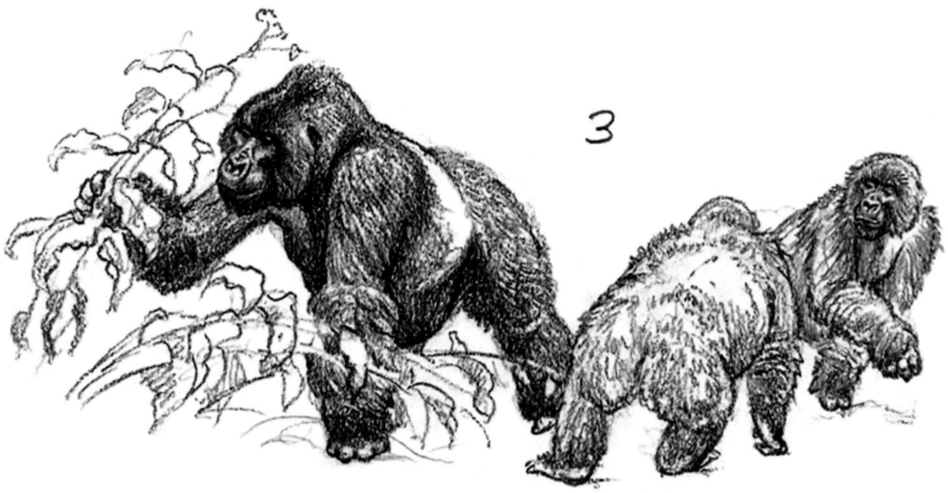

add details to the crude sketches. Some of them were not even made in the field, but were memory sketches made later in my cabin while they were still fresh.

"One day, I did not go out to view the gorillas, but at Dian's suggestion, with my guide Nmeye, I climbed the volcanic peak Mt. Visoke and descended down to the shores of the beautiful lake in its crater, an experience so memorable that it would have justified the entire trip, even if there had been no gorillas. Some years before, I had spent two weeks in the Paramos, the above-timberline tropical alpine zone of the Andes of Colombia, at about the same 11,000–12,000 ft. elevation, and I was struck by the similarities of the dominant vegetation of these frigid regions of two continents.

"Mountain gorillas came within a hair's breadth of extinction within less than two hundred years after they became known to poachers and Western trophy hunters. Happily, their numbers are slowly rebounding, thanks largely to Dian's lifelong dedication to promoting understanding and protection for this remarkable living relic of our shared primate past."

Dian's Secret Daring "Experiment"

From Dian Fossey, July 25, 1972:

> I have tried out another "experiment" with Group 4—and this is to remain please private information. For several years I have wondered how they would react to a naked human form. When the weather improved, a few weeks ago, and I made contact with them within 15' with a minimum of thistles and nettles, I took off my clothes very slowly and "self-groomed" while taking off several layers of thermo underwear, etc. Needless to say I was completely alone. By the time I'd peeled off everything the entire group was all bright eyed and "bushy-tailed." The orphan juvenile, Simba, came directly up to me without any hesitation to try to nurse—as a rule she is very cautious about her proximity to humans. She pinched my breasts and then lay down on top of me to suck, then immediately retreated to smell her fingers before advancing again for more sucking.
>
> She was frightened away by the advance of all of the animals who wanted to pinch and smell my skin. Uncle Bert, the silverback, came within two feet to pinch and smell only and this type of behaviour is quite unusual for him. Of interest to me is the fact that the orphan, Simba, recognized a maternal structure and lost her fear or apprehension of "humans" so that she allowed me to cuddle her, etc. as her own

An adolescent called Tiger munches on a stinging nettle plant, a favorite food. He rolls up a stalk and gobbles it up quickly, showing no concern for the multiple sharp thorns.

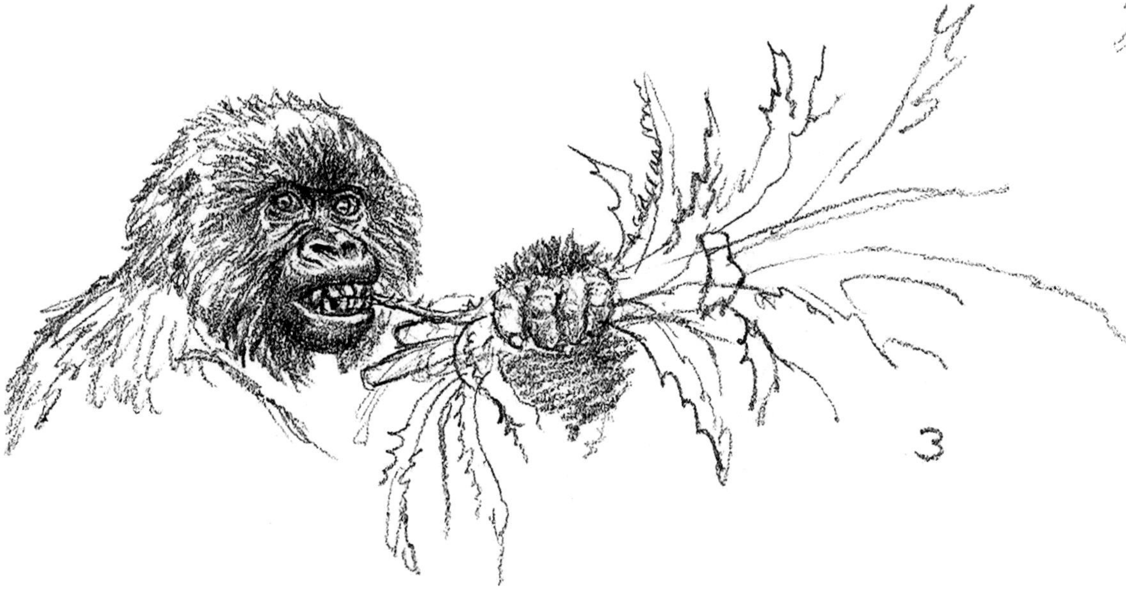

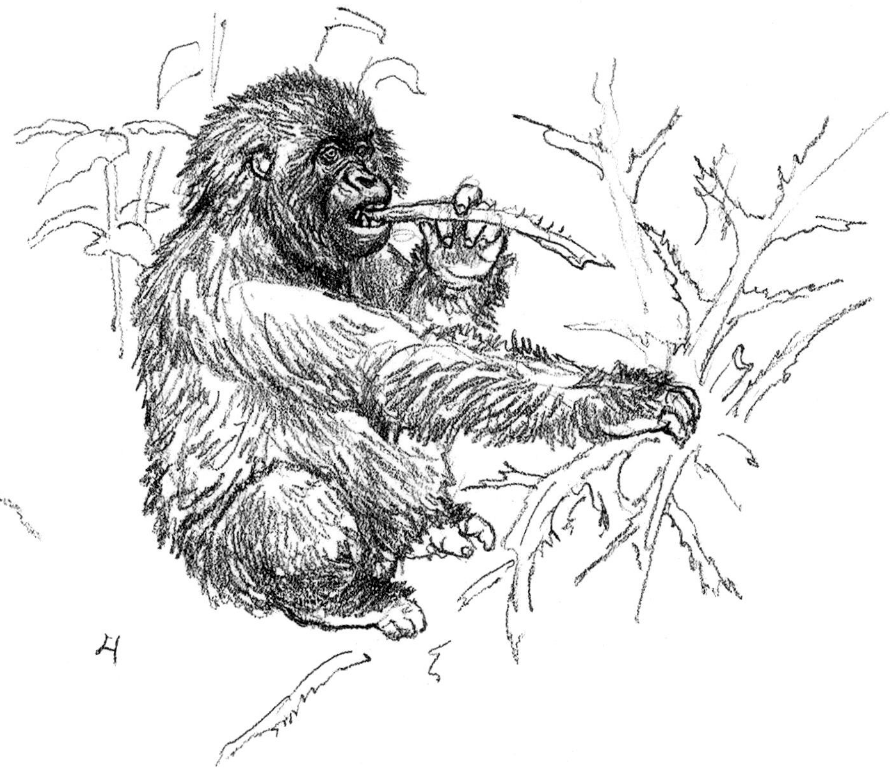

mother, Mrs. X, might still be doing (Mrs. X disappeared last June). The other animals only wanted to stare or pinch my skin.

Unfortunately my next contact with the group was in the presence of one of the students so I couldn't very well strip down. Of interest though is the fact that again Simba came up to my side and attempted to pull up my shirt without hesitation, but after a few seconds she retreated. I would very much like to try this "experiment" a few more times when alone, however as I do not want my animals "people" or object orientated, I will not continue such so as not to interfere with their normal behaviour. I doubt very much that I will try to explain this to *National Geographic* for the time being as it might seem a bit strange to them and obviously they will not publish anything like this despite its scientific interest. Another reason that I feel it is not good to repeat this type of "experiment" is the fact that I might convey germs to the animals via this close proximity.

Reply from Jay, no date, 1972:

Del and I were fascinated with your disrobing experiment, and the extraordinary reaction of Group 4. No doubt, they regard you as a very strange gorilla who can take off your "furry coat," and who has white, hairless skin underneath. Of equal interest, I'm sure, is the fact that you lack the heavy pungent smell so characteristic of themselves. We were especially amused trying to imagine their amazed expressions as they sat around you, staring, or coming forward to feel and pinch you (Lord, that must have been painful, as their skin is rough, and their nails are so hard and strong—all endured in the name of science!) I agree that this all has considerable scientific interest, although obviously it cannot be made public at this time…

From Dian Fossey, June 13, 1972:

Sorry that [a particular scientist] did not visit camp. "Perhaps he was warned away…or has heard that I am a very 'irascible' person. For sure I am that when it comes to hordes of tourists or hippy kids who want free lodging and think this is the place to get it. When it comes to scientists and people such as yourself whose knowledge, interest, & enthusiasm about gorillas matches my own, I am only too glad to have them as guests. So, you see, you and Del will just have to pack up your rain clothes & your palette and get up here!

From Jay, August 4, 1972:

What is this about your "wrangling days in Montana?" [A prize-winning equestrian as a teenager, young Dian spent a summer working at a horse ranch in Montana. —RM]

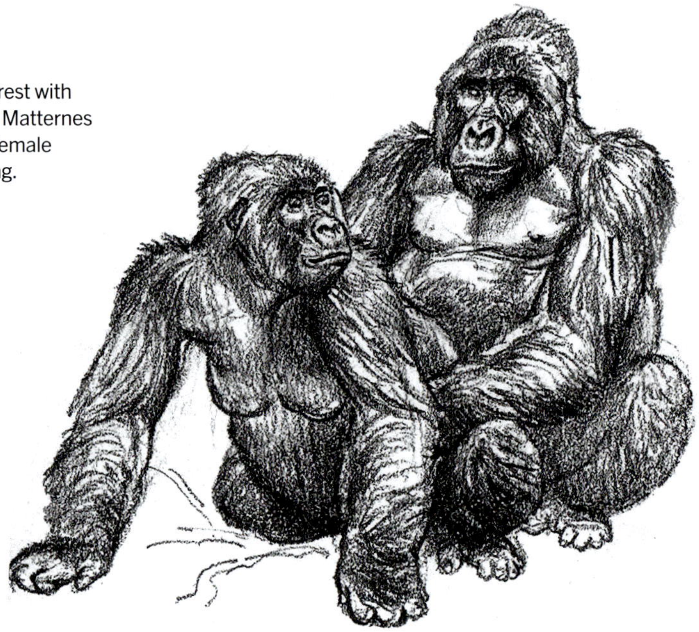

During a walk through the forest with grad student Kelley Stewart, Matternes spotted Uncle Bert and the female known as Old Goat copulating.

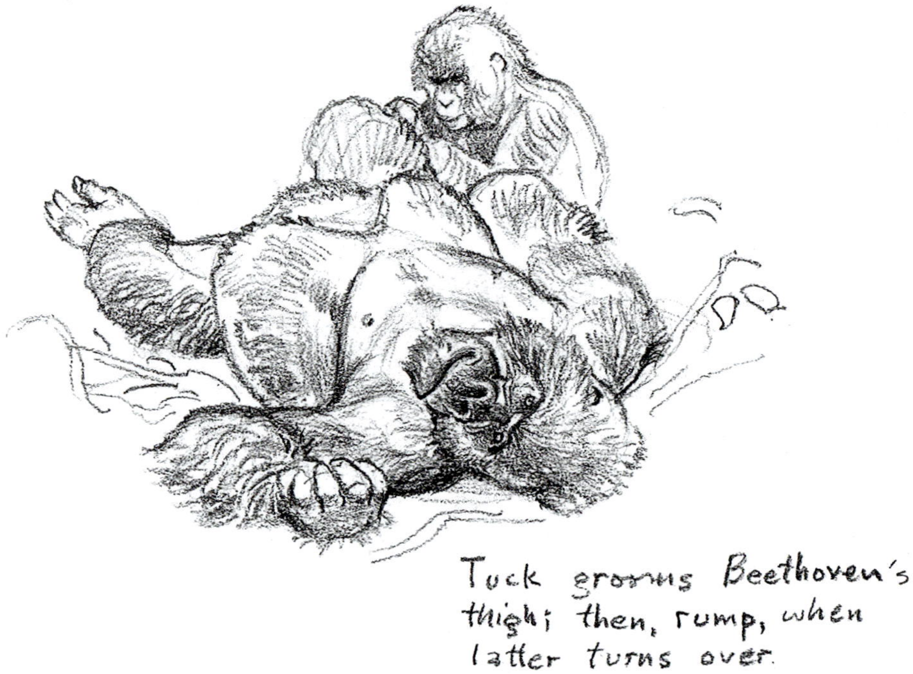

From Dian Fossey, August 8, 1972:

Mutilated gorilla body found out killed by poachers for "black magic" remedies—toes hacked off, ear cut off—distal tarsals of left foot, scrotum and testicles have been cut out; the tongue has been cut out.

From Dian Fossey, March 30, 1973:

Have you seen the April issue of the *National Geographic Magazine*?... There's a blip with 2 photos of a gal and her husband which the National Geographic Society is supporting in a year's study of the behavior of orangs in Borneo—the woman, one Birute Galdikas Brindamour if you please, was recommended by the late Dr Louis Leakey, according to the spiel. Now, actually, Dr. Leakey told me of this about 2 years ago (he volunteered the information, because he knew of my interest in apes), but he didn't know at that time whether this woman he had recommended would make her study in Borneo or in Sumatra, when I asked him. He said that she wouldn't make her observations from the ground like other previous observers, but would actually get up into the trees to observe her subjects from their own level. When I asked him what qualification his candidate had for the work, he replied "she has magnificent deltoids!" (I am not kidding, that's exactly what he said!) and that Birute would be able to get up into the trees to study the orangs at close range, etc. If Birute ever reads that, she'll be greatly amused, and at the time that Leakey told me this, I considered it extremely fanciful, which of course, it was. [It turned out that Galdikas found it impossible to follow orangs through their arboreal travels, and had to make most of her observations looking up from the treacherous, swampy ground. —RM]

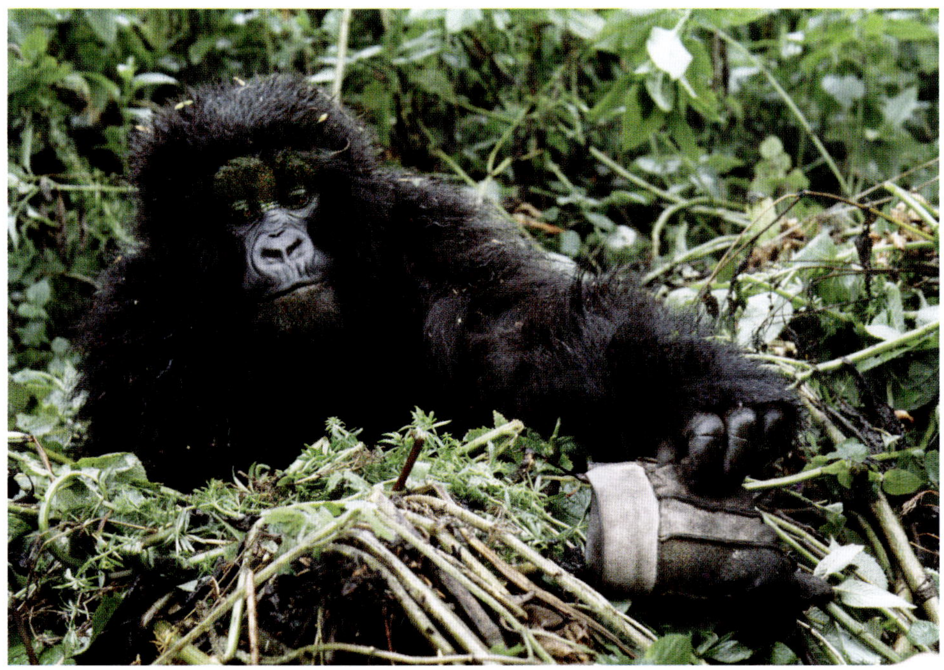

Jay's only gorilla photo was this snap of Tiger, taken during his first visit to Karisoke in the 1970s, when the youngster was four or five years old.

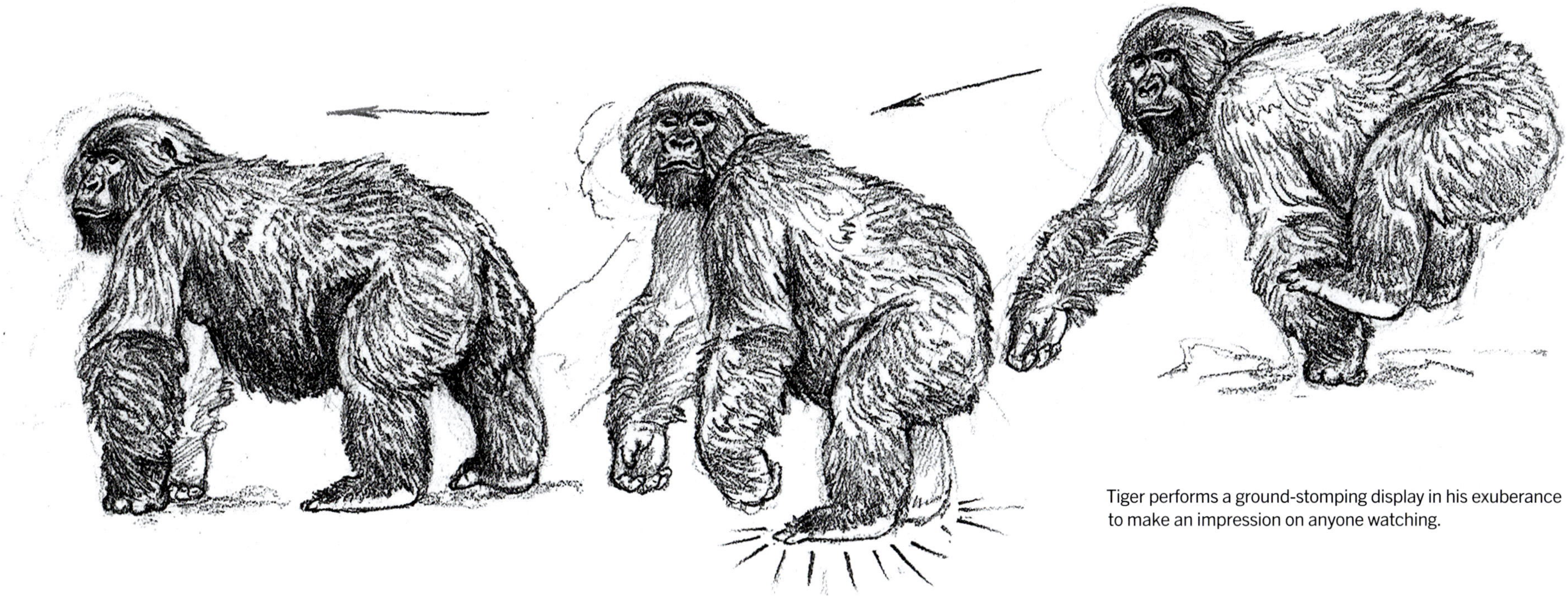

Tiger performs a ground-stomping display in his exuberance to make an impression on anyone watching.

From news article "Poachers Slay 'Gentle Gorilla'" (*Washington Star*, February 3, 1978):

> MORGES, Switzerland (AP)—Digit, the gentle gorilla who starred in a U.S. zoologist's film, has been killed by poachers who sliced off his head and hands to sell to tourists, the World Wildlife Fund said here yesterday.
>
> Digit's impish portrait stared down from posters beckoning tourists to "Come and See Me in Rwanda," his Central African homeland where the gorilla was a legend for his gentleness and approachability.
>
> Peter Jackson, a World Wildlife Fund spokesman, said "at least two" native poachers killed Digit, inflicting "multiple spear wounds to his lungs, heart, stomach and intestines."
>
> Word of Digit's death came from the gorilla's long-time companion, U.S. zoologist Dr. Dian Fossey, who had been studying his movements for 11 years in a project funded by the Washington-based National Geographic Society.
>
> Although Fossey did not see the incidents, signs of a struggle in nearby vegetation made it clear the gorilla has "put up one hell of a fight," the spokesman said, relating details of a letter from Fossey.

From Dian Fossey, August 7, 1978:

> [Describes that on July 24, 1977 (?)] Uncle Bert was killed by bullet to heart, and seventeen year old female also killed—they found the poachers who killed Digit; they have been caught and will be punished.

"You are going to be an artist in history, Jay; I would like to see you enjoying more fame while you're still able to hobble about."

—Letter to Jay Matternes from Dian Fossey, Camp Karisoke, Rwanda, December 8, 1980

A POIGNANT PONGID PORTRAIT

At Dian Fossey's request, Matternes made this extraordinary portrait of a gorilla family, which she kept framed in her Rwandan forest cabin. On October 11, 1974, she wrote him: "Well Jay, you just can't believe how much that pencil sketch of Flossie, Titus, and Uncle Bert means to me. I can now sit at my desk and look up at [it] instead of just looking out of a window. Until now I never realized to the full extent exactly what an artist could convey that a camera couldn't…I've only seen one photograph out of thousands…which typifies the gorilla as it really is. Your sketch goes beyond [all the] photographs. I remain tremendously grateful to you for having taken the time to do it for me."

Little Titus grew up to become a strong adult leader of the largest gorilla group on the mountain. He was also famous to millions for a filmed high-spirited romp with Sir David Attenborough when he was five years old.

Later, in 1983, Fossey used Jay's sketch as the frontispiece for her classic *Gorillas in the Mist*, and wrote in the acknowledgments: "Words fail me when it comes to thanking you for being able to use that drawing [in my book.] I have yet to see a photograph that does the gorilla the justice of this piece of artwork...thank you again so much for the use of your masterful drawing."

From Dian Fossey, December 8, 1980:

> "You are going to be an artist in history, Jay; I would like to see you enjoying more fame while you're still able to hobble about."

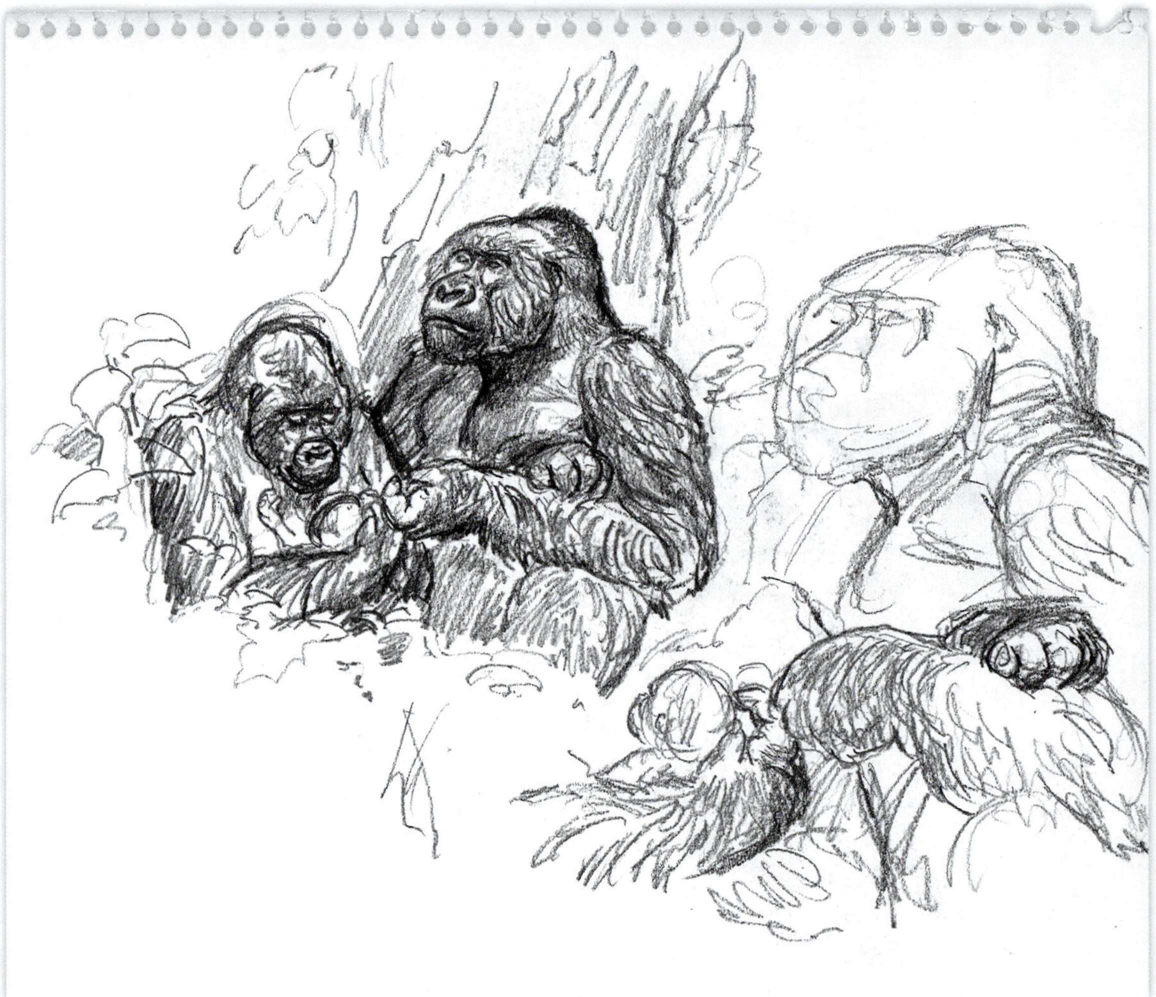

An idealized family portrait of Uncle Bert, Flossie, and their infant Titus, which Dian requested Matternes to draw. It hung on the wall of her cabin and became the frontispiece of her classic 1983 book *Gorillas in the Mist*.

> "I never realized exactly what an artist could convey that a camera couldn't."
>
> —Dian Fossey

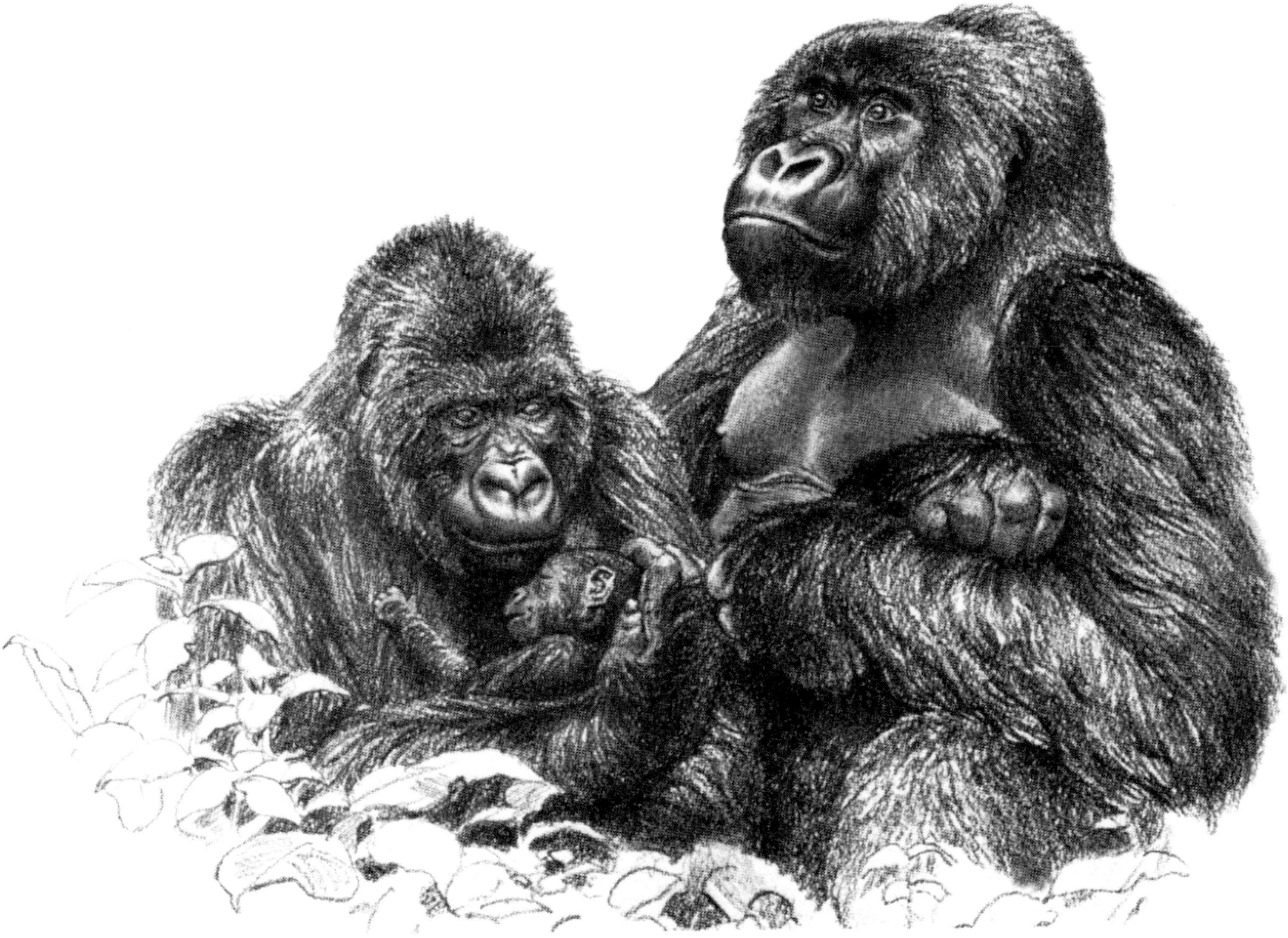

Flossie, Titus, and Uncle Bert, 1974

ENDING THE DREAM OF KONG

An undated photo was tipped into a folder of Dian's letters in Jay's collection—an iconic publicity still from the original 1933 film *King Kong*. Against a stormy, dramatic nightscape of New York City, the giant ape rises defiantly, holding a woman in his huge hairy paw—demonstrating Fossey's capacity for delicious self-satire of her lifelong romance with gorillas.

On the back of the photo, written in Fossey's hand, are a few musical notes followed by the phrase "I had the craziest dream last night…" from a 1940s hit record by Frank Sinatra:

I had the craziest dream, last night, yes I did
I never dreamed it could be
Yet there you were, in love with me

Dian visited the States again in 1983, and returned to Rwanda on November 20 of that year. Notation on her final letter by Jay: "She called to tell us of this [imminent departure] the night before from Ithaca, New York. This was the last time Del and I would see her."

Fossey was brutally murdered in her forest cabin December 26, 1985, aged 53, by an unknown assailant, possibly one of the poachers. She is buried among the slaughtered members of her gorilla family in the little animal graveyard at Camp Karisoke.

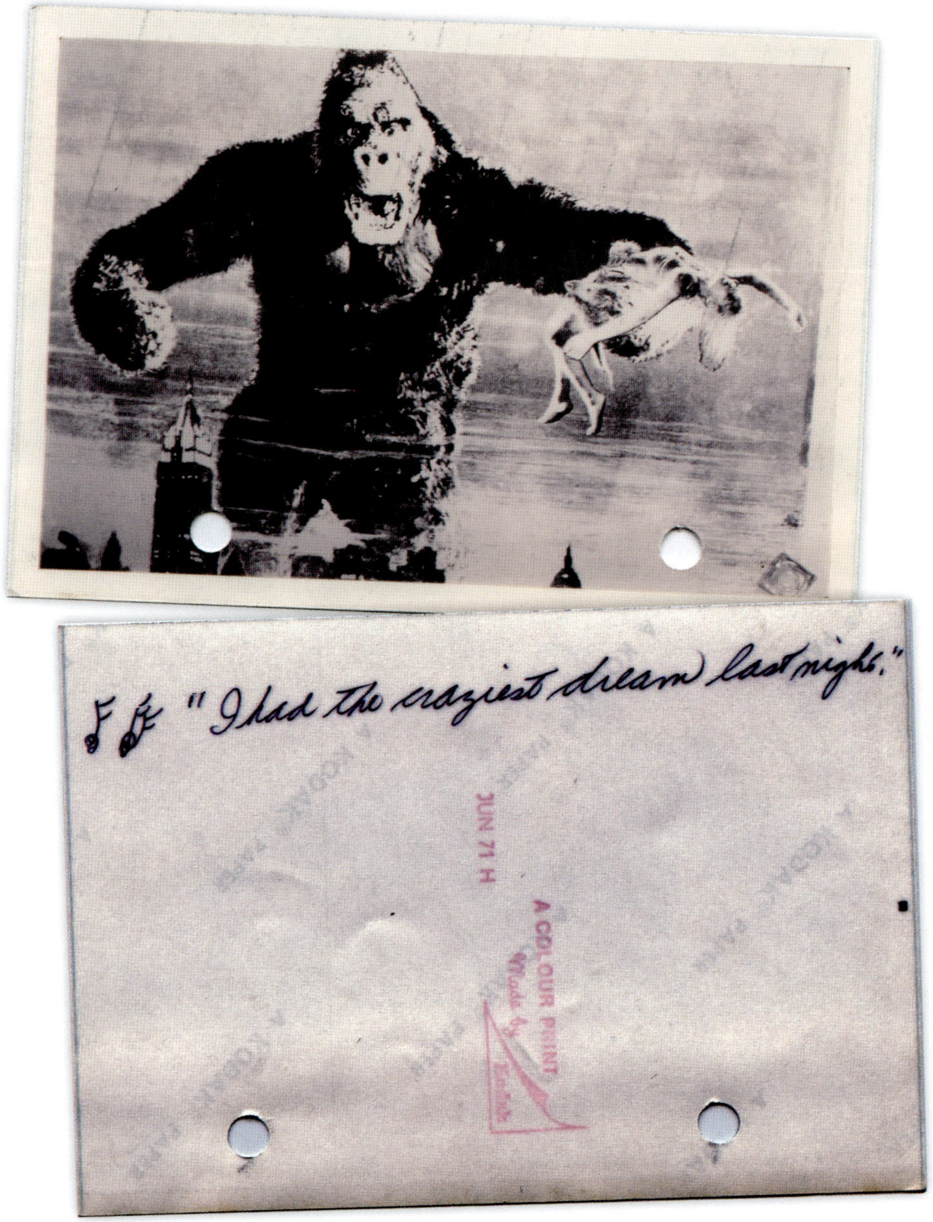

A promotional still from the 1933 classic film *King Kong* was found among Dian Fossey's letters to Jay, which he had squirreled away for more than 40 years. On the back she had handwritten a line from a popular song: "I had the craziest dream last night."

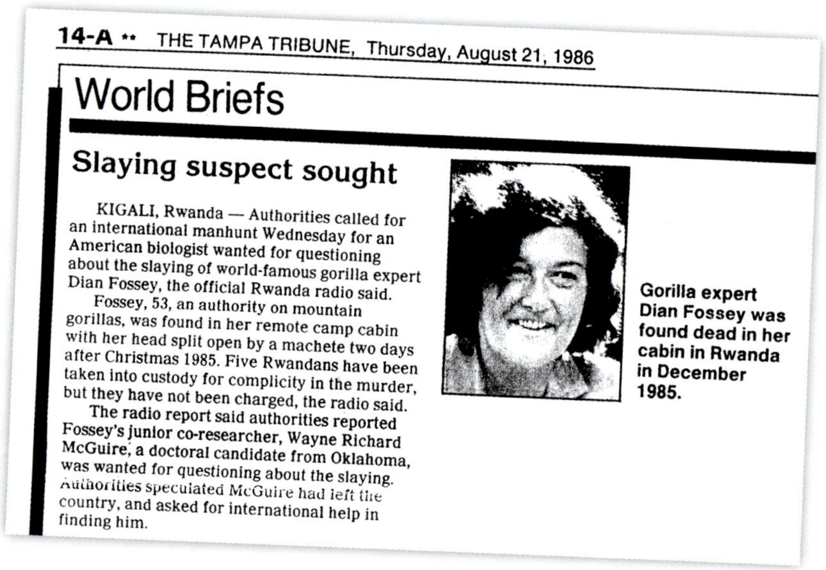

14-A •• THE TAMPA TRIBUNE, Thursday, August 21, 1986

World Briefs

Slaying suspect sought

KIGALI, Rwanda — Authorities called for an international manhunt Wednesday for an American biologist wanted for questioning about the slaying of world-famous gorilla expert Dian Fossey, the official Rwanda radio said.

Fossey, 53, an authority on mountain gorillas, was found in her remote camp cabin with her head split open by a machete two days after Christmas 1985. Five Rwandans have been taken into custody for complicity in the murder, but they have not been charged, the radio said.

The radio report said authorities reported Fossey's junior co-researcher, Wayne Richard McGuire, a doctoral candidate from Oklahoma, was wanted for questioning about the slaying. Authorities speculated McGuire had left the country, and asked for international help in finding him.

Gorilla expert Dian Fossey was found dead in her cabin in Rwanda in December 1985.

DINOSAURS

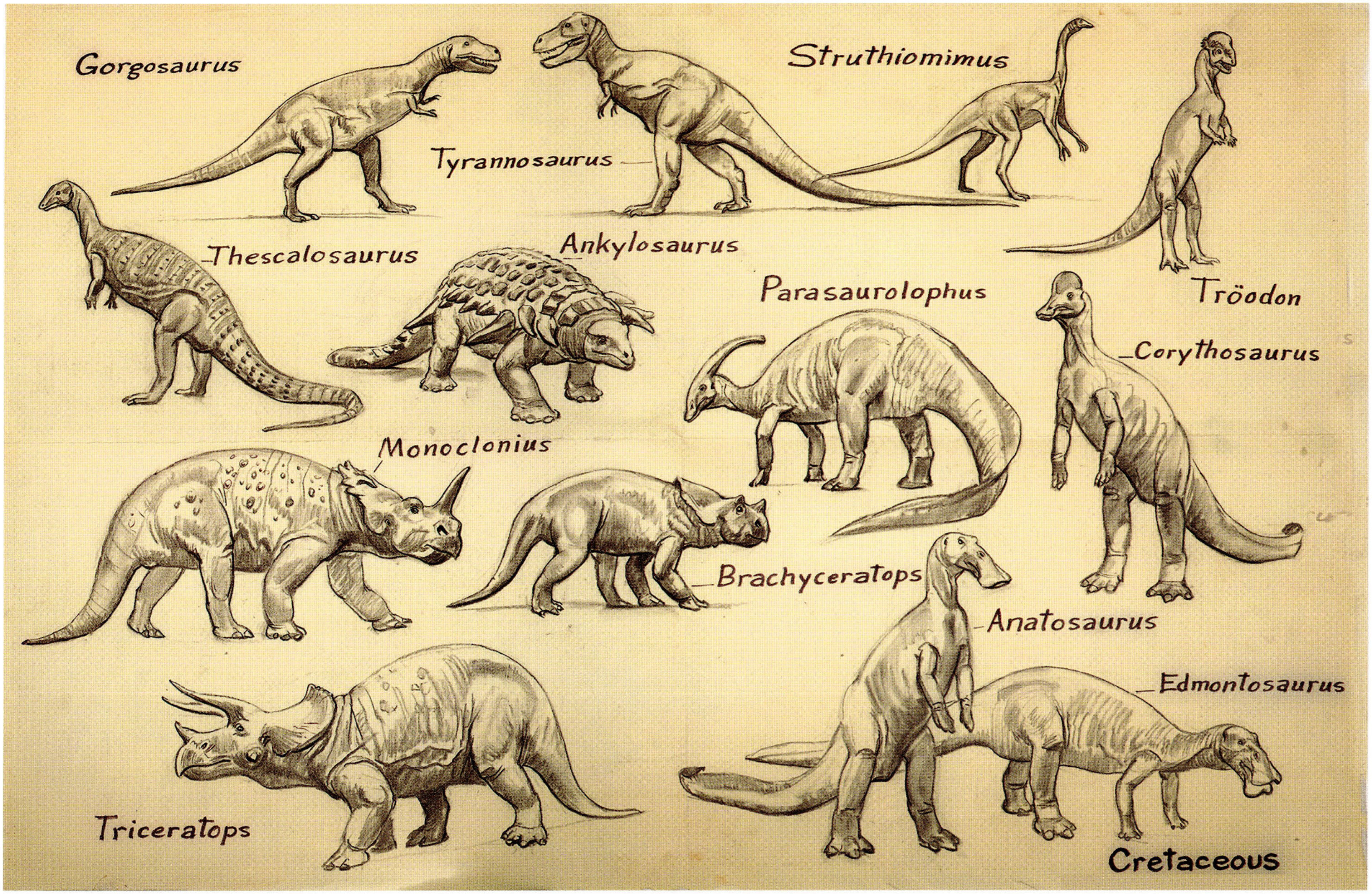

Dino Jubilee!

Matternes created this "cast of characters" chart to introduce the Smithsonian's model makers to Cretaceous dinosaurs (145–65 mya) for miniature "Age of Dinosaurs" dioramas. Sculptors followed Matternes's drawings faithfully in depicting the dinosaurs' postures, shapes, and skin textures.

Brachiosaurus with *Camarasaurus*

Brachiosaurus, one of the largest North American dinosaurs, with *Camarasaurus*, the continent's most common large sauropod, at a lakeside in ancient Colorado. Their fossil skeletons, which date from the Late Jurassic period, 150 million years ago, are frequently found in the same deposits. Although *Brachiosaurus* could grow to 90 feet long, the much smaller *Camarasaurus* was no lightweight either at 44,000 pounds! Matternes created this small pastel just for the joy of it.

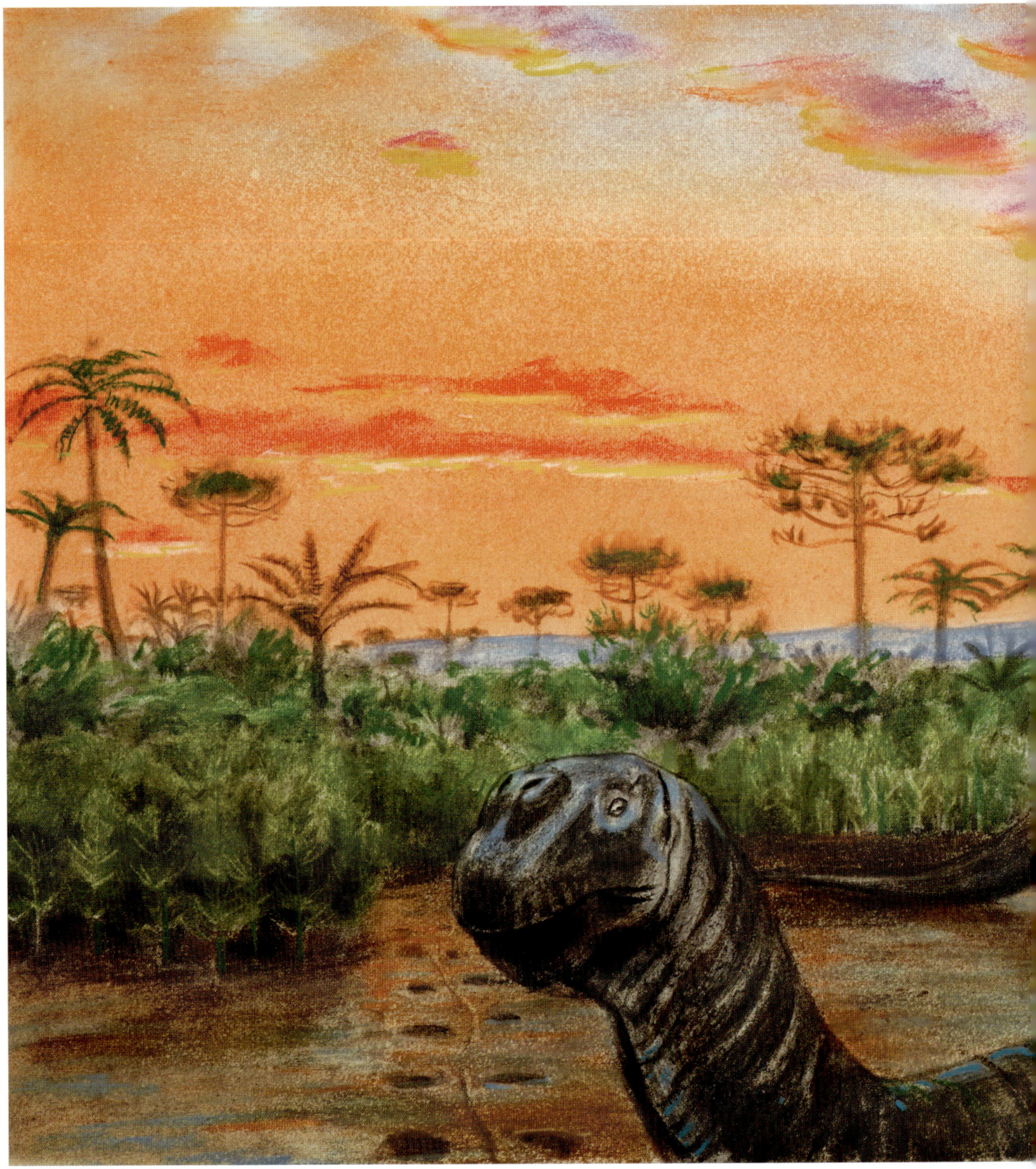

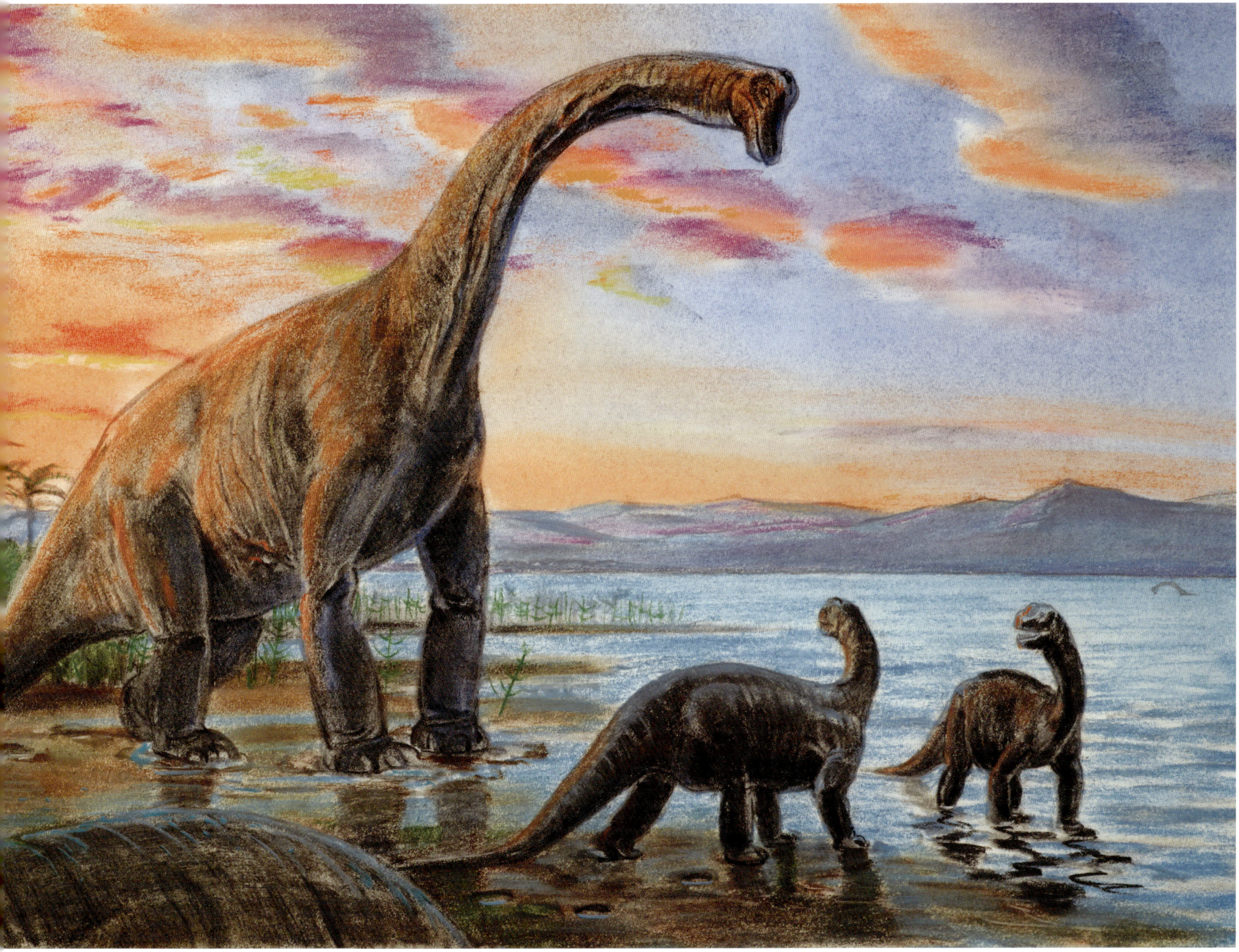

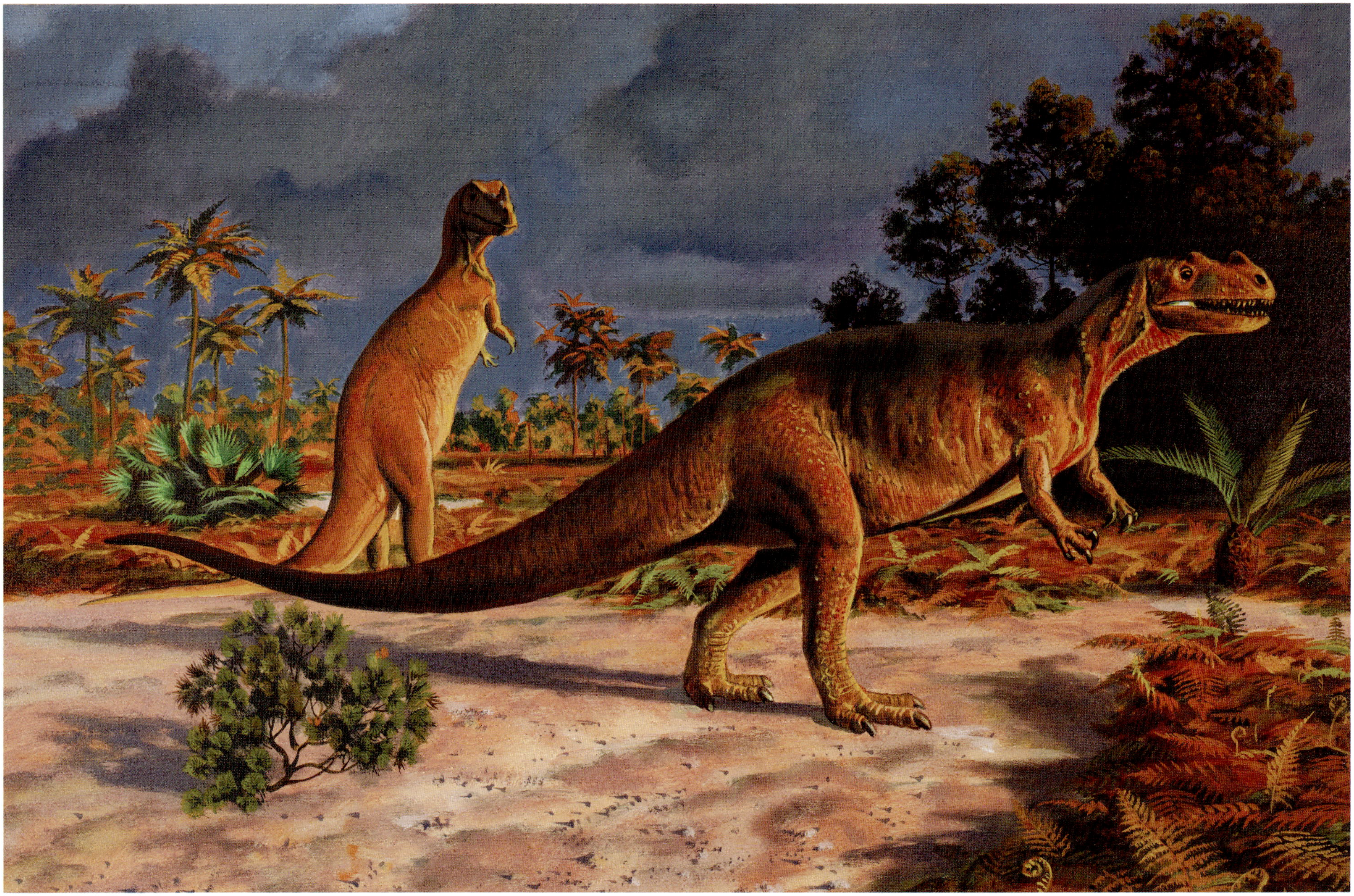

Ceratosaurus

Ceratosaurus was a medium-sized carnivorous dinosaur from the Late Jurassic period, about 150 mya, named for the horn (*ceras*) on the tip of its snout. Shaped much like its gigantic cousins the tyrannosaurs, *Ceratosaurus* had elongated hind legs and very short forelimbs. This colorful acrylic was made to relieve the monotony of the many black-and-white sketches Matternes was making for the diorama model makers.

Trachodons

A pair of "duck-billed" dinosaurs, one of the most prevalent herbivores of the Cretaceous. Apparently, it was frequently preyed upon by large carnosaurs.

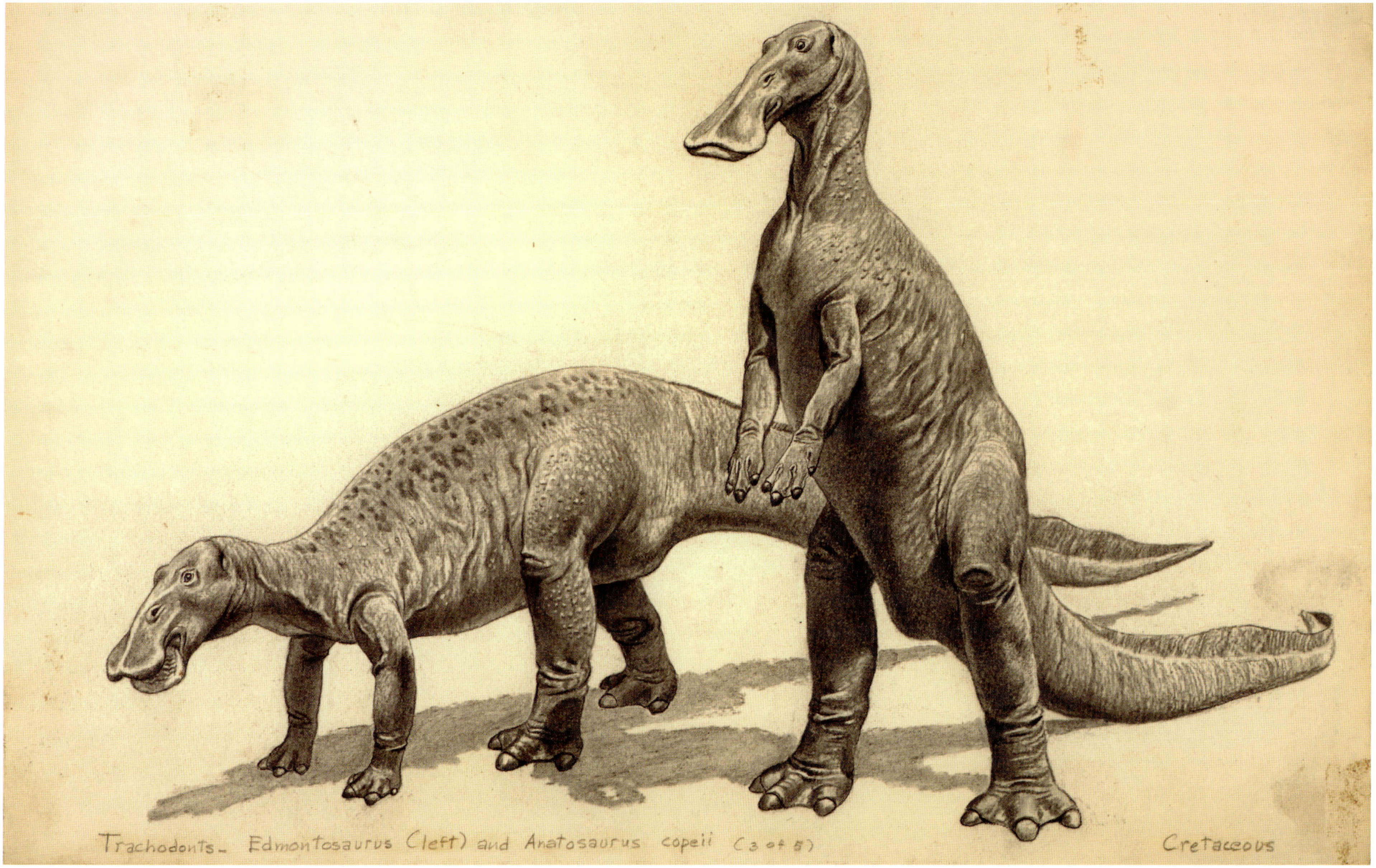

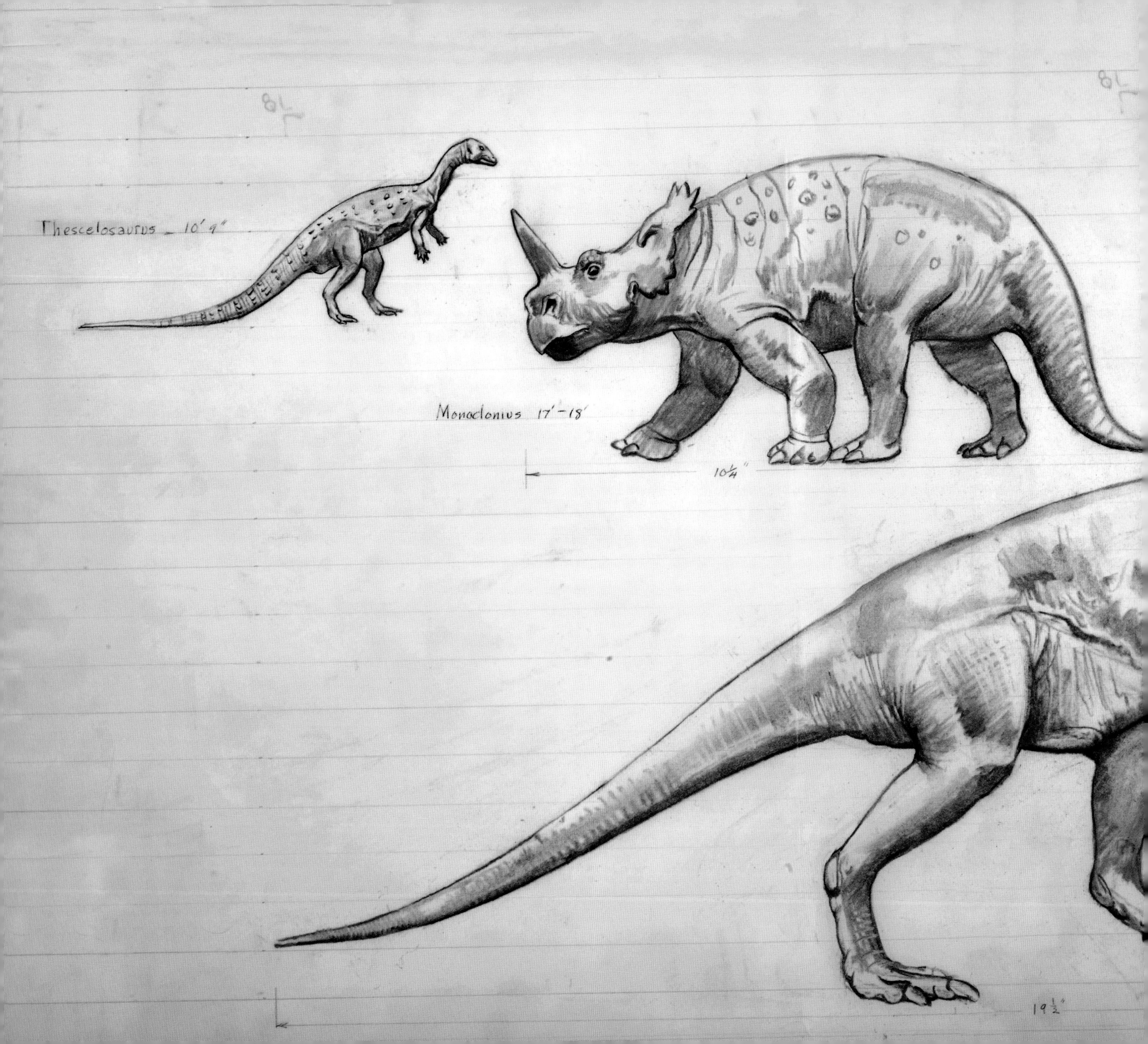
Thescelosaurus _ 10' 9"
Monoclonius 17'-18'
10¼"
19½"

Thescelosaurus, Monoclonius, Gorgosaurus, Triceratops

Thescelosaurus, Monoclonius, Gorgosaurus, Triceratops and other dinosaurs drawn in correct relative scale to each other on lined composition paper. Matternes made these charcoal sketches in 1962 to assist preparation of the Smithsonian's dioramas of Mesozoic reptiles.

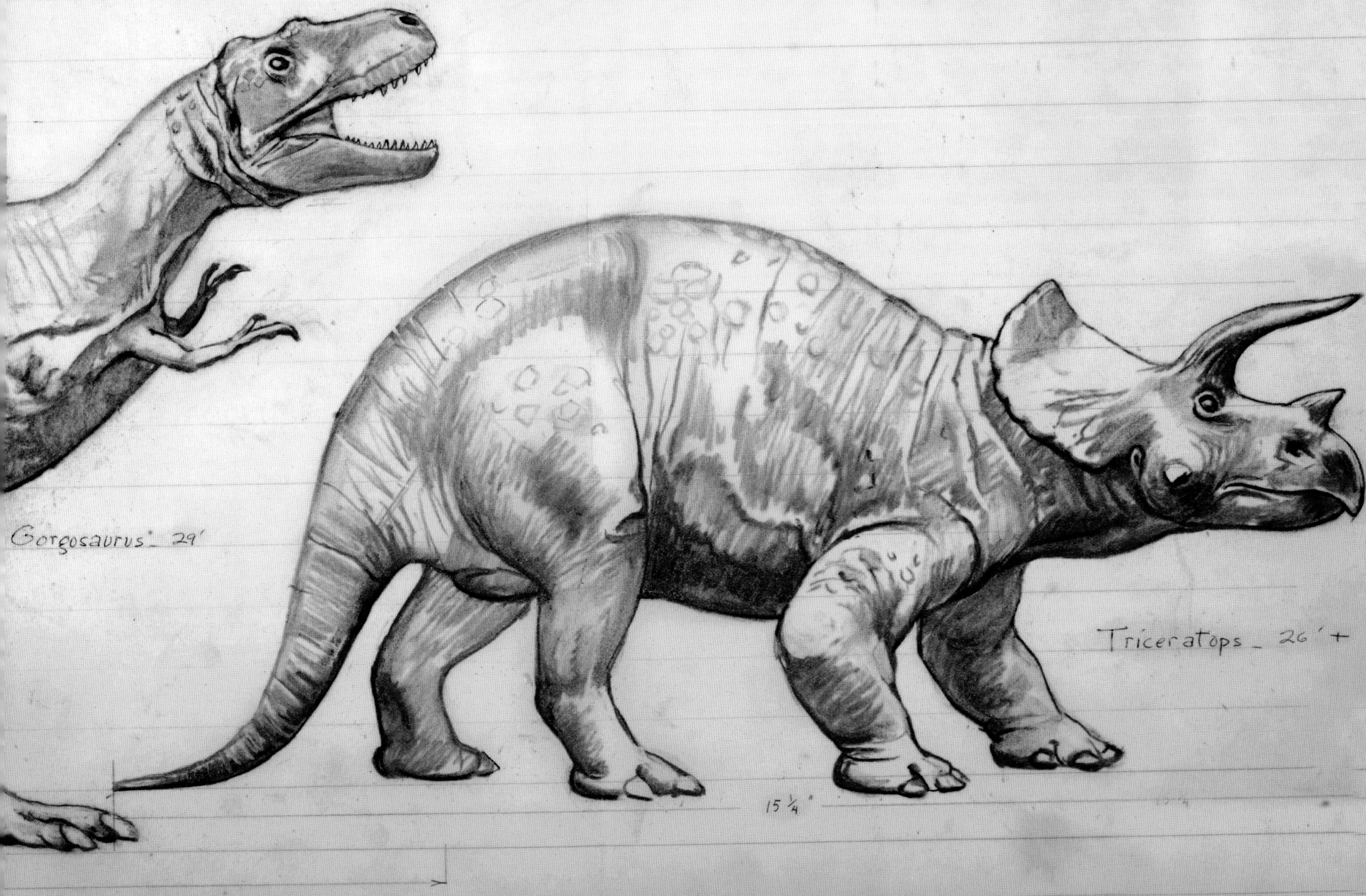

Ornitholestes

Ornitholestes ("bird robber") is a small theropod dinosaur of the Late Jurassic (154 mya) of the part of Laurasia that was to become North America. It is known only from a single partial skeleton found at the Bone Cabin Quarry near Medicine Bow, Wyoming, in 1900.

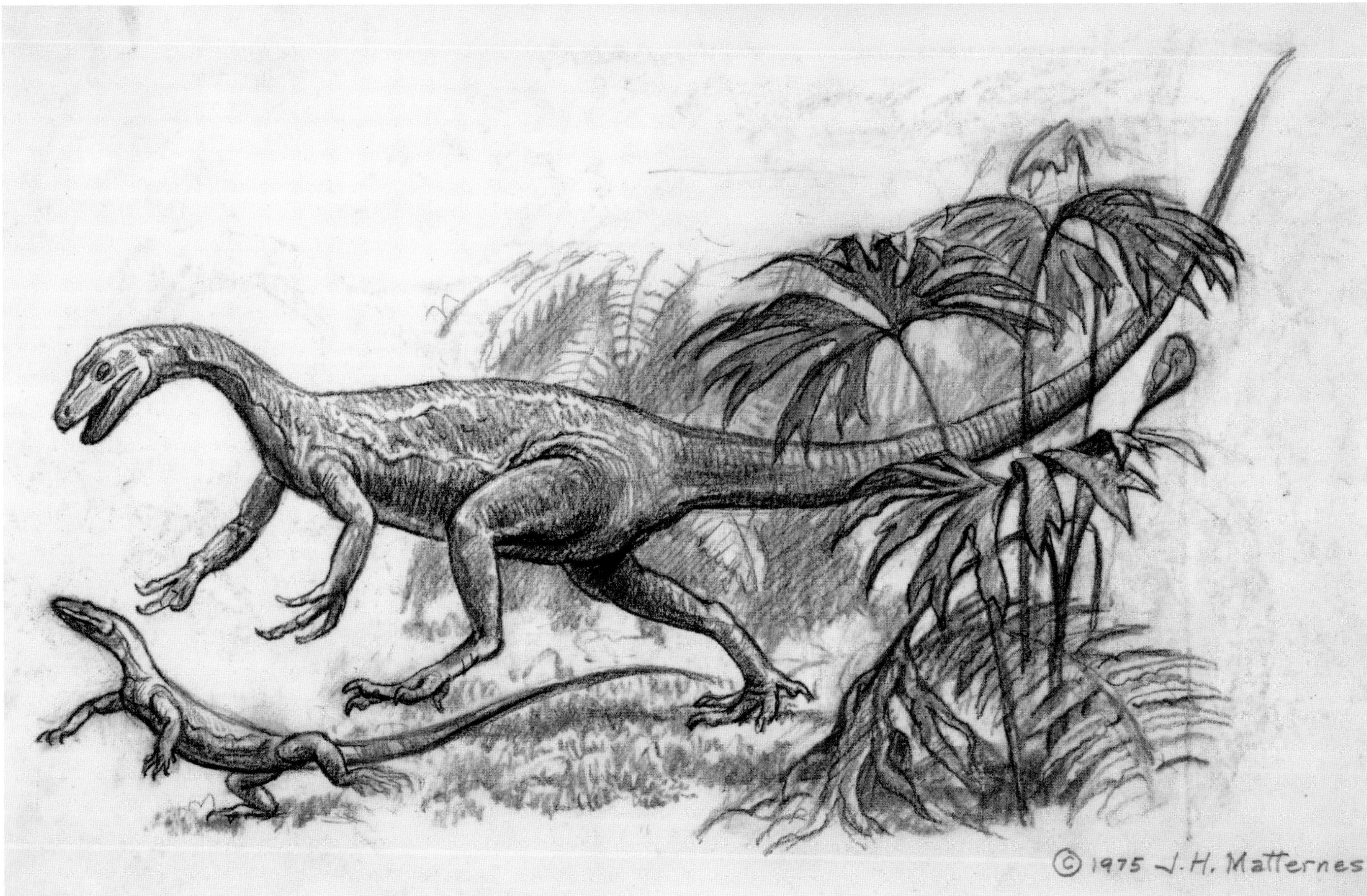

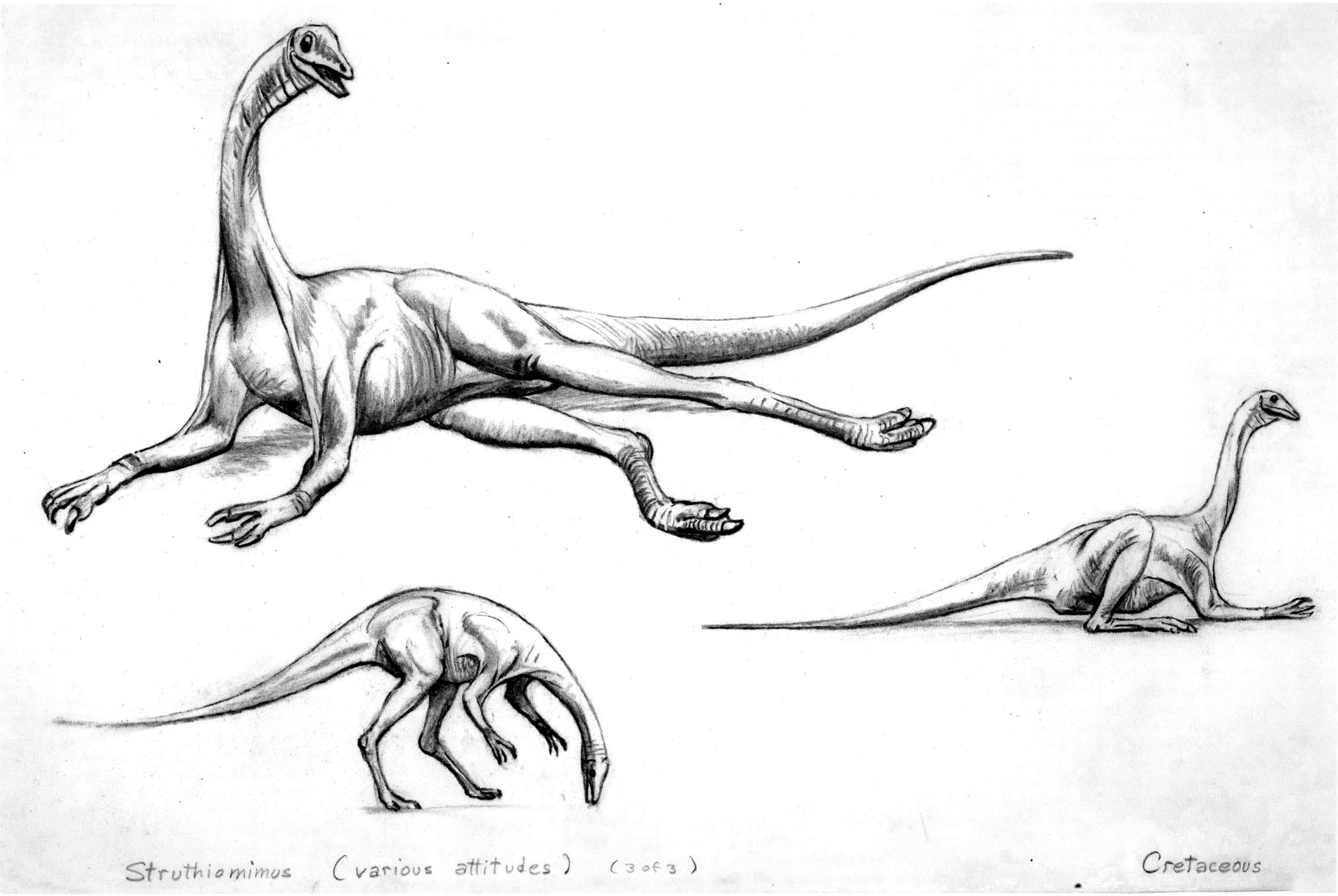

Struthiomimus

Here's another swift-moving animal depicted by Matternes in the relaxed posture commonly seen among bored zoo animals. It had a toothless beak and the proportions of an ostrich, so was named *Struthiomimus* ("ostrich-like"). It lived in North America during Late Cretaceous times and weighed about 330 pounds. This depiction includes a hypothetical flap of skin between elbow and side, supposedly to assist speedy running, but that has since been dropped from subsequent restorations.

Plesiosaurs

Plesiosaurs were Cretaceous marine reptiles with snake-like necks and oar-like paddles, known from the fossiliferous Cornwall Coast in England and the long-gone seas of Kansas. In this lyrical underwater composition, Matternes depicts the short-necked plesiosaur *Brachauchenius* and its long-necked cousin *Elasmosaurus* as they pursue giant herringfish dinners.

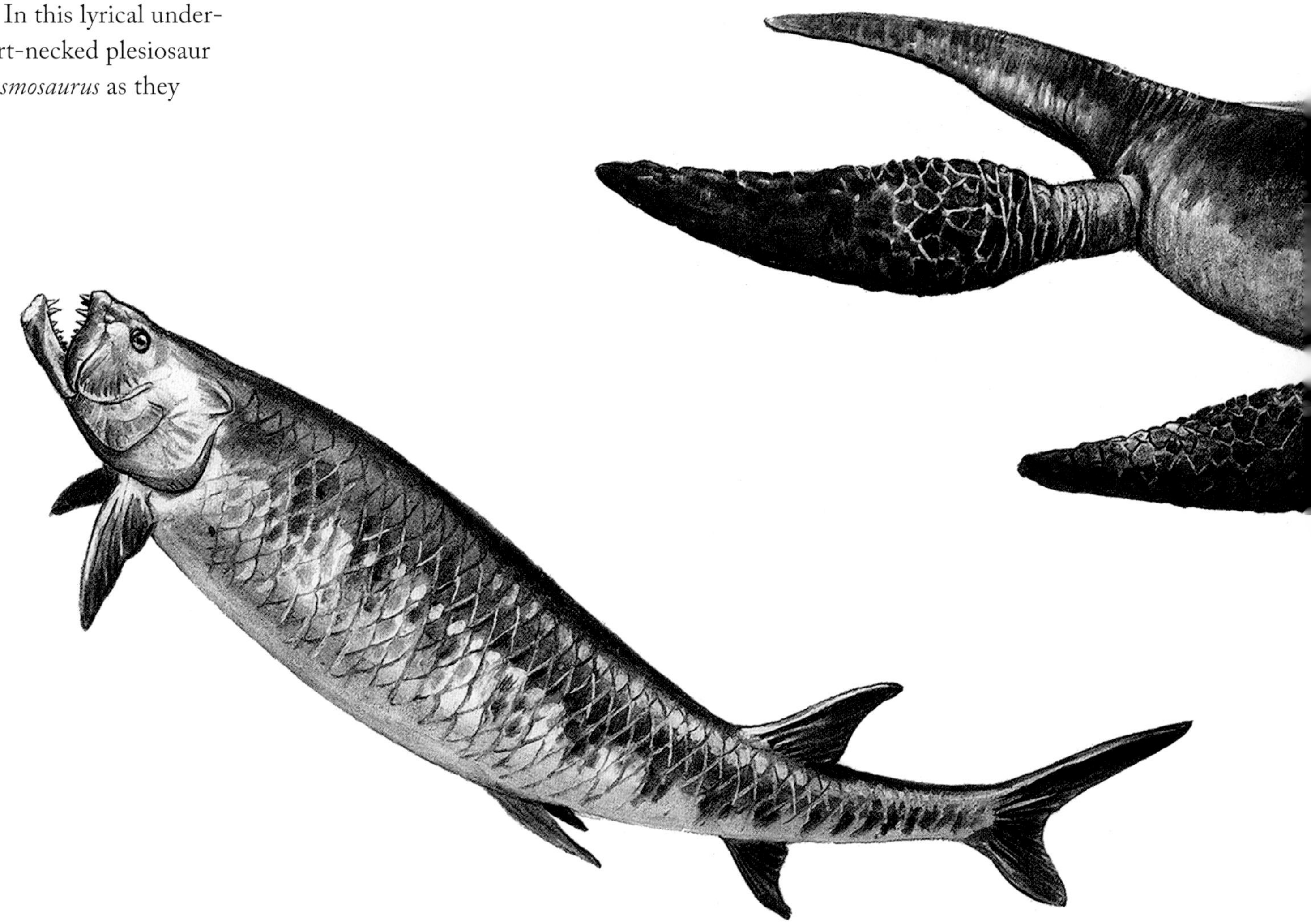

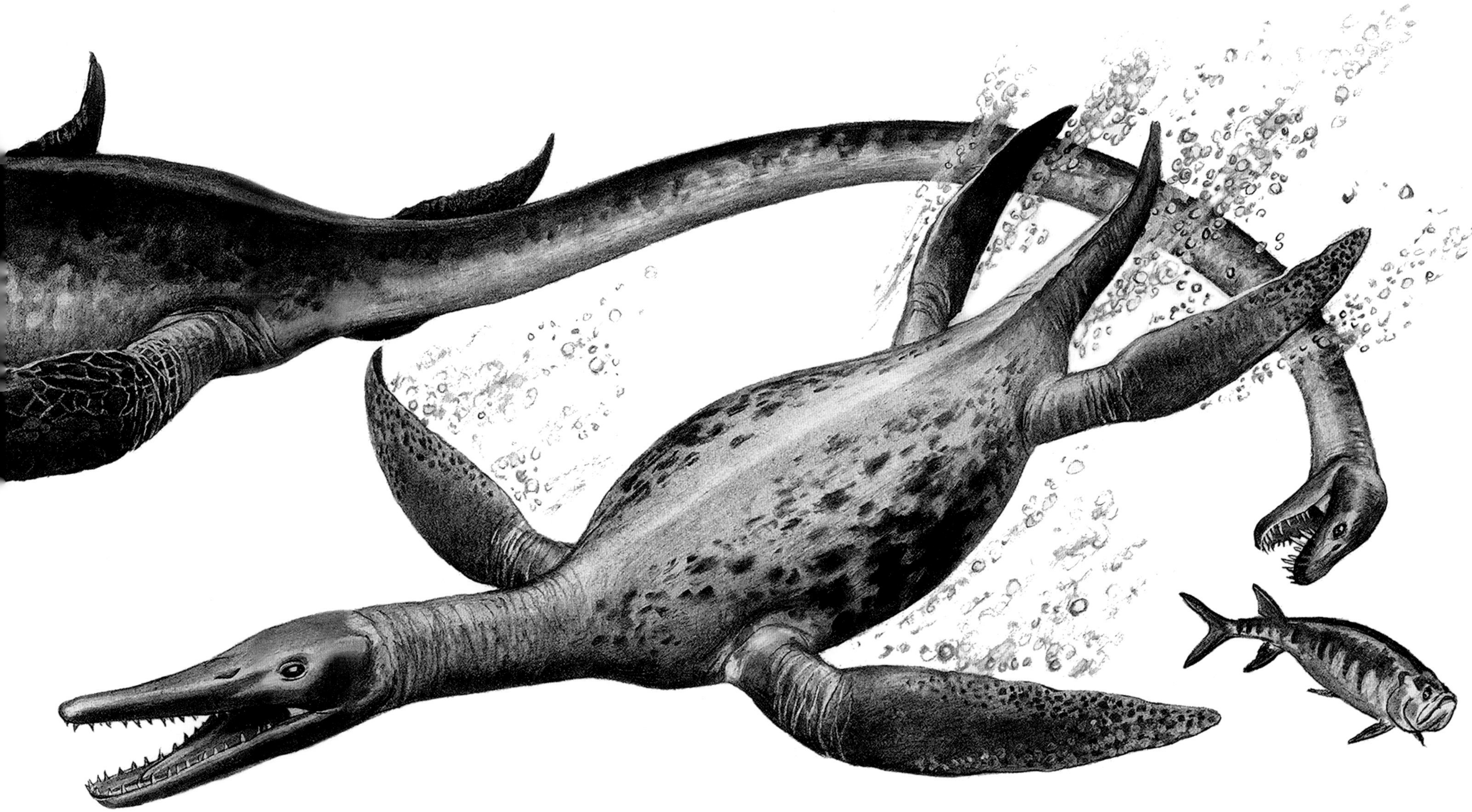

RESTORING SOME EARLY MAMMALS

These exquisite pencil drawings of early North American mammals from the Eocene epoch demonstrate Matternes' stages of reconstructing extinct animals from the inside out—first piecing together a complete skeleton from fossil bone fragments, then adding each muscle, and finally covering the body with skin and representing conjectural coat texture, facial expression, and locomotor behavior. All are on permanent display in the Lila Acheson Wing of the American Museum of Natural History, where Denis Finnin photographed them for this book in a temporary darkroom he rigged with long black curtains to eliminate glare from sunlight on the glass.

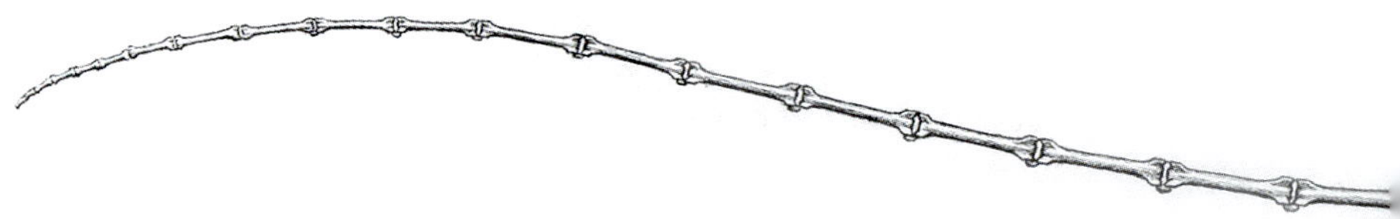

SMILODECTES GRACILIS, a primitive Eocene primate, had the limb proportions of an early arboreal leaper and vertical clinger, much like lemurs of today. It inhabited the tropical forests of Egypt near the Tethys or early Mediterranean Sea.

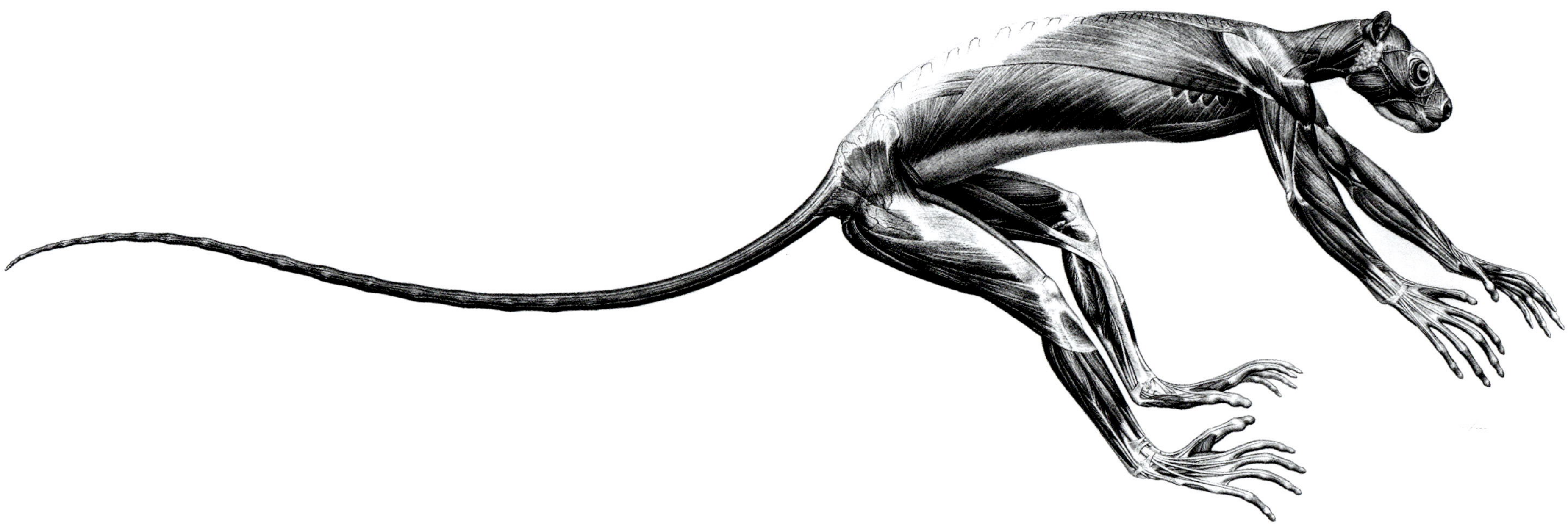

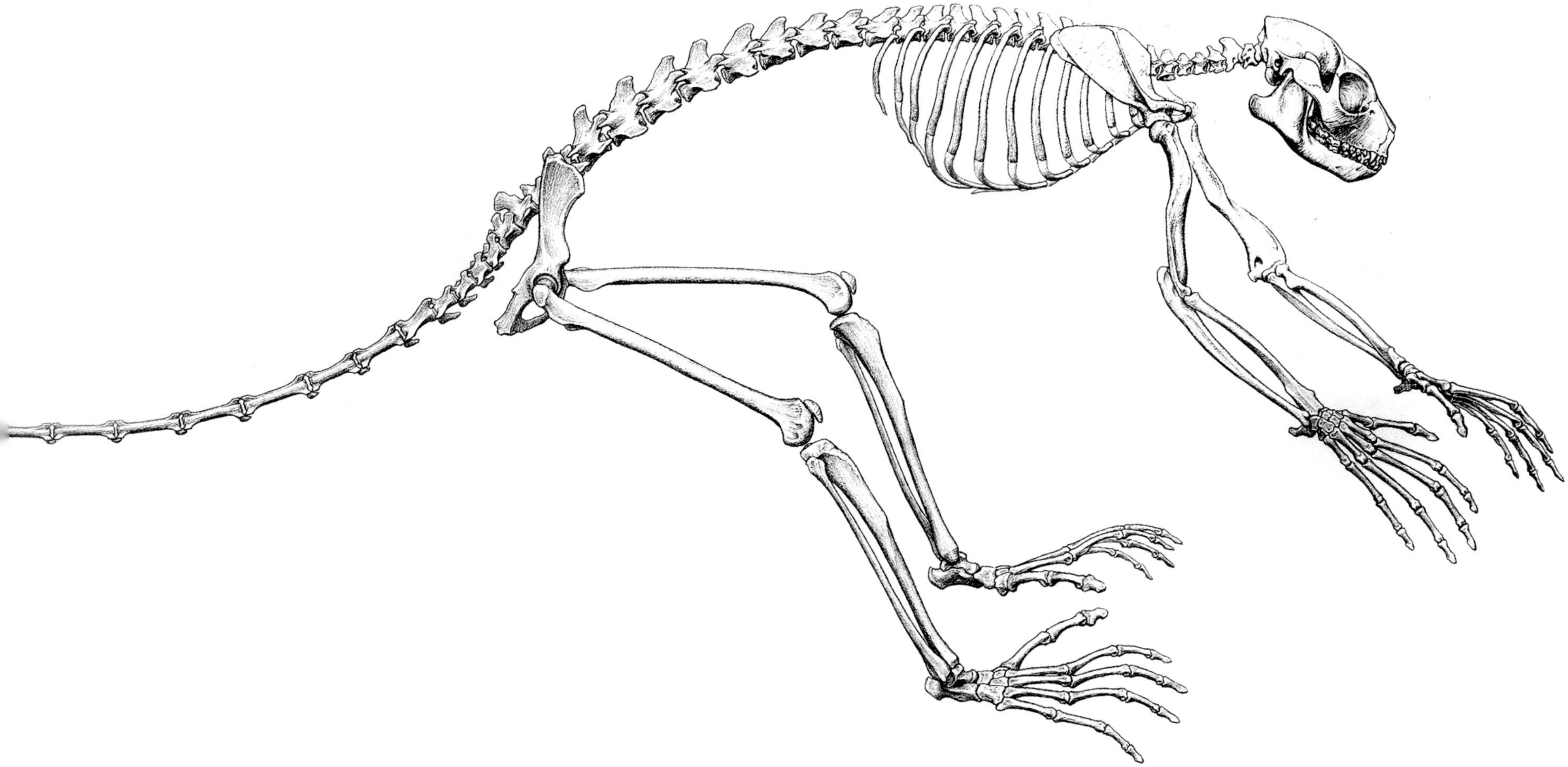

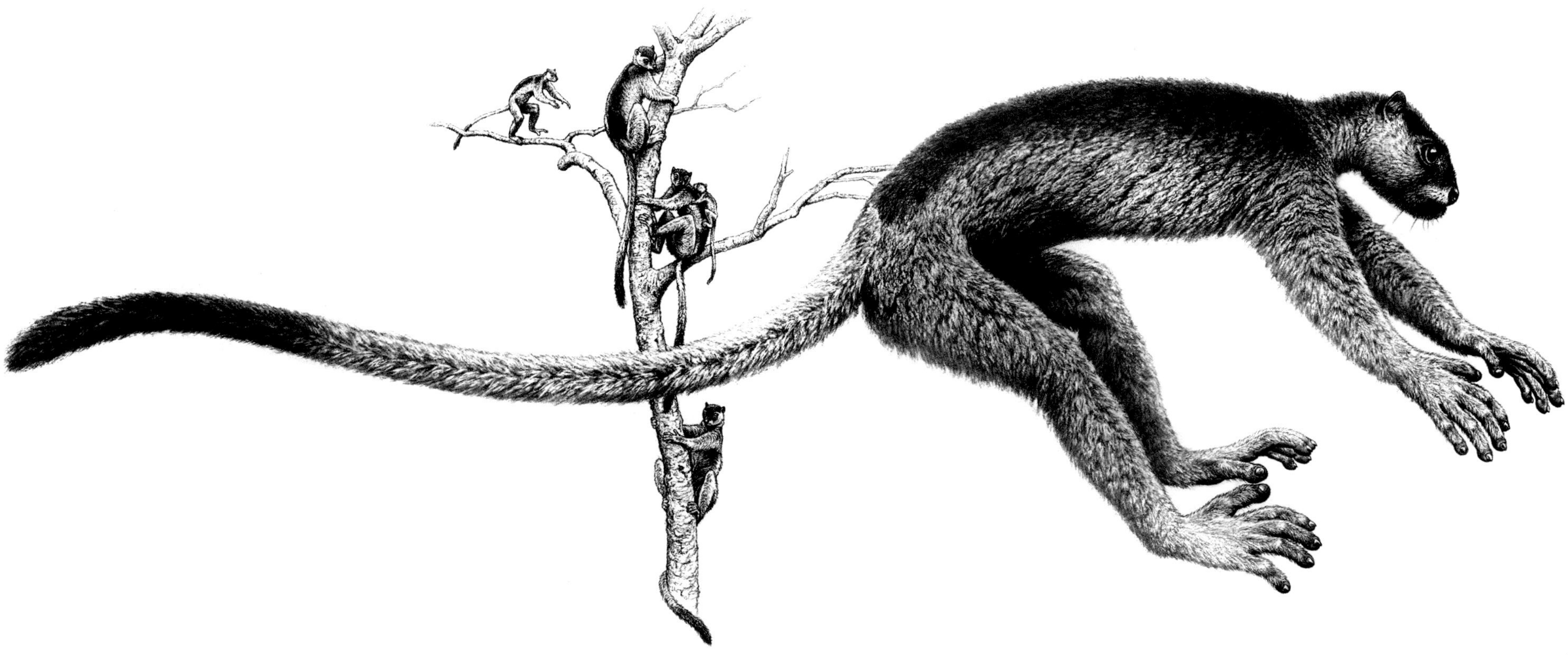

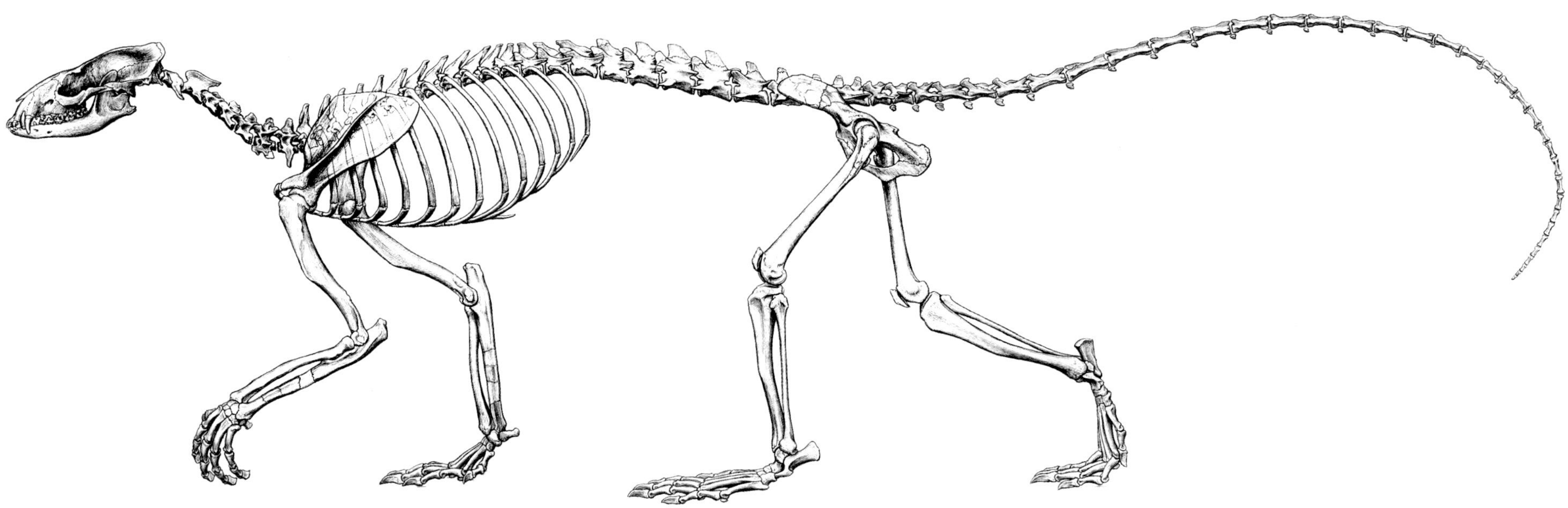

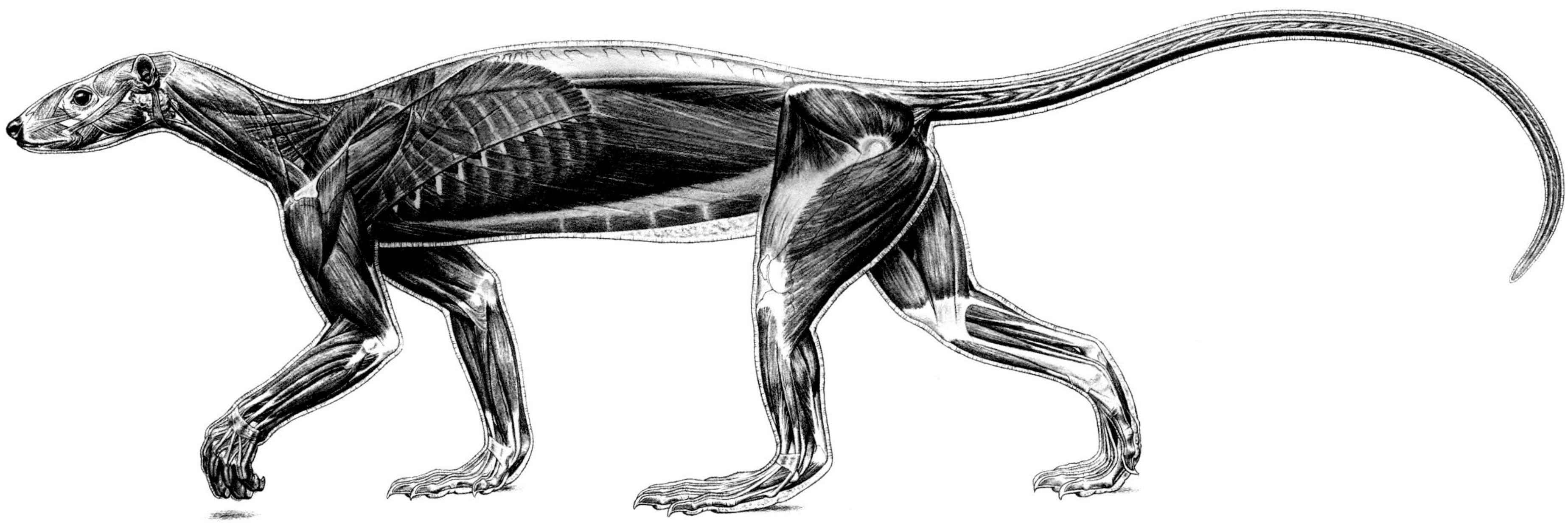

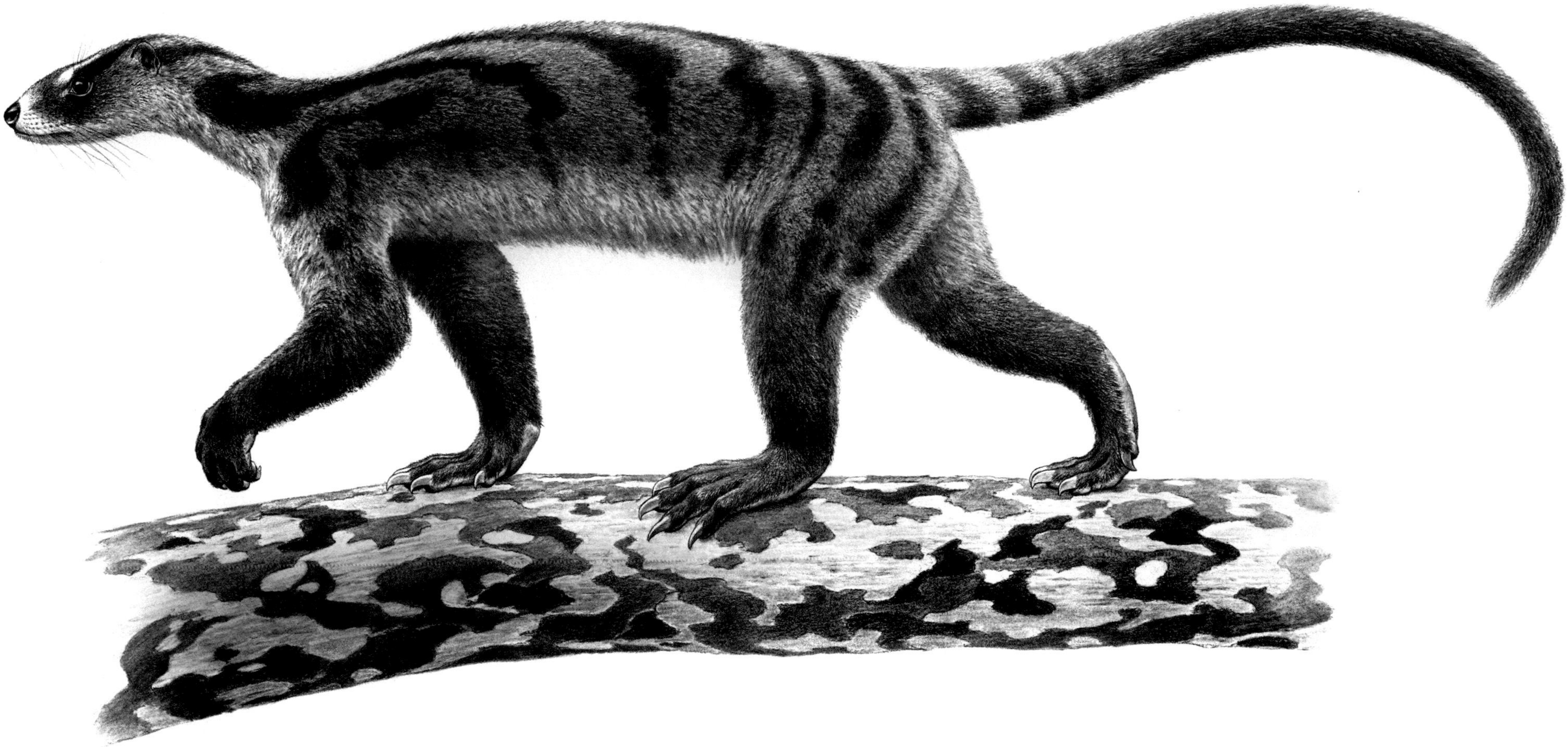

VULPAVUS OVATUS, a cat-sized Eocene carnivore from North America, appears very similar in its lithe body to a modern genet or to a pine marten, and was probably a good climber and ate birds, bird's eggs, reptiles and small mammals. Matternes shows this little predator on the branch of a sycamore or plane tree (*Platanus*), which was very common in its habitat.

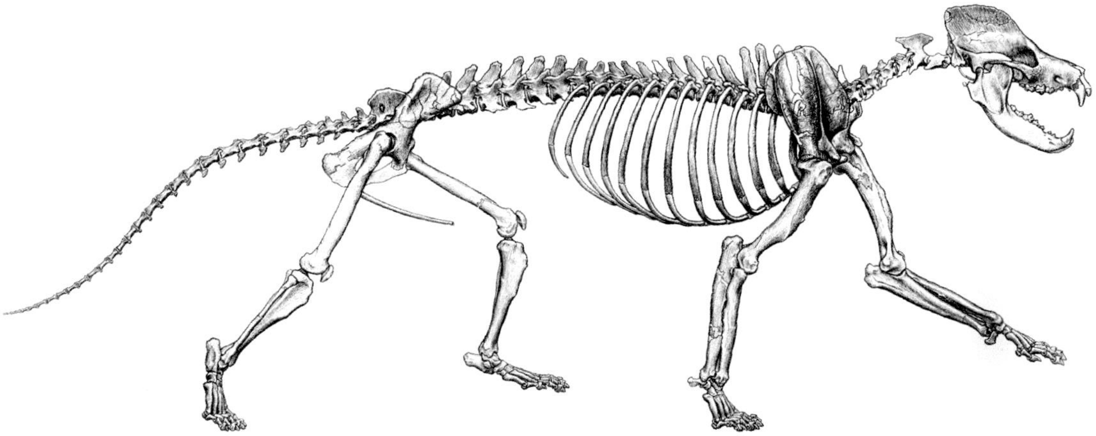

AMPHICYON INGENS, a formidable bear/dog, was probably plantigrade, or flat-footed, in its hind legs and sported a large tail. While this species is known from the Miocene of North America, its genus *Amphicyon* had a wide distribution in Eurasia and Africa as well. As large as today's biggest brown bears, they were powerful carnivores and probably the top predators of their time.

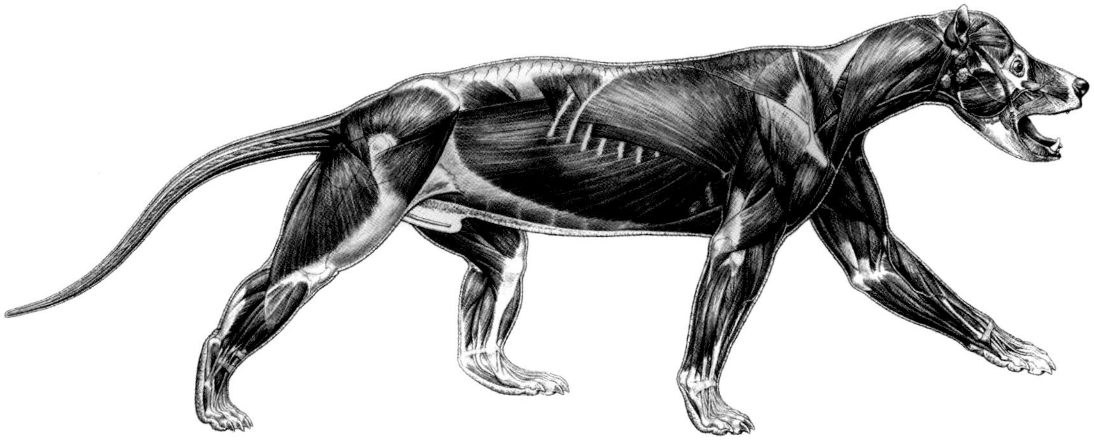

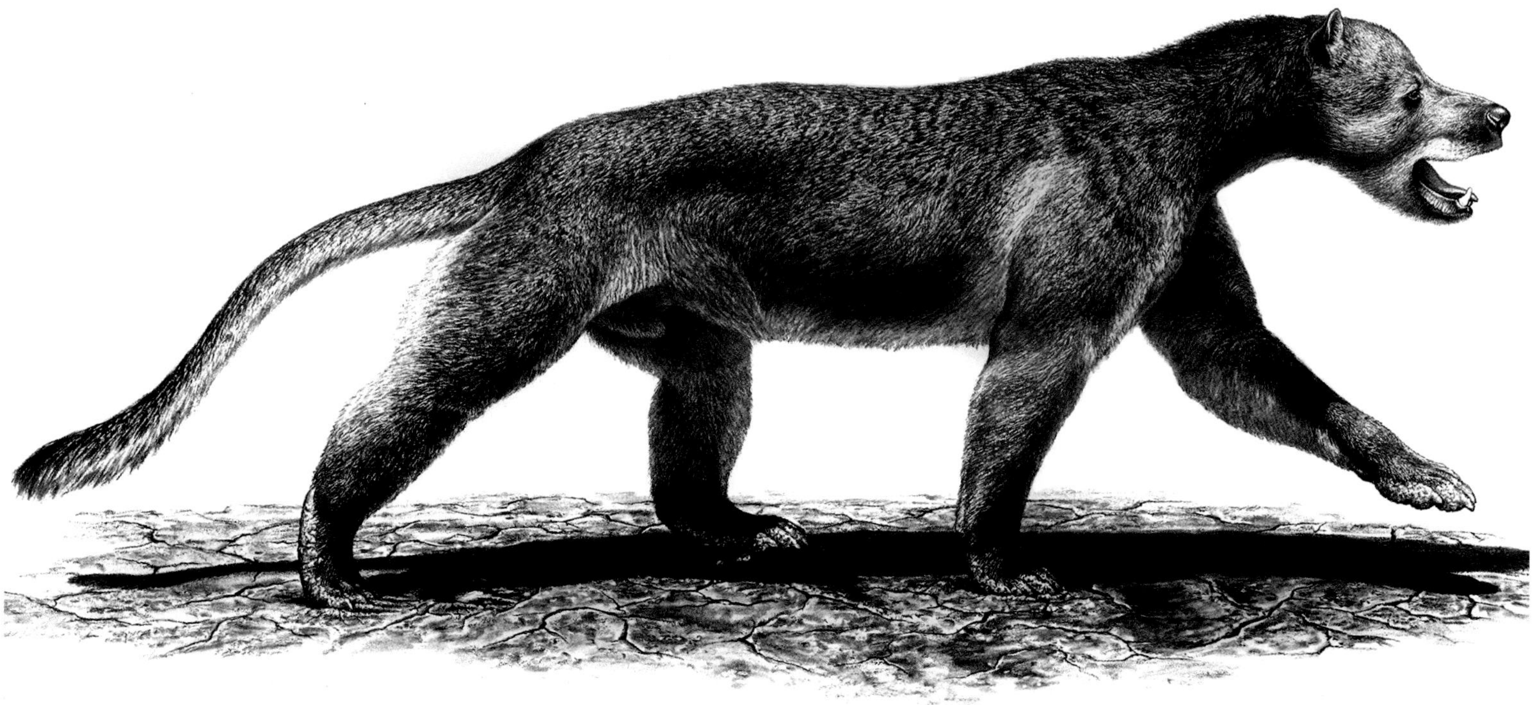

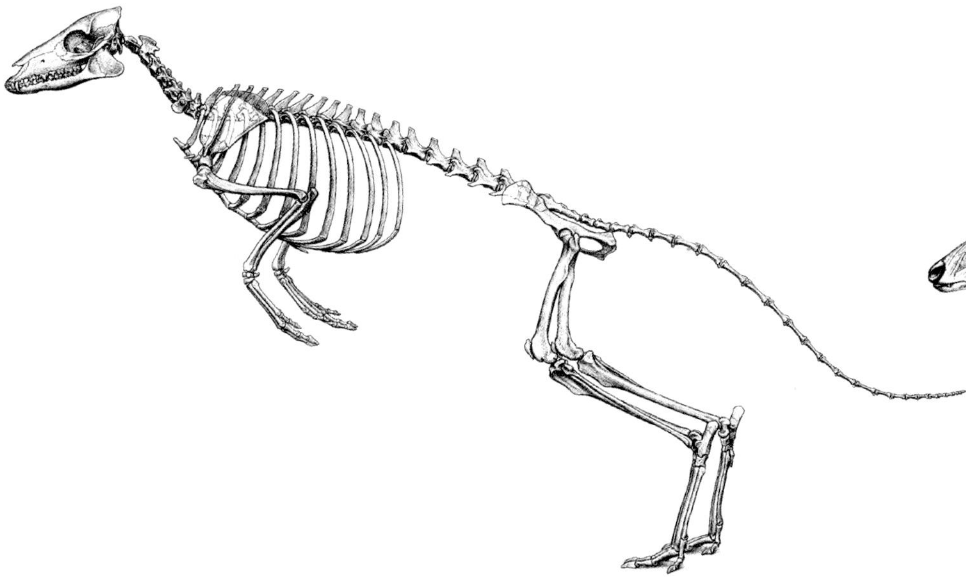

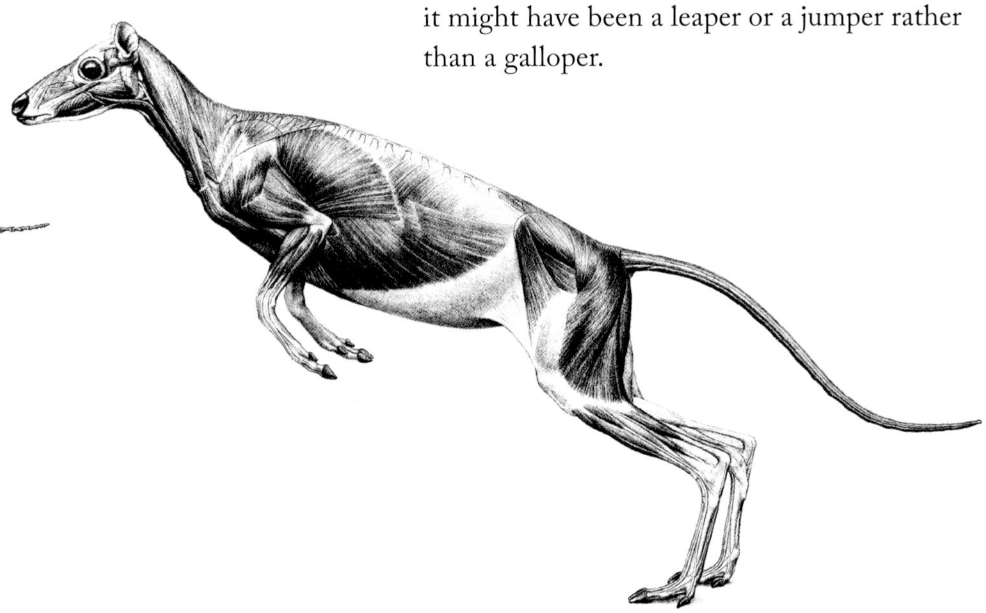

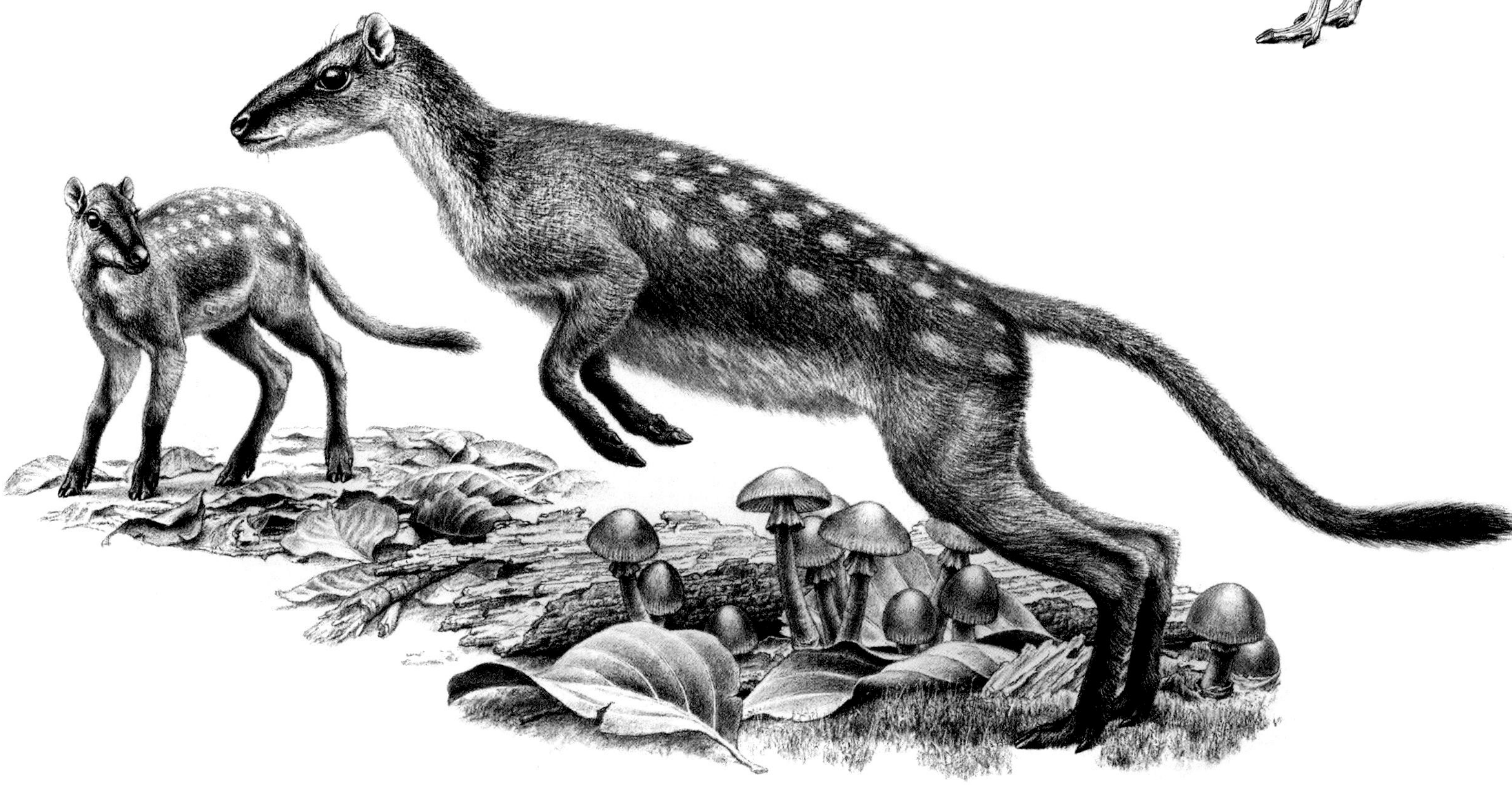

DIACODEXIS METSIACUS, from the lower Eocene of North America, pictured in mid-leap, was a beautiful little antelope-like creature with tiny hooves. Matternes wanted to show that it might have been a leaper or a jumper rather than a galloper.

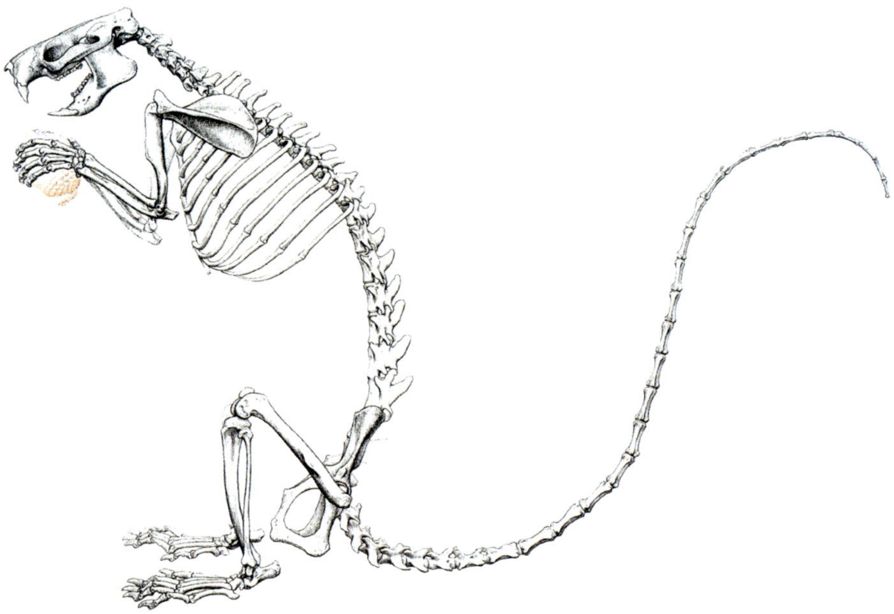

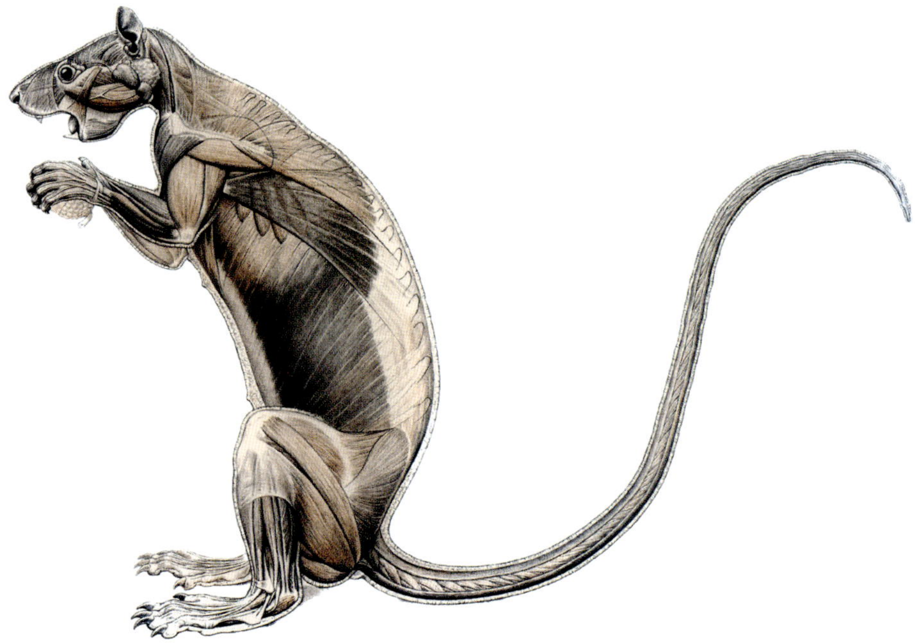

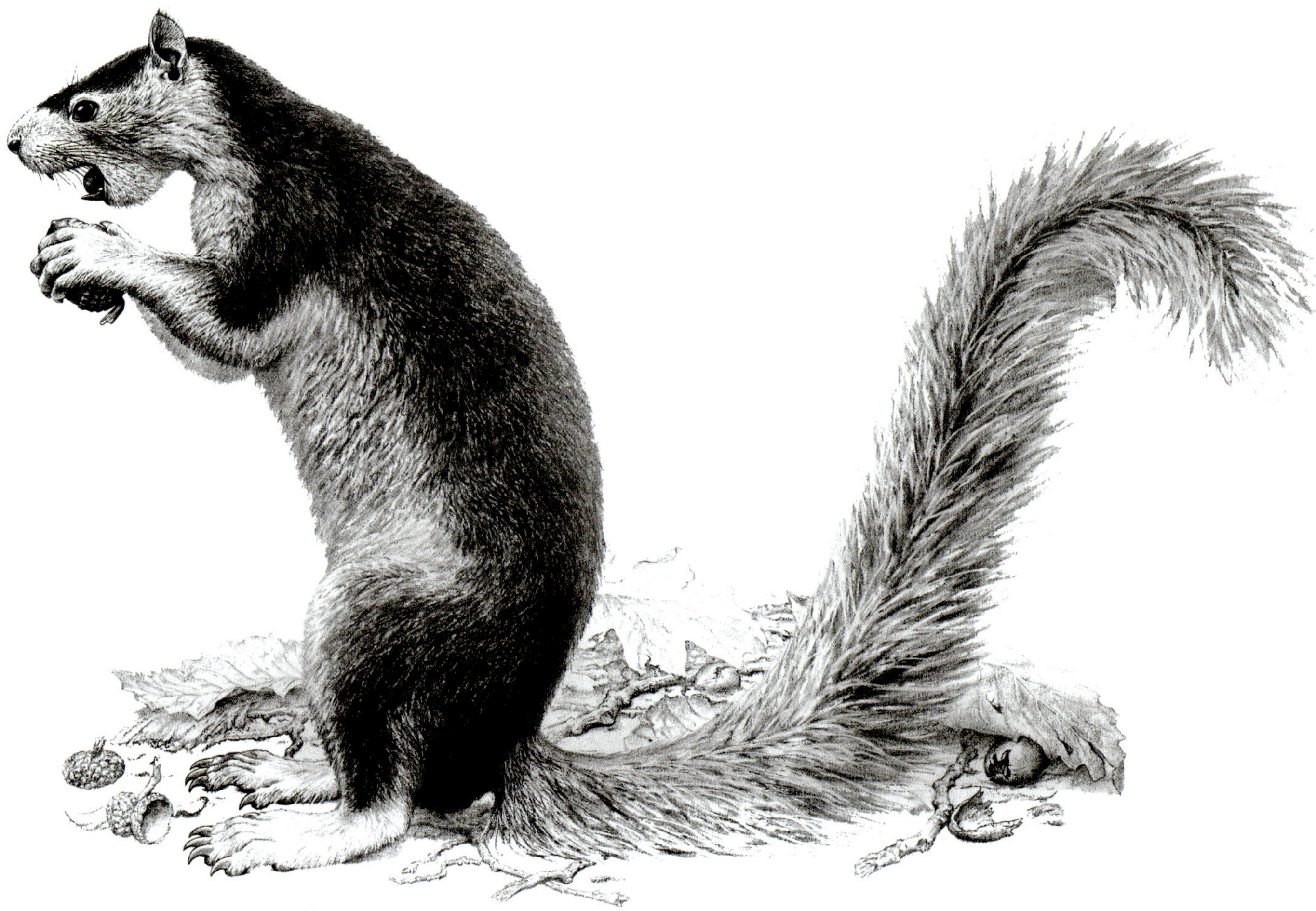

PARAMYS DELICATUS, a large North American rodent of the Eocene, is so anatomically similar to modern squirrels that the artist chose to depict it in a familiar upright pose. *Paramys* holds a food item in its forepaws, a behavior resembling its modern counterparts.

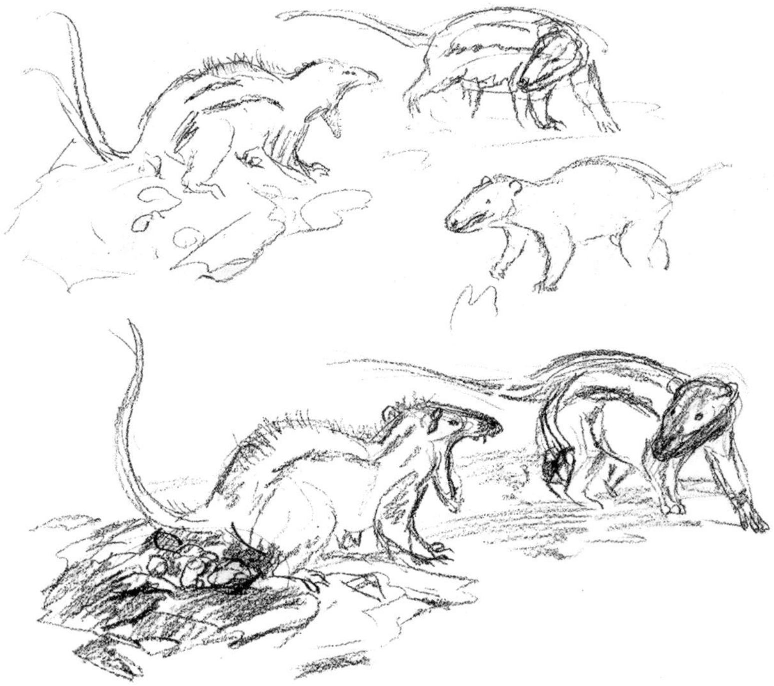

Earliest North American Mammals

Evidence for early mammals in North America, mainly from teeth, goes back 100 million years. There were no known skeletons until 1974, when the first one was discovered in the shale-and-sandstone badlands of Montana. It was an opossum-sized meateater from the Cretaceous of the triconodont group, so called from the triple cusps on its teeth. Matternes reconstructed the animal using his usual method: from skeleton to musculature to full external appearance. But in this drawing, he goes a step further to create an imaginative reconstruction of behavior. A female is defending her young against an intruder amid a litter of ginkgo leaves and sequoia needles. Were their litters born or hatched, and were they sequestered in a nest or an earthen burrow? The artist makes his best guess.

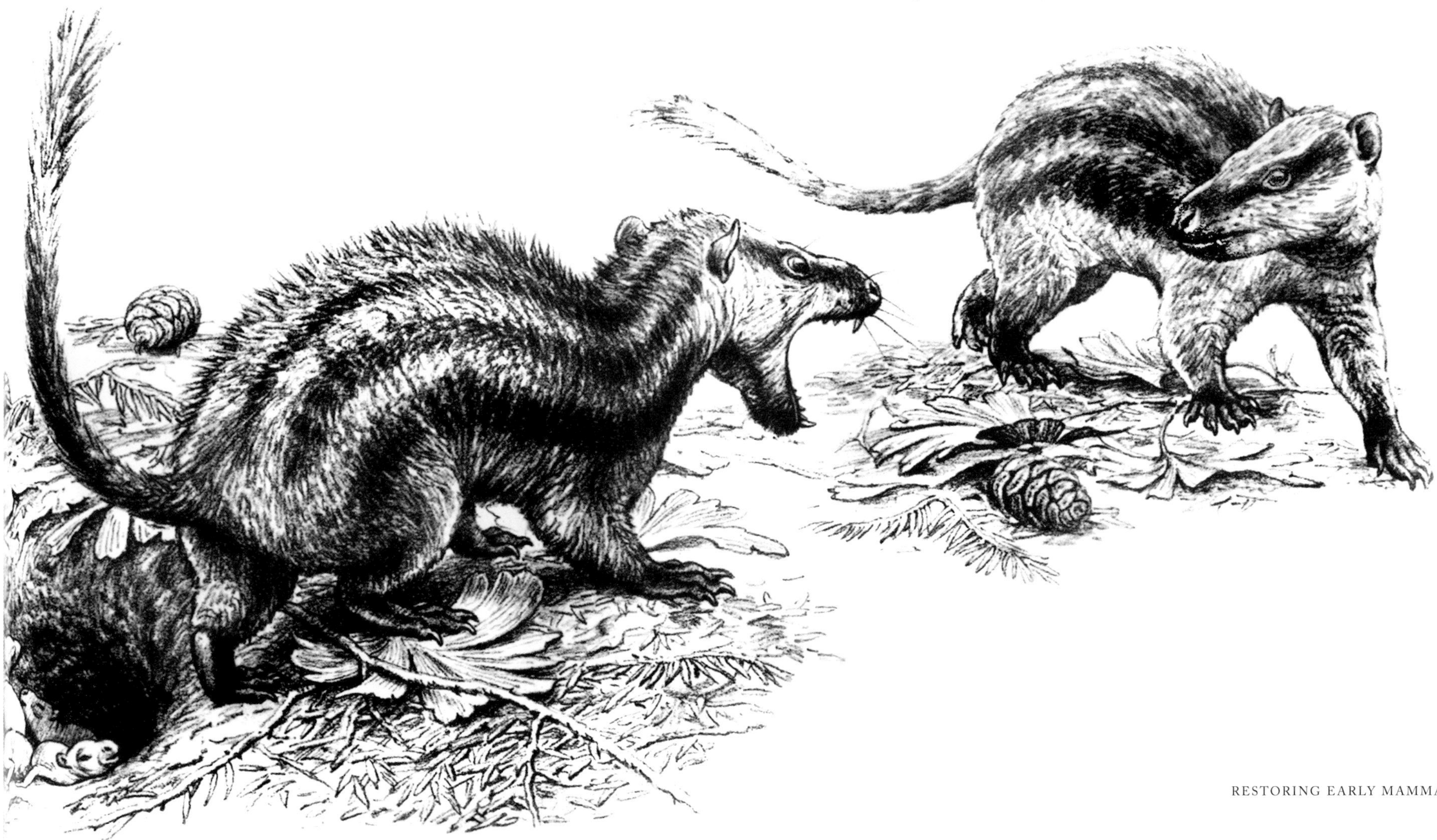

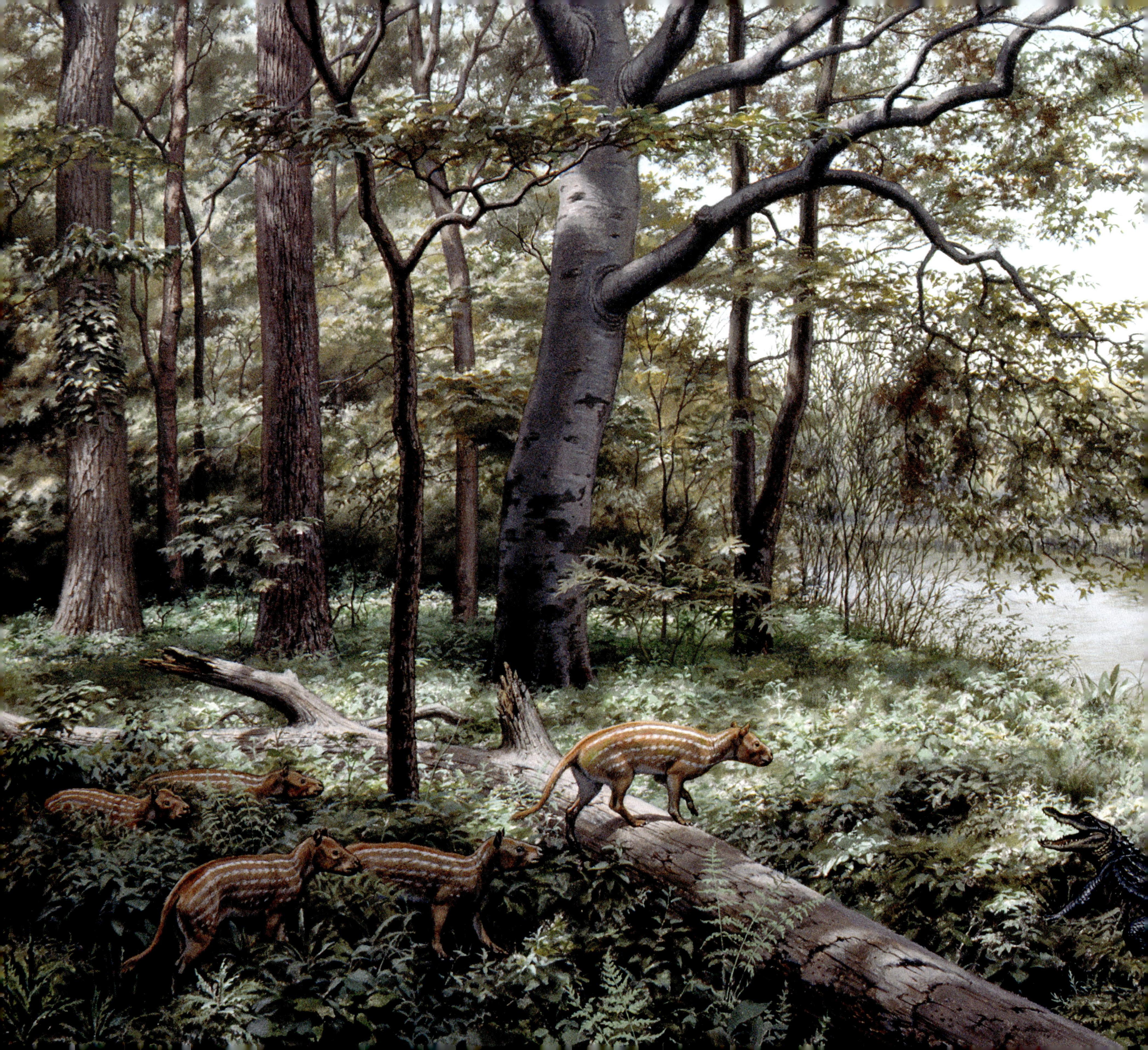

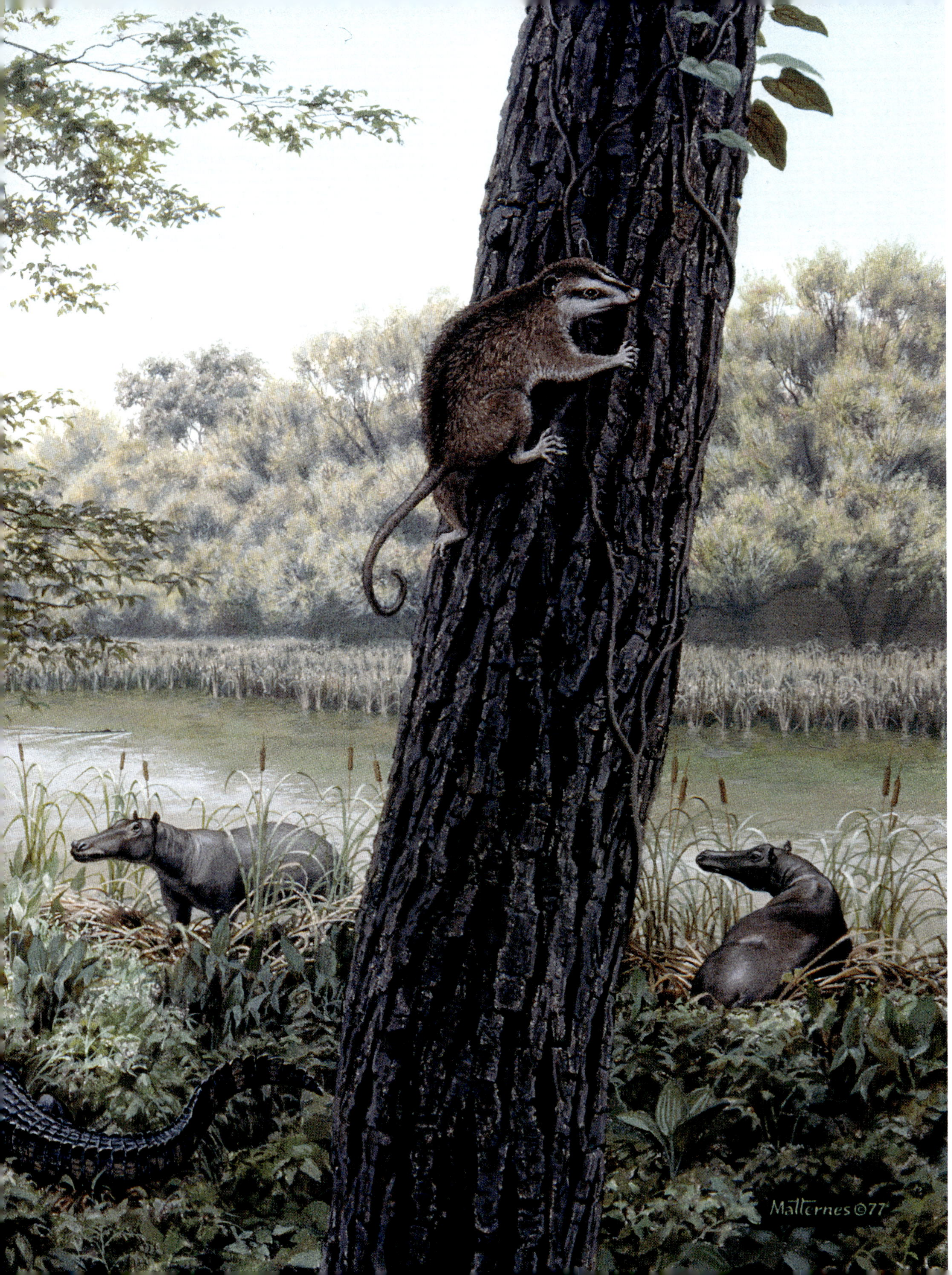

EARLY MAMMALS along the Niobrara River, near Agate Springs, Nebraska, some 35 million years ago. A group of oreodonts have startled an alligator and two hippo-like *Aepinacodon*. In a tree is an opossum, one of the oldest New World mammals still extant. Familiar vegetation, including cottonwoods, willows, beech, and dogwood, as well as cattails, are also present. A 1977 restoration of the paleoecology of Nebraska's fossil-rich Agate deposits during the Eocene epoch, commissioned by the U.S. National Park Service for its guidebook to the area.

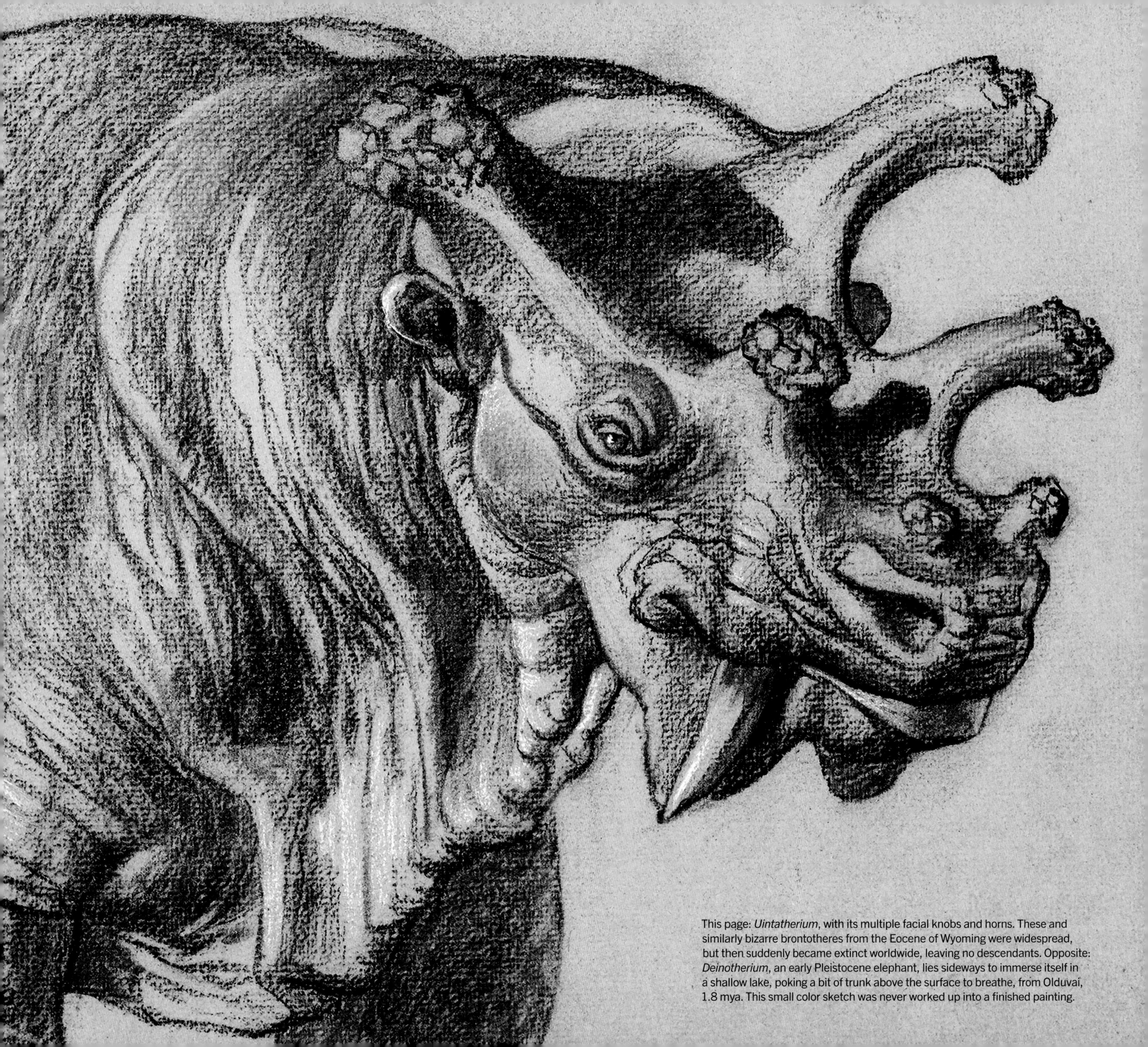

This page: *Uintatherium*, with its multiple facial knobs and horns. These and similarly bizarre brontotheres from the Eocene of Wyoming were widespread, but then suddenly became extinct worldwide, leaving no descendants. Opposite: *Deinotherium*, an early Pleistocene elephant, lies sideways to immerse itself in a shallow lake, poking a bit of trunk above the surface to breathe, from Olduvai, 1.8 mya. This small color sketch was never worked up into a finished painting.

LARGE MAMMALS

Cenozoic Murals

Around 1938, the curators at the Smithsonian's National Museum of Natural History had begun to discuss a massive renovation of their fossil halls, but World War II delayed progress until twenty years later. To assist public understanding of the fossil record, the curators decided to create murals portraying life in the Cenozoic era—the Age of Mammals, 66 million years ago until today—and sought a talented wildlife artist to work with the museum's paleontologists.

The first of four large paintings would focus on the 48-million-year-old middle Eocene deposits from the arid Bridger Basin in Wyoming, which anciently supported lush subtropical forests like those in present-day Florida. In 1958, young Pvt. Jay Matternes was a working artist at the Army Exhibit Unit at Cameron Station, near Washington, DC. Also a part-time freelancer, he had designed and painted backgrounds of two dioramas on North American mammals at the Smithsonian. While he was still in the army, Smithsonian curator C. Lewis Gazin invited him to a meeting at the museum's Division of Vertebrate Paleontology, where he first heard about the murals. Gazin encouraged him to apply for the project, and suggested that he paint

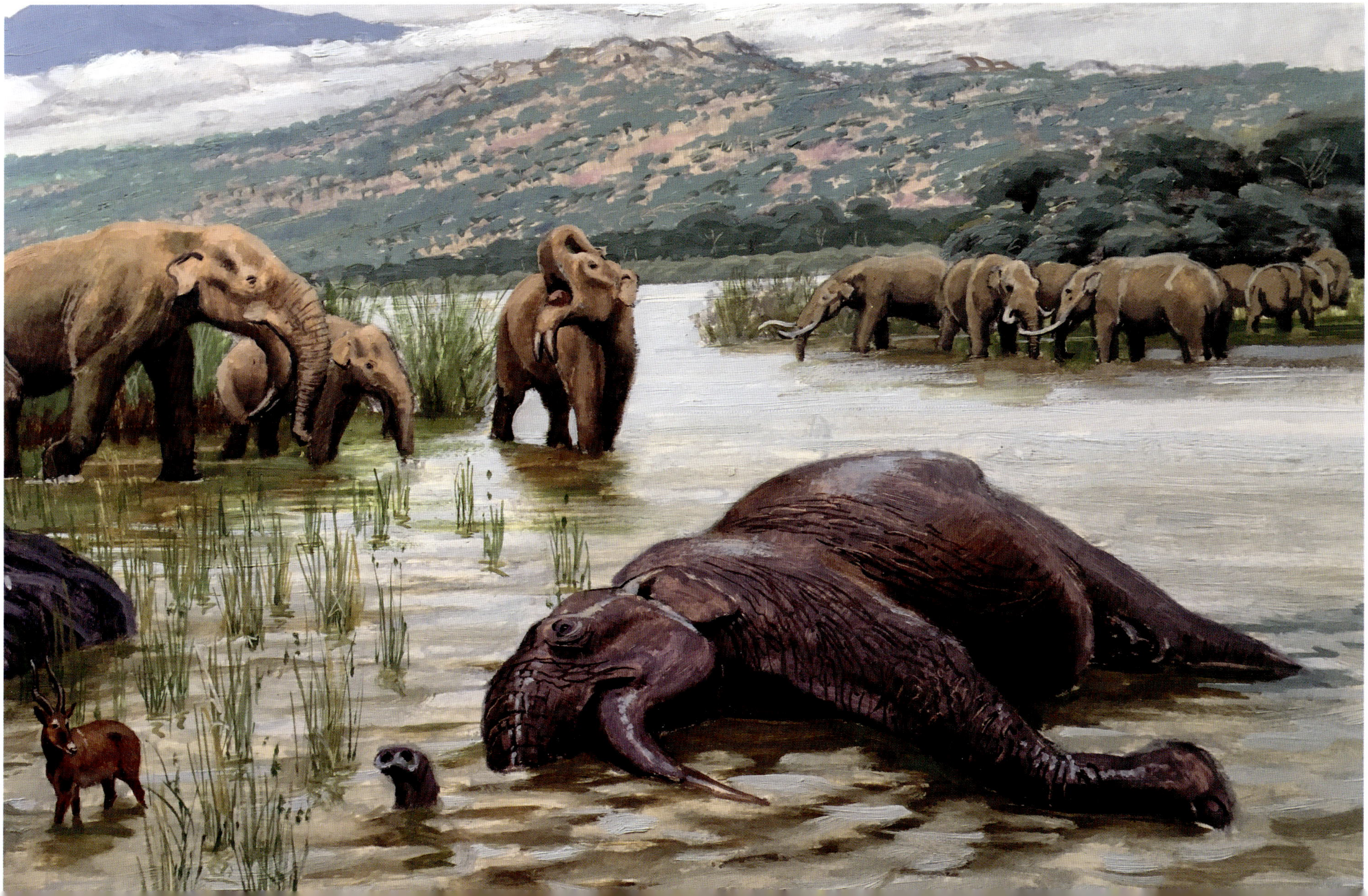

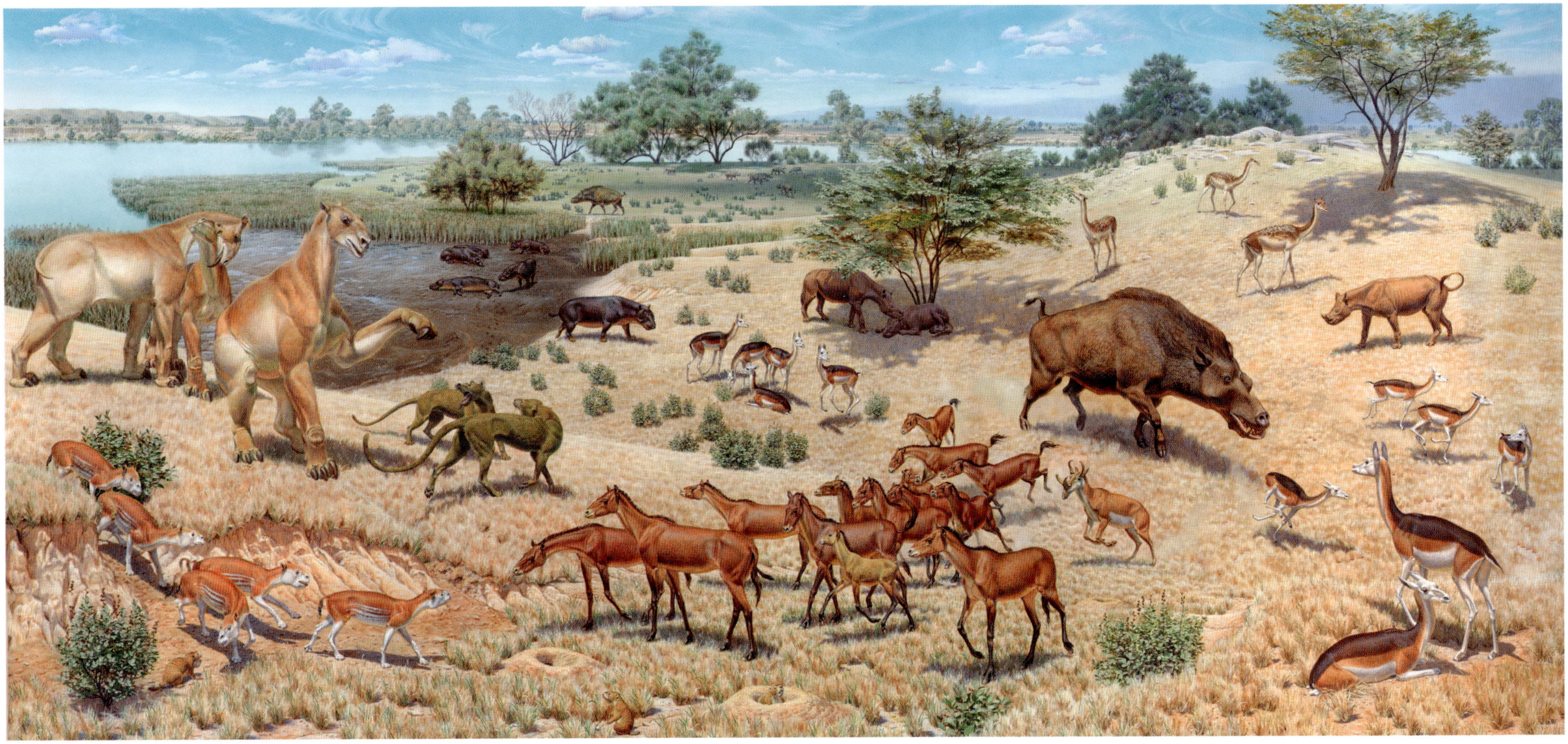

NEBRASKA SAVANNA, late Oligocene–early Miocene, Harrison Formation, 24.8–20.6 mya. Painted 1961.

a restoration of a primitive tapir, *Helaletes*, as an audition. As Jay worked on it during the evenings, the curator brought him fossils and journals as references, and he won the job. A few months after completing his army hitch, he began full-time as a "scientific muralist" at the Smithsonian.

Near the beginning of his work, young Jay became embroiled in an argument with a senior scientist over his reconstruction of the six-horned ungulate *Uintatherium*. Its skeleton had been mounted with a bent-armed stance, more like crawling reptiles, which Matternes thought impossible for a heavy mammal. He painted it with straight legs, like a modern elephant, which irritated a museum curator. "Damn it, Matternes," he said, "it would be much better if you made your paintings correspond to the skeletal mounts in the hall, rather than require expensive repositioning of the skeletons to match your paintings!" Nevertheless, he was rehired for the next 40 years to create the other five murals. As his hero, pioneering paleoartist Charles R. Knight of the American Museum, had done, he turned down repeated offers to join the museum staff, remaining an independent freelancer throughout his career.

Eocene animals preserved in the Bridger Formation of southwestern Wyoming (not shown), evolved from indigenous fauna that developed in North America but were soon to be outcompeted by arrivals from Eurasia (rodents, primates, horses, tapirs, brontotheres, early cloven-hoofed ungulates), most of which have persisted to the present. One group that appears in all six Cenozoic murals is the horse family, which originally evolved in North America but emigrated to Asia during the Pleistocene, leaving no indigenous wild horses or asses in the Western Hemisphere. All the American varieties are *Equus caballus*, the domesticated horse, which was reintroduced from Europe by Spanish conquistadores.

In the early Eocene mural, only one primitive cloven-hoofed ungulate appears, whereas in the later Eocene (36 mya) there were at least seven kinds, including pigs (peccaries), camels, and sheep. They thrived in the new grasslands, as did horses, tapirs and rhinos.

Matternes' mural of the Oligocene–Miocene era (above), based on fossils from the Harrison Formation in Nebraska, depicts a time when grasslands expanded, though they did not yet resemble prairies like the Great Plains. The large fawn-colored animals at left with the strange clawed feet are *Moropus*, which may have used their forefeet to dig up edible roots and tubers, or for defense against predators, like the bear-dogs that are confronting them.

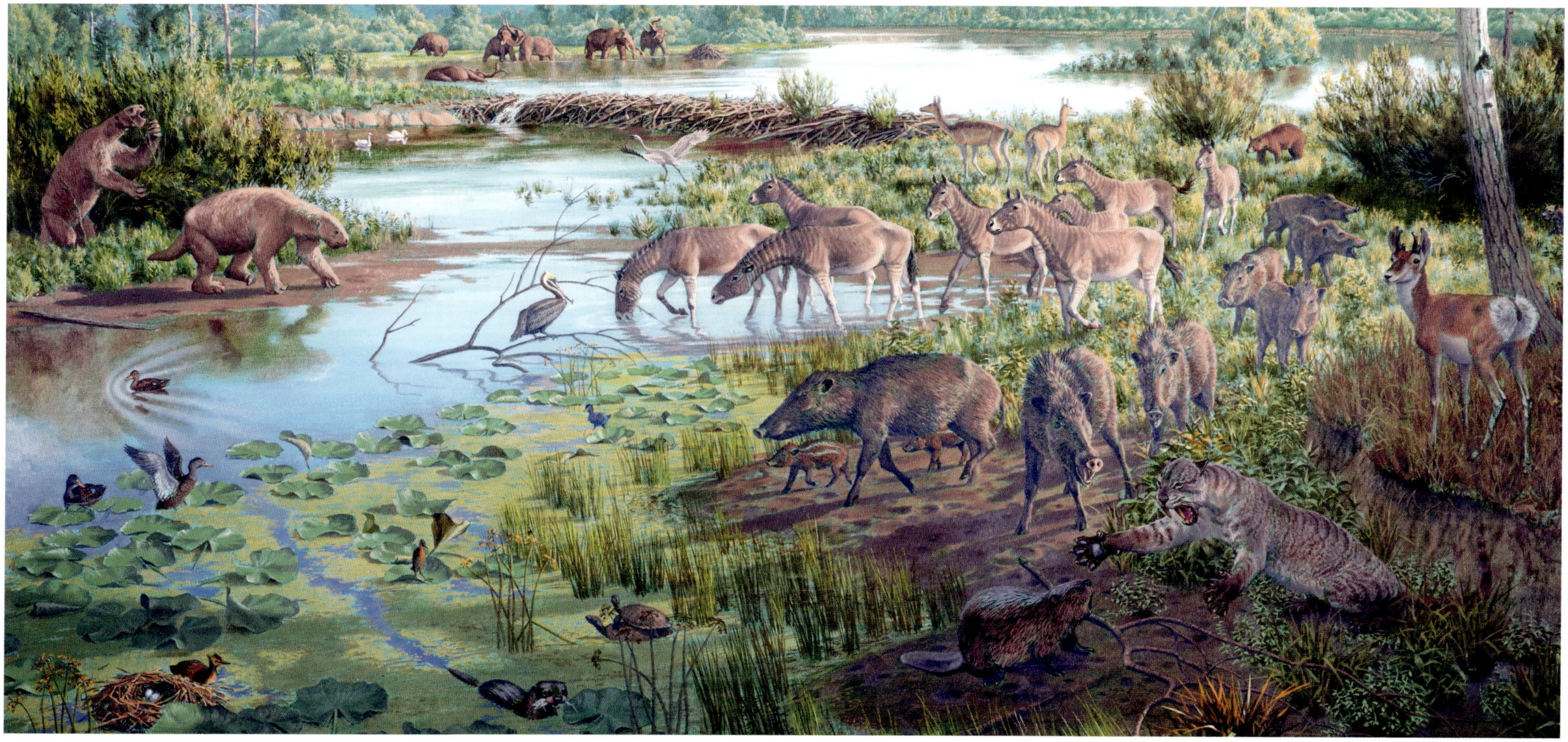

IDAHO RIVERBANK, late Pliocene, Hagerman Formation, 3.4 mya. Painted 1969.

However, they would soon bow out and disappear from North America, leaving only horses, tapirs, and rhinos to represent the odd-toed ungulates, while such cloven-hoofed species as cattle, sheep, goats, bison, hippos, boars, giraffes, and camels continued and are still with us. The exquisite little horses in the mid-foreground are *Parahippus*, a relative of the famous *Eohippus*, or "Dawn horse" of the Eocene. The fearsome extinct predator *Daeodon* right midground, is notorious for its huge jaws and teeth, and a head almost a third as large as its body. Paleontologists call it the "Demon Pig."

The final mural in the old Ice Age Hall (above) depicted a south Idaho riverbank about 3.5 million years ago with a long beaver dam in the distance. During this transitional period, the habitat of what is now the plains states east of the Rockies was a true grassland, very much like the present. The wildlife of this Pliocene continent, like that of the Miocene, was mostly adapted to being either swift runners or burrowers, like rodents. Bizarre shovel-tusked mastodons and the short-legged, barrel-shaped rhinos were browsers rather than grazers, and favored vegetation near bodies of water. Mixtures of modern North American mammals and ancient types persisted in the Hagerman "Horse Quarry."

Together with an additional mural based on the "Gold-Bearing Mucks" fossils near Fairbanks, Alaska, these six grand artworks formed a magnificent, imaginative yet scientifically-based chronicle of mammal evolution over more than 52 million years. Each mural took Matternes more than 10 months to complete; they remained on exhibition for more than 40 years and were beloved by several generations of museum visitors. After decades of new fossil discoveries, and consequent scientific revisions, they were retired from exhibition in 2018, then digitized full-size and archived; two of the replicas were incorporated in the Smithsonian's renovated "Deep Time" Halls.

In their definitive 2019 book on the Cenozoic murals (*Visions of Lost Worlds: The Paleoart of Jay Matternes*) Smithsonian scientists Kirk Johnson and Matthew Carrano describe the Matternes murals as admirably "set apart from other works of their time by their density of animals, intense attention to details of anatomy and behavior, and realistic depictions of the environment."

At the black-tie gala opening of the new halls in 2019, a woman enthused to Matternes, "Thank you for all the beauty and joy that your murals have given to so many people." Jay's sincere, understated response: "Thanks, just doing my job."

Pelorovis

Sweeping horns adorn the extinct African ungulate *Pelorovis oldowayensis*, an immense Pleistocene relative of the present-day Cape buffalo. Stone Age hunters may have trapped and killed these bovids by driving them into quicksand-like mires where they became helplessly bogged. Probably, if the local ecology was similar to today's, they attracted oxpeckers and cattle egrets, which fed on parasites and insects drawn to their fragrant mud wallow.

This restoration appeared in Louis S. B. Leakey's 1969 book *Animals of East Africa*, based on fossils his field team recovered from 1.75-million-year-old strata at Tanzania's Olduvai Gorge. Fossils of *Pelorovis*'s skull and horn cores, first discovered in 1913, were originally thought to belong to a giant sheep, which it superficially resembles. However, the skull and horns are three-and-a-half times the size of a ram's, and one-third larger than a Cape buffalo's.

From modern bovids, Jay estimated the full size of their immense headgear, and placed them at the pinnacle of hoofed animals of their time, capable of dominating the most coveted mud patches or waterholes. Matternes' resting herd vividly suggests what must have been an overpowering experience for any creature approaching them, with the lowing of the huge animals combined with pungent aromas of dung and urine mixed with mud.

Pelorovis would probably have favored open country, where their horns would not be an impediment to moving about, and would serve them well by intimidating would-be predators. The setting for this painting was based on the present-day Lake Manyara habitat, which is a mixed bag of grassland, patchy scrub, and widely spaced trees of a tropical lakefront woodland—similar to the African habitat in which ancient hominins also flourished.

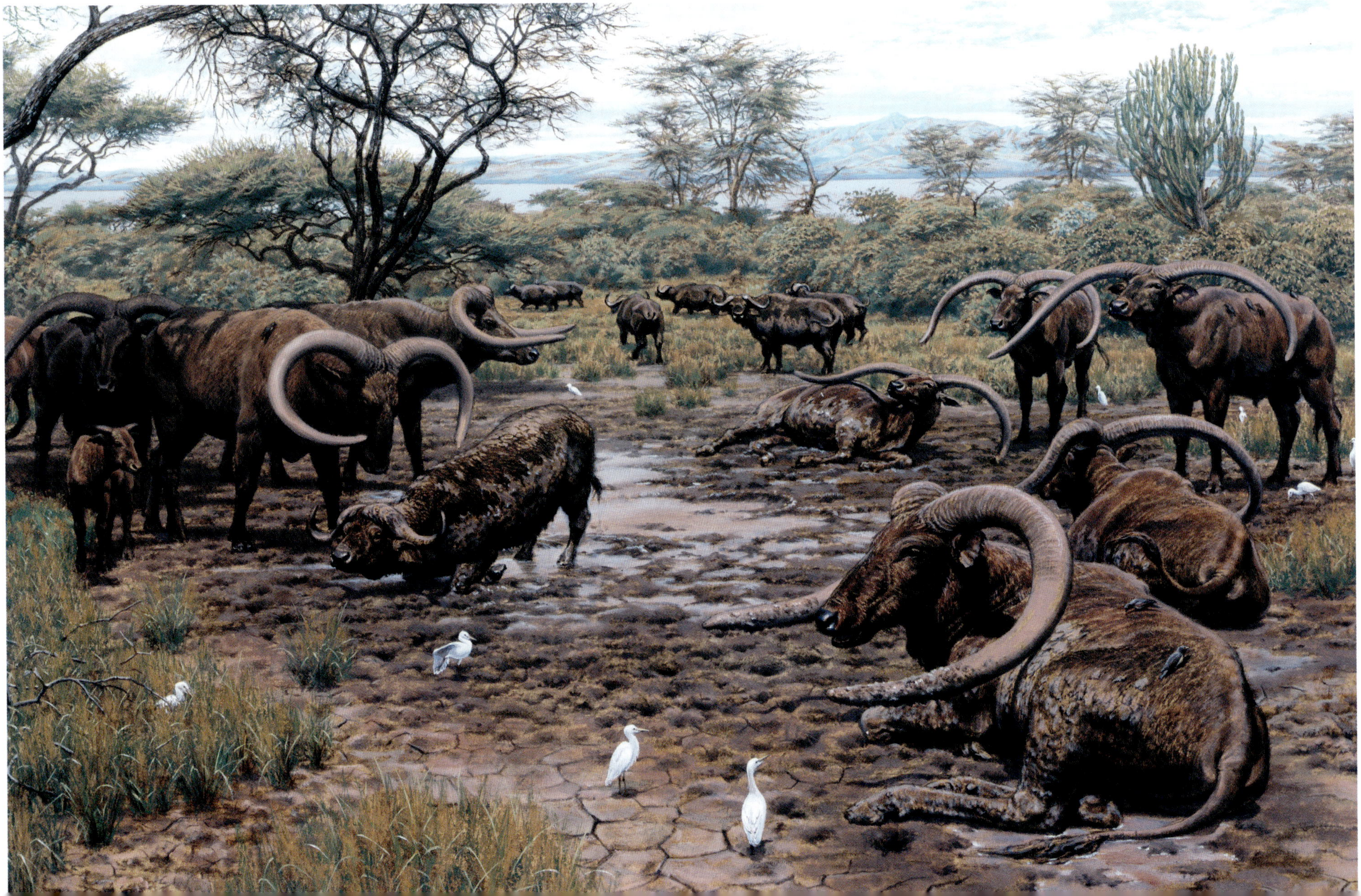

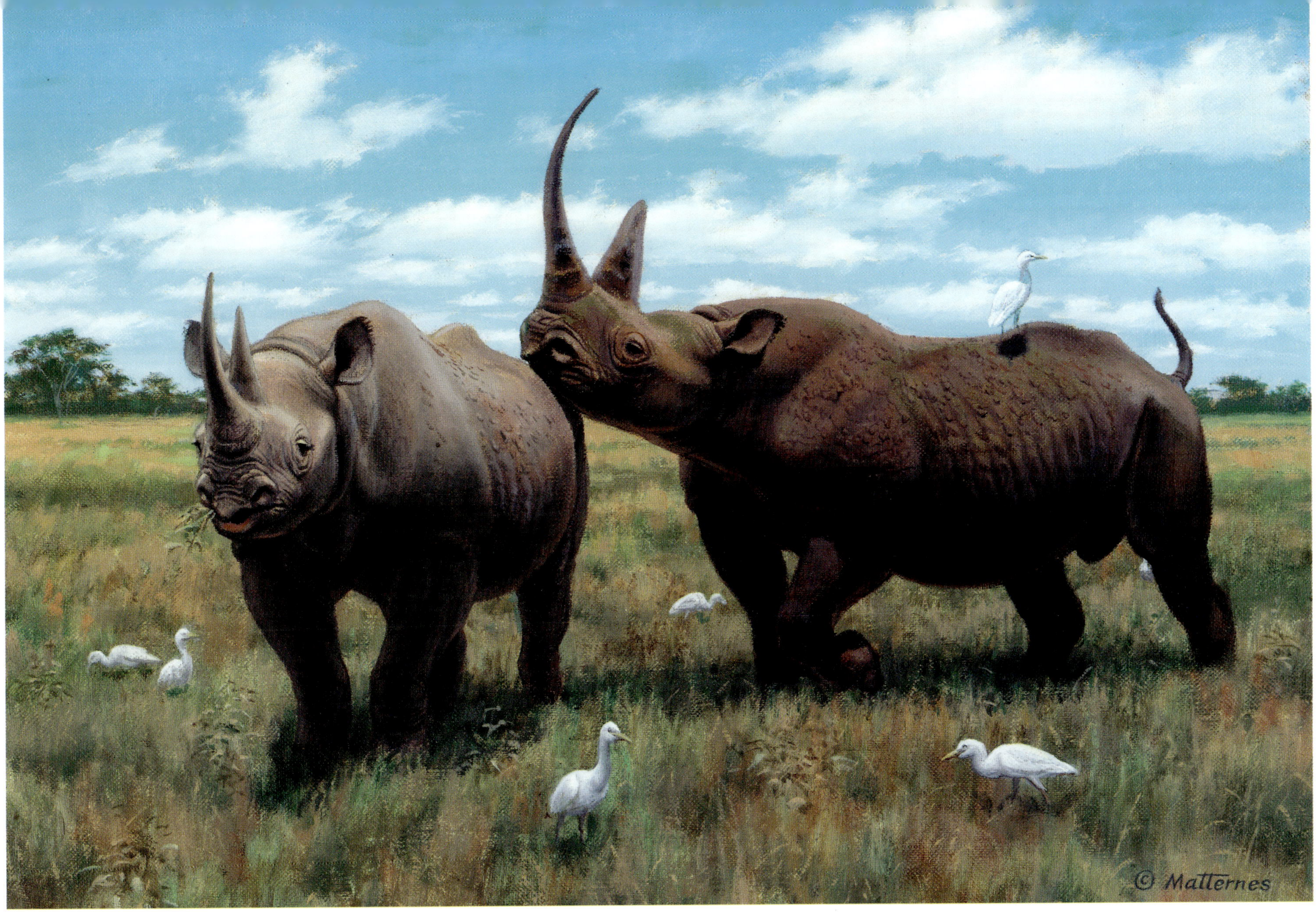

African Black Rhinos

"Rhinos of all living species have always appealed to my imagination, since they are so exotic and primitive in appearance, reminding me of the many fossil forms from which they descended.

"This painting was inspired by an amusing incident I witnessed in Tanzania's Ngorongoro Crater in 1968. A cow black rhino with her half-grown calf were approached by a male rhino who was evidently in rut. He trotted up to the cow, who was clearly not receptive, as she and her calf moved away. The amorous male followed close behind with tail raised, uttering high-pitched whinnying sounds that seemed incongruous coming from such a big and powerful animal. He appeared to be begging for her attention. Whenever the cow was moving slowly, the bull attempted to rest his head on her rump or flank. When his persistent advances began to annoy her she backed away, bellowing, and, with lowered head, charged the unwelcome suitor. He blocked her attack head on, to avoid serious injury from her two long, sharp horns. When the two collided with a thud, a cloud of dust rose from both. However, her aggressive defense didn't deter him; the last I saw of them, the lovesick male was still in tow, accompanied by hitchhiking white cattle egrets which feed on insects stirred up by the behemoths as they plow through the savanna grasses."

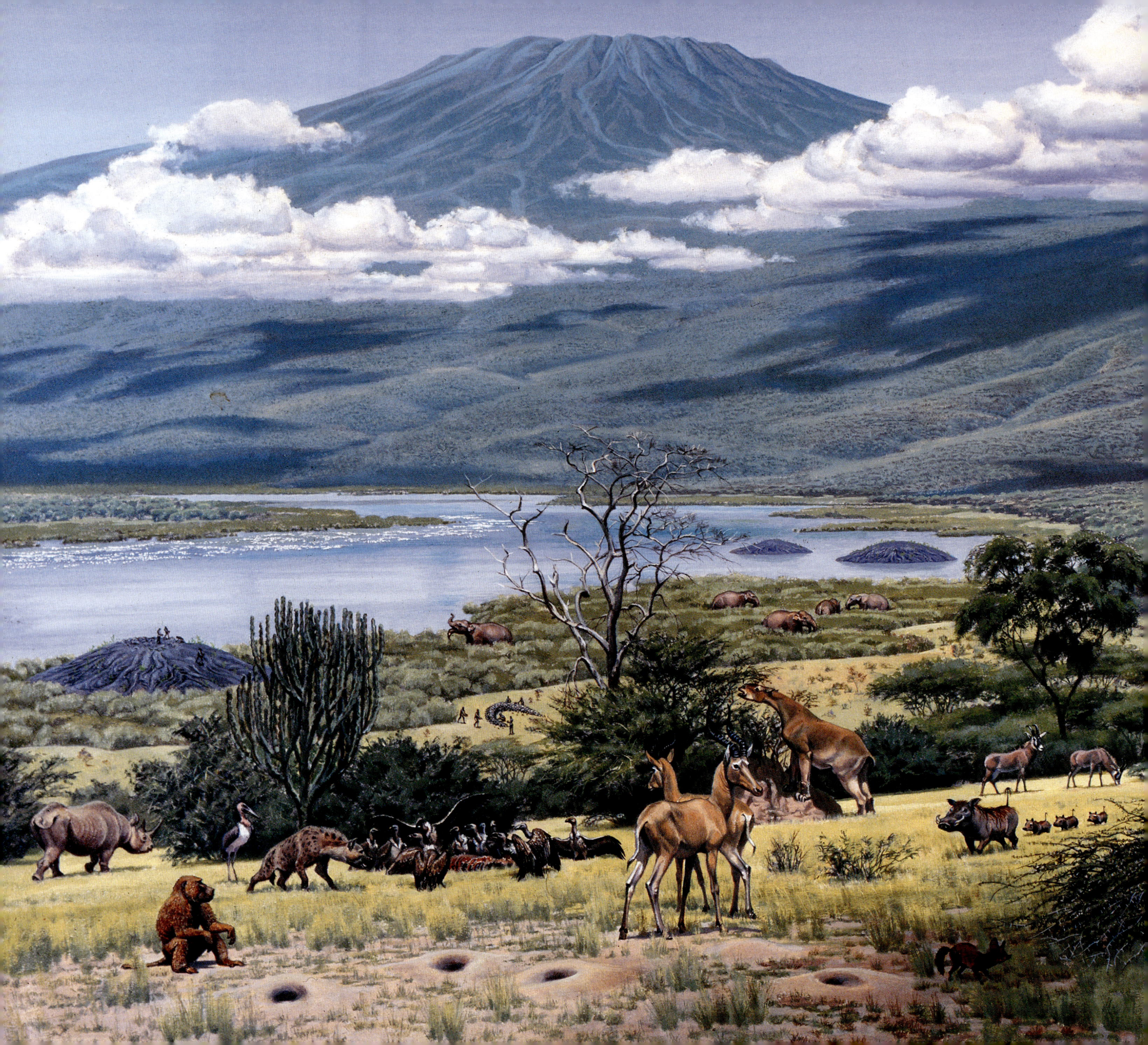

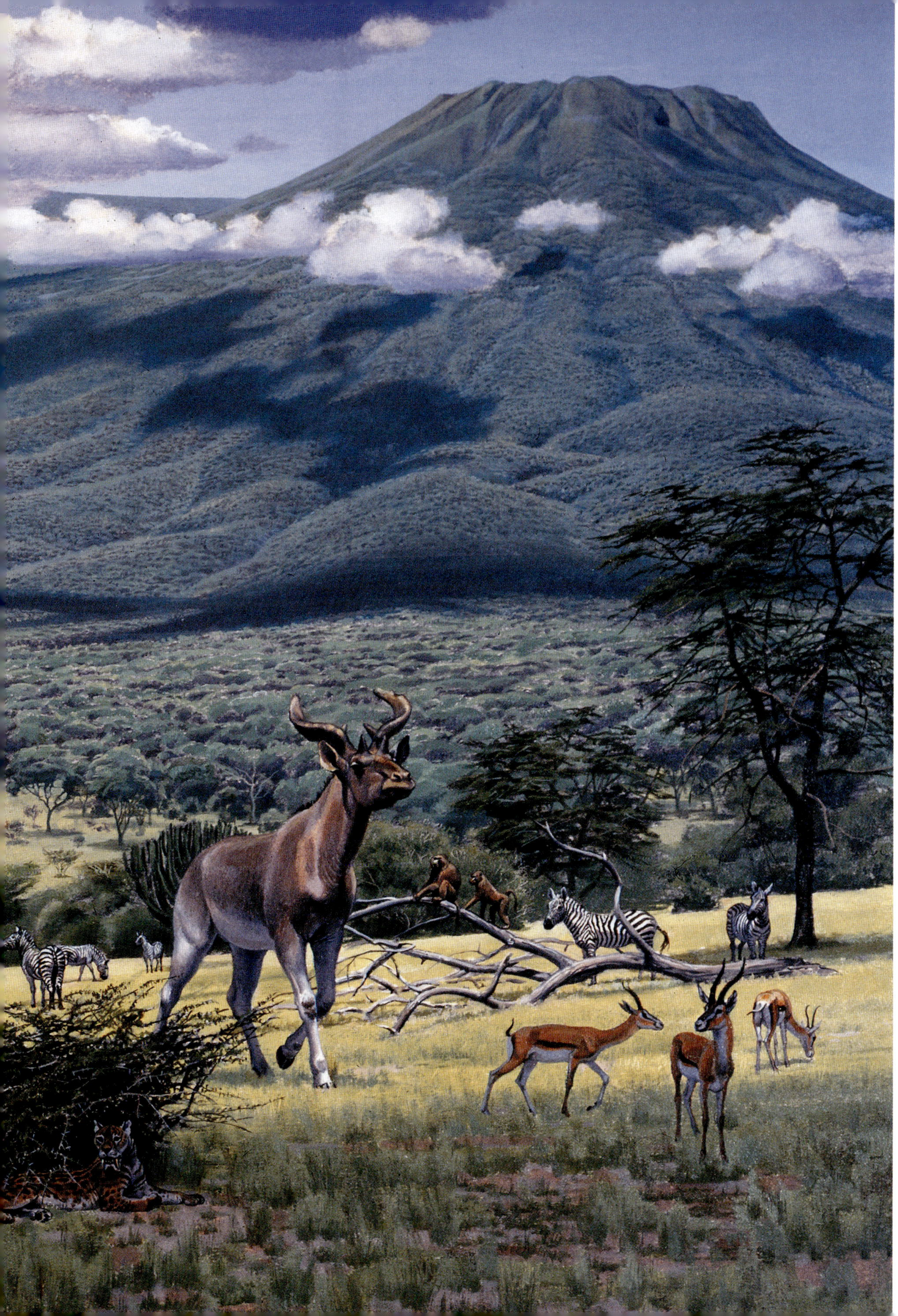

When Olduvai Was Green

After consulting with geologists, Matternes reconstructed the lush temperate ecosystem of Olduvai 1.34 million years ago for Time-Life's *Early Man* (1965), showing how the area appeared long before it became today's arid, steep-sided valley. Two prominent volcanic mountains, Ngorongoro and Lemagrut, dominate the terrain, surrounded by small lakes and swampy marshes.

Olduvai's climate was much wetter than it is today, supporting many species of plants and animals. Among those shown are baboons (left), wild pigs, ancestral antelopes and gazelles, vulture-like birds, and flamingos.

Hares and foxes similar to those living in East Africa today were present, along with several larger beasts, including rhinos (opposite, mid-background), the elephant-like marsh browser *Deinotherium*, the tree-cropping chalicothere, and the oddly antlered relative of the giraffe *Sivatherium* (right foreground), towering over some grazing Thompson's gazelles. Carnivores included lions, hyenas, and sabertooth cats.

The early hominins *Australopithecus*, *Paranthropus*, and *Homo habilis* shared this ecosystem as well. Evidence of their presence has been found throughout Olduwan strata, including a mysterious arrangement of carefully placed lava rocks (left page, distance) forming a large crescent whose purpose remains unknown. Exercising his unique ability to paint extremely small figures, Matternes has placed seven conjectural hominins near it, but it is suggested that the reader use a magnifying glass to see them.

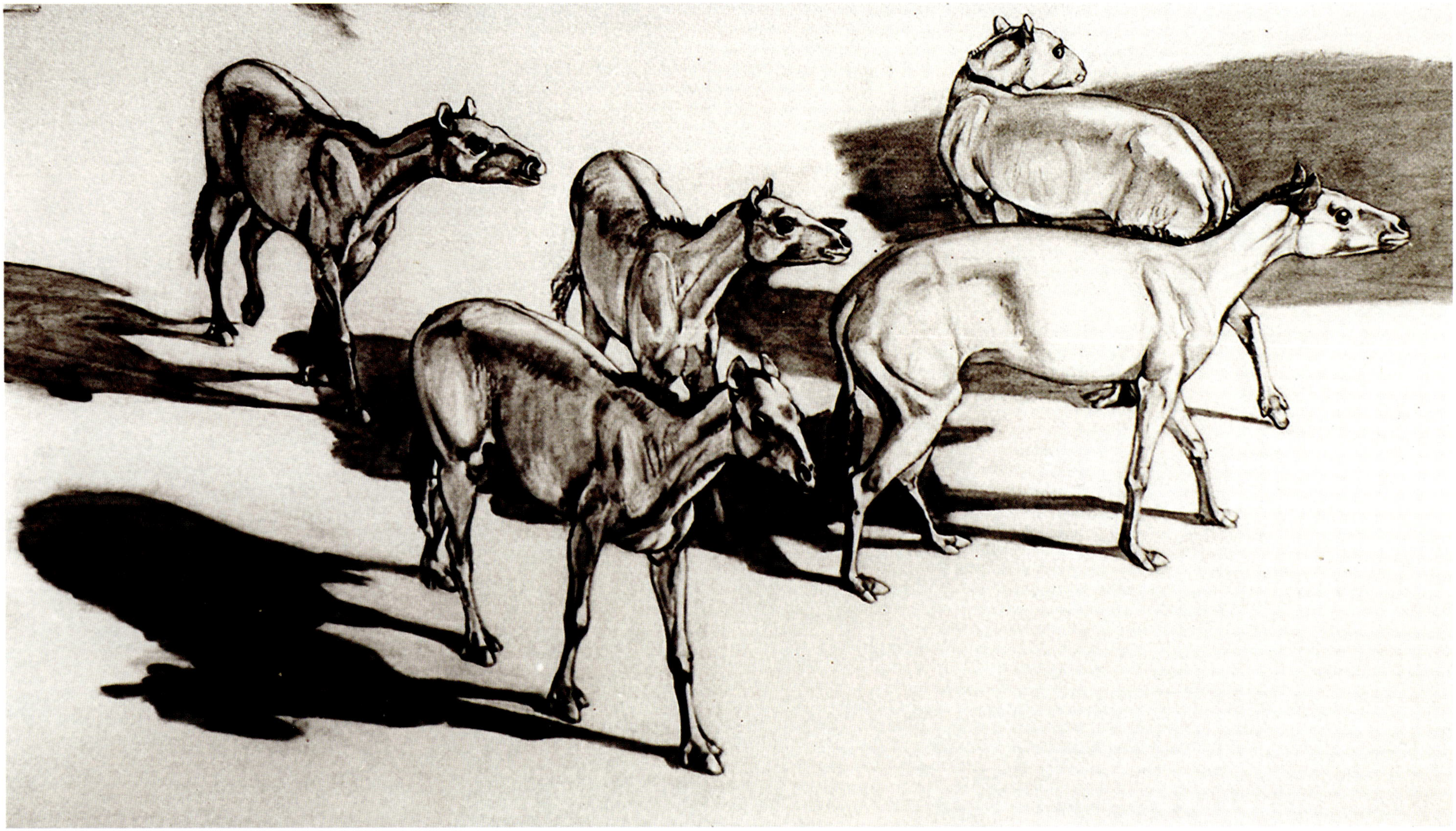

HOLD YOUR HORSES!

The Matternes Guide to Equine Evolution

"Horse evolution, according to the fossil record, is primarily a tale of teeth and toes. Because foot structure is so important to the environmental adaptation of fossil animals, I show their feet clearly in my paintings, unobstructed by ground cover vegetation. As for teeth, horses and a few other hoofed animals started out with flat molars that were easily worn down. As the climate cooled and vegetation became tougher, their molars developed vertical shafts that continually replace the enamel as it wears down. Thanks to these innovations, the unobtrusive horse family (and a few other ungulates) were able to survive for millions of years, while some more imposing-looking contemporaries—such as the magnificent titanotheres—all died out.

"Over millennia, the horse's feet, frame, and dentition adapted to the drier, cooler landscape. The earliest little horse, *Hyracotherium*, known from Wyoming's semitropical Eocene and elsewhere, was originally no larger than the proverbial fox terrier. About 55 million years ago, this 'dawn horse' (originally called *Eohippus*) foraged along the forest floors on its four-toed feet. By Oligocene times, its successor, the larger, collie-sized *Mesohippus*, had lost a digit. Next appeared the pony-sized *Merychippus*, equipped with a single central toe

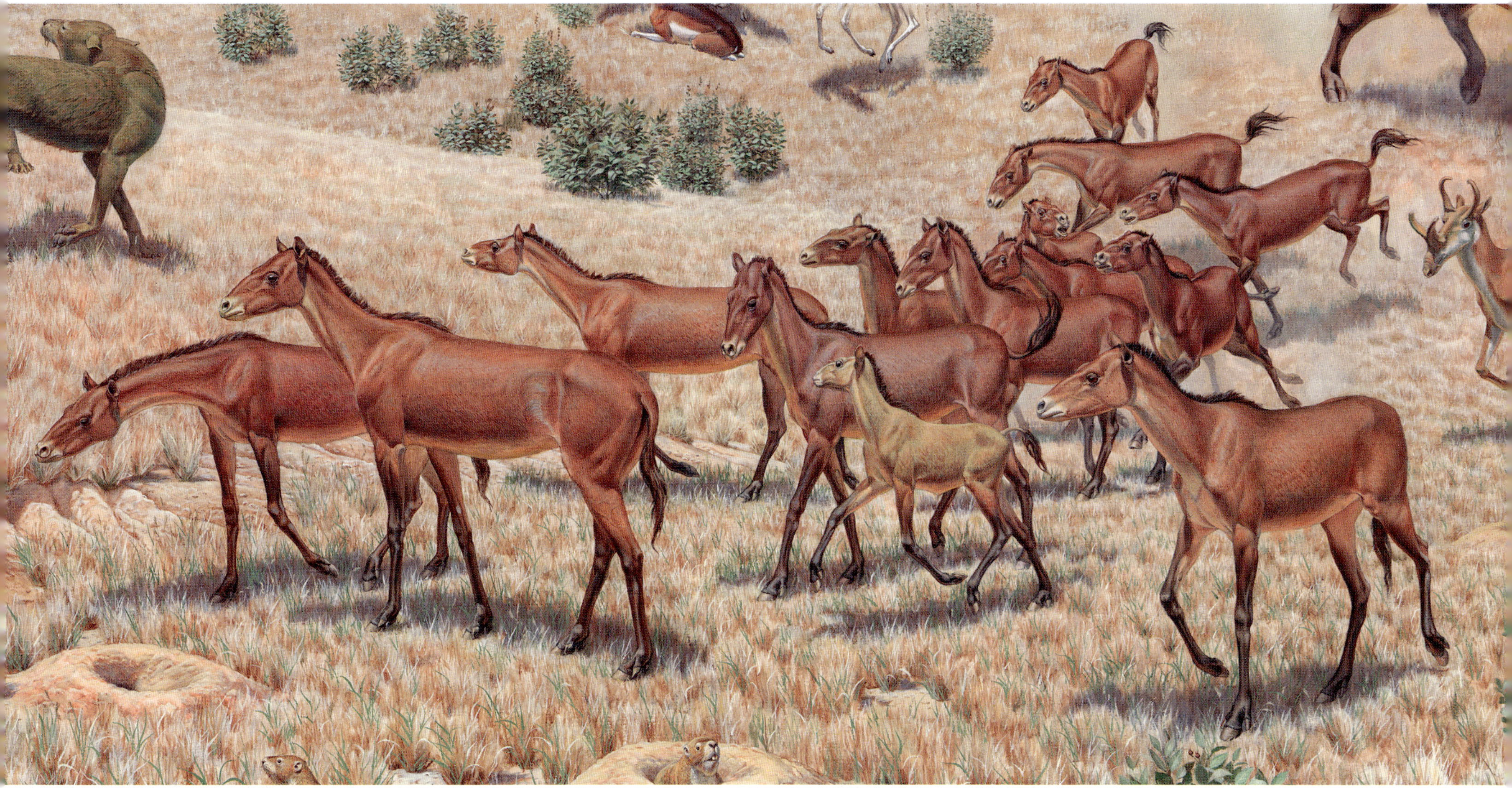

Above: An exquisite herd of elegant little horses, *Desmatippus*, from the early Miocene mural set in the Nebraska Plains, circa 20 mya. At upper left is a bear-dog, *Daphoenodon*, one of their frequent predators. At middle right is the head of strangely horned *Syndyoceras*. Opposite: *Mesohippus* ("middle horse"), late Eocene to early Oligocene, 38–33 mya, from South Dakota and Nebraska. They had three toes on both their forefeet and hind feet, with hooves evolved from the middle digit.

on each foot, flanked by two vestigial ancillary toes, which further enhanced its running ability. *Merychippus* was joined during the Pliocene by *Pliohippus*, the direct ancestor of the modern horse and the first to evolve a hoof on all four feet—a unique feature of horses and part of their basic adaptation as a runner on the plains.

"I gave the little *Orohippus* horses (a descendant of *Eohippus*) a spotted coat, which may have conferred some protective coloration in their Eocene forests, like the pelage on such small hoofed animals as chevrotains and duikers today. Actually, the genes for spotted or striped coats are common in the horse lineage, even in domestic breeds, emerging as dappled, piebald ('pinto') or reticulated, and striped, in zebras. However, the relict population of Przewalski's wild horse from the Mongolian steppes seems a reasonable color model for Pleistocene horses: solid beige, with dark legs, neck, and head. Similarly, the little *Parahippus* in my Miocene mural is uniformly colored, with dark 'stockings.' Blending into a monochrome beige landscape on the grasslands may have conferred some protection from predators, which often lack color vision.

"Other even-toed hoofed animals since the Eocene have also lost digits. Cattle, deer, giraffes, antelope, sheep, goats, and swine have lost all but two primary

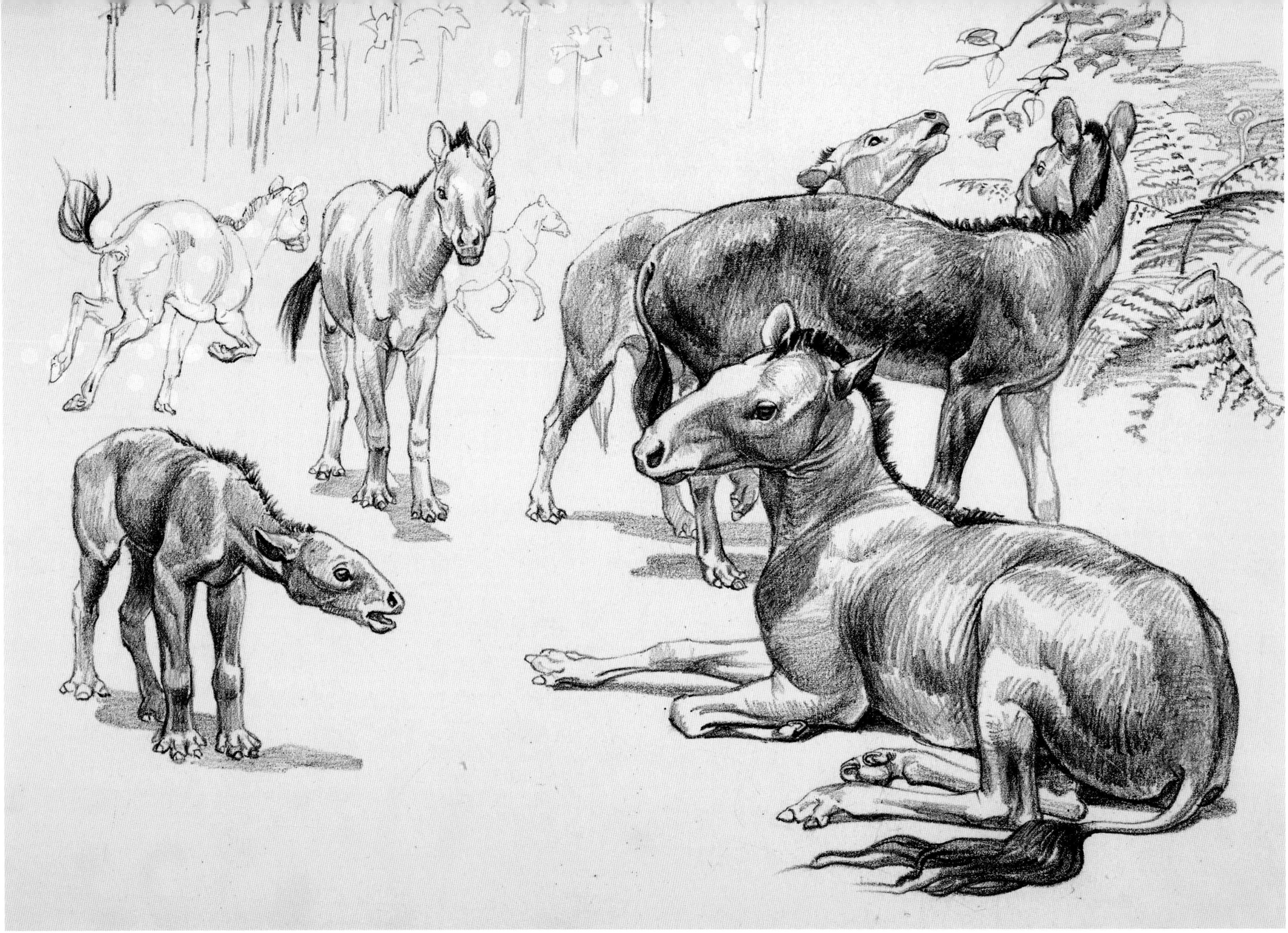

Seen here in a preliminary drawing for the Eocene mural, *Orohippus*, from the Bridger Formation of eastern Wyoming, was a browser, not a grazer, that inhabited semi-tropical forests. It had four toes on its front feet and three on its back legs.

functional hooves. The only member of the even-toed hoofed animal group to retain multiple fore and rear digits is the hippopotamus, which inhabits rivers or lakes with soft muddy shores, where a splay-toed foot is advantageous.

"After prospering in North America for millions of years, horses disappeared from the continent less than 15,000 years ago for unknown reasons. By the late Pleistocene, there were no longer any native horses in North America. When the Spanish conquistadores brought domesticated horses (*Equus caballus*) with them in 1519, they quickly became reestablished in the Western Hemisphere, but this time as working animals for farm labor, construction, transportation, and warfare—until their primacy was displaced overnight in the late 1920s by the automobile. Nevertheless, there are still 86,000 free-range wild mustangs in the western United States, with more than half of them in Nevada."

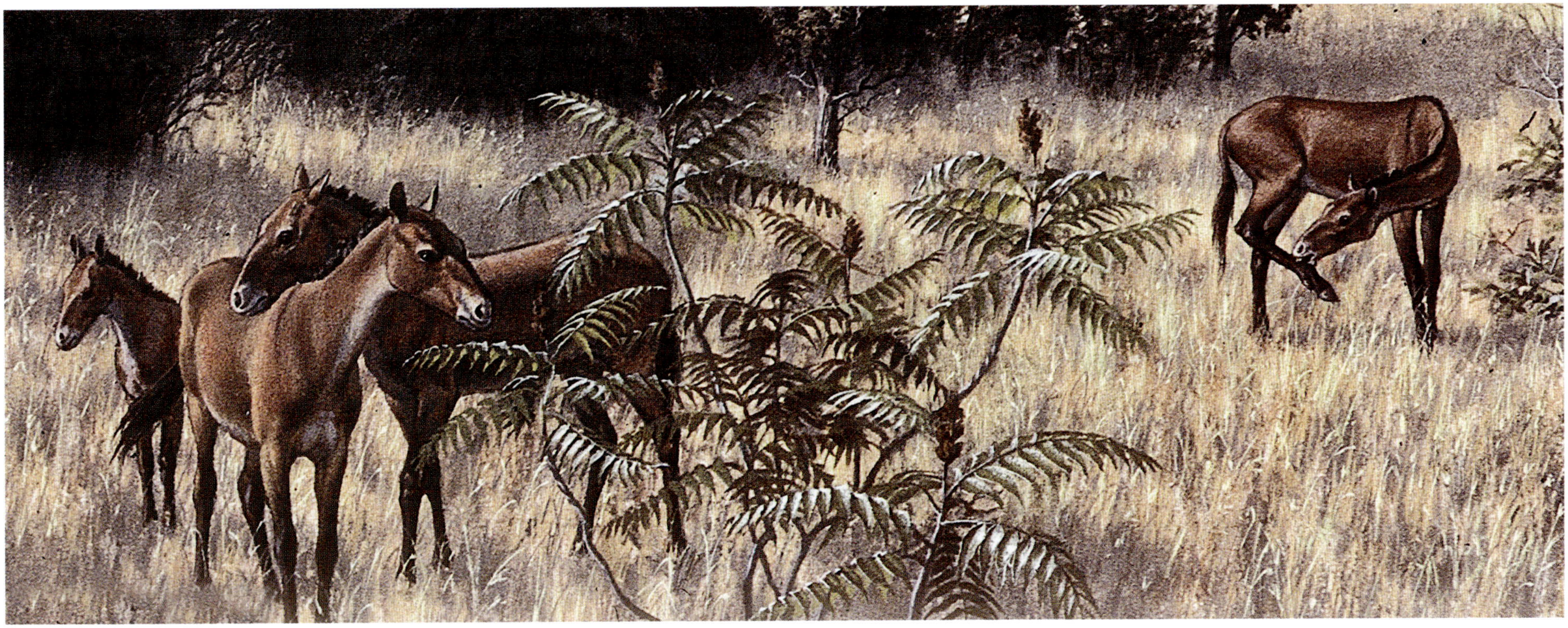

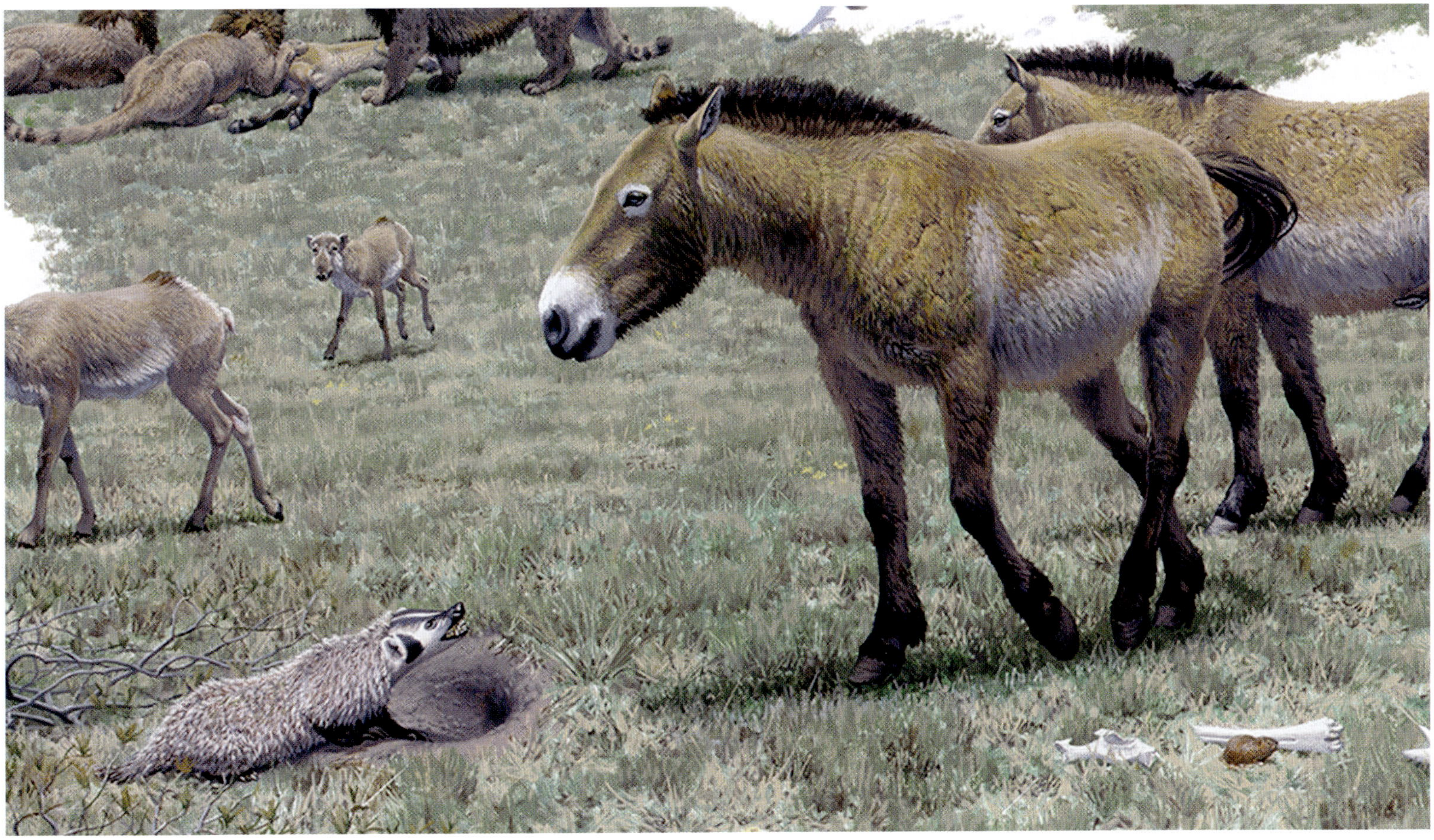

Above: *Desmatippus*, early Miocene horses, from Matternes' illustration for the National Park Service guide to the Agate Springs fossil beds in Nebraska. Left: A Late Pleistocene modern horse, *Equus*, is confronted by a feisty badger, a detail from his Fairbanks, Alaska, mural.

Overleaf: A small herd of *Merychippus*, Shetland pony–sized three-toed horses, gallops across a plains landscape that includes cottonwood, elms, sycamores, willows, and cedars. Local climate had become drier and cooler in the middle to late Miocene of Nebraska, 15 mya, producing true prairies, which were a boon to ungulate populations. The equine lineage developed longer limbs and greater lung capacity for sustained running. Although its usual hunting grounds are woodlands along waterholes and streams, a jaguar-like cat, *Pseudaelurus*, tries to bring down a passing horse.

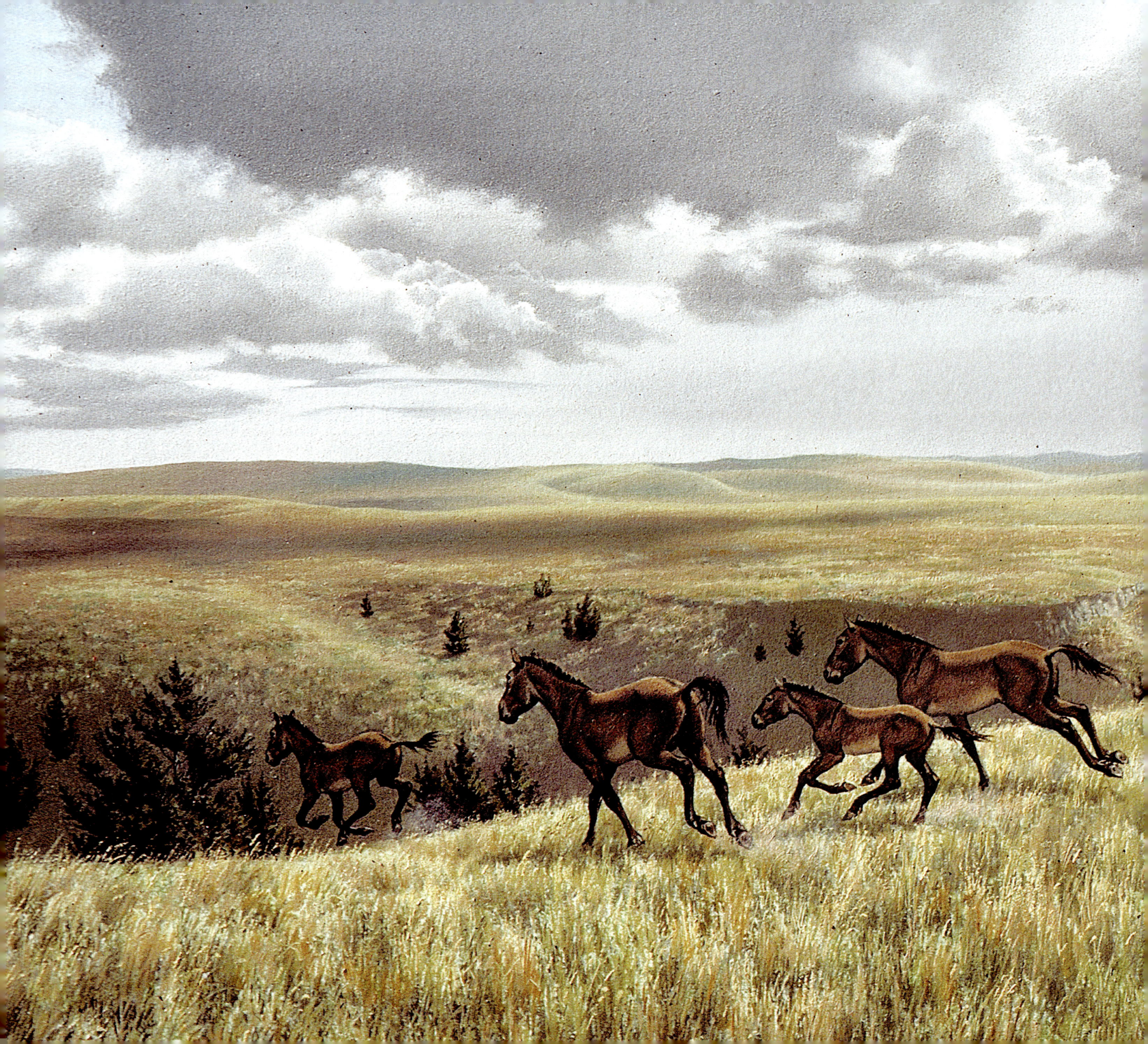

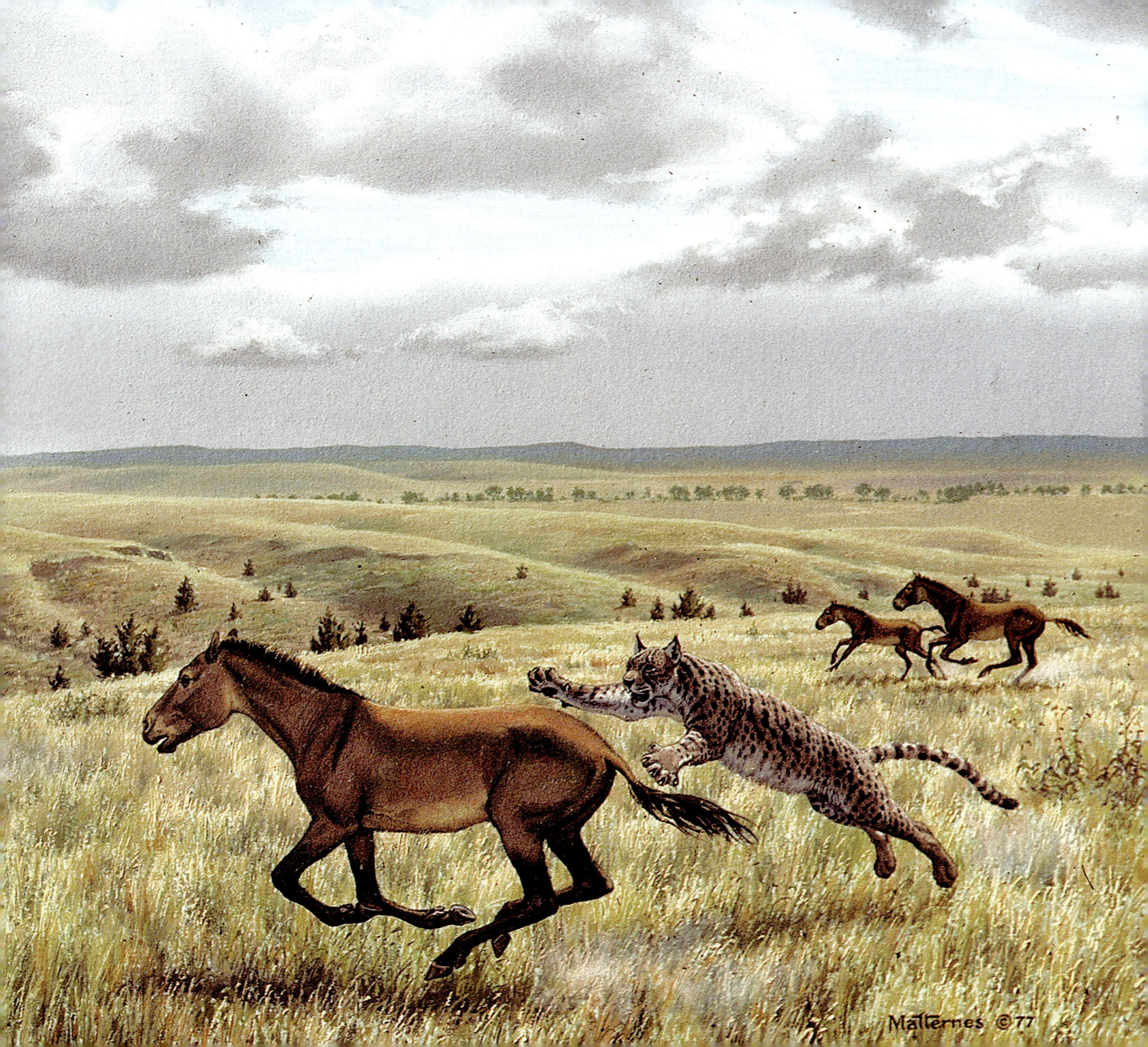
Matternes ©77

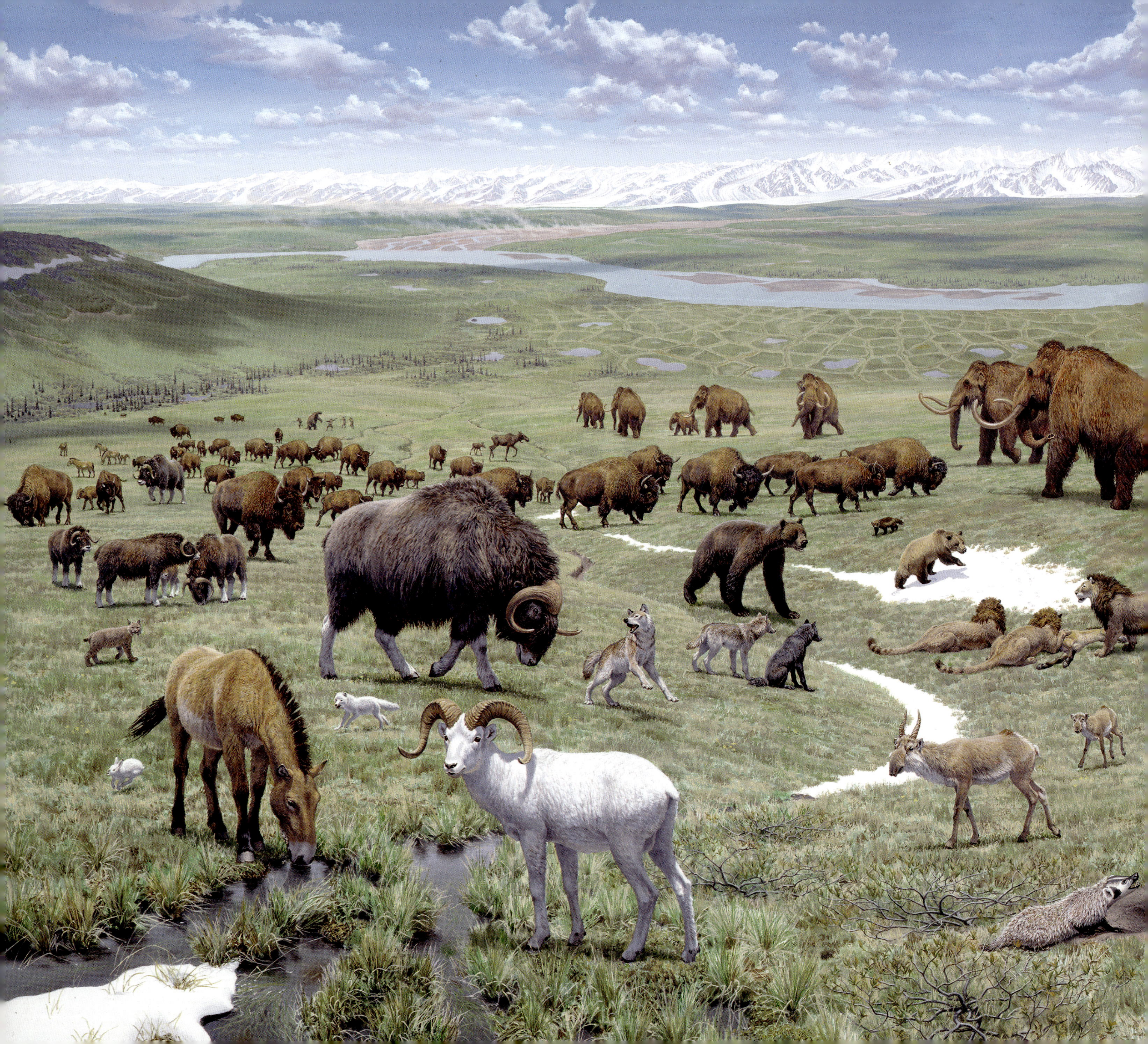

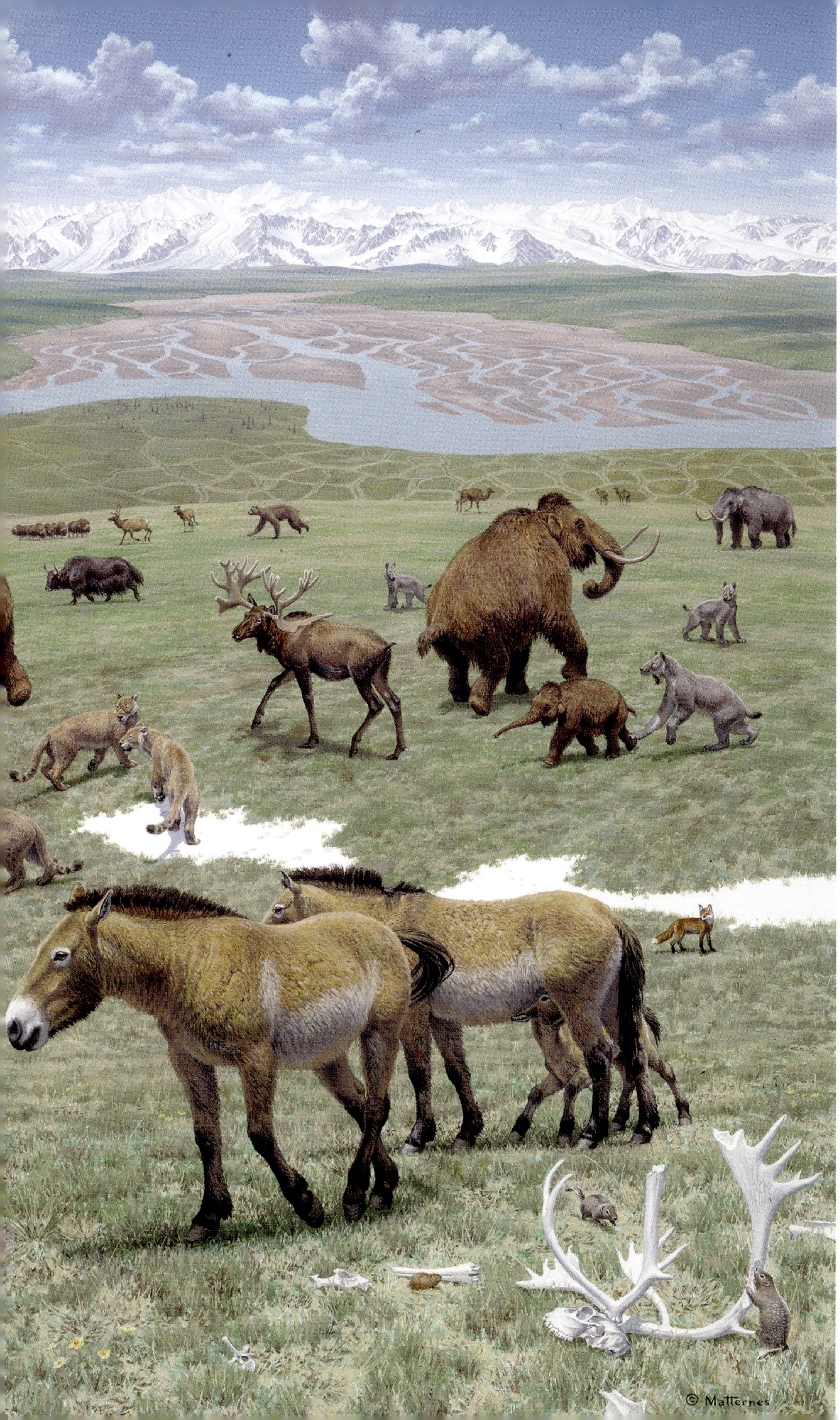

FAIRBANKS MURAL

An Ice-Free Enclave in Ice Age Central Alaska

In 1975, Matternes completed a 12 by 20 foot panorama of North American animals from the 14,000-year-old "Gold-Bearing Mucks" of Fairbanks, Alaska. It depicts 31 different mammals that thrived in the central Alaskan steppes during the latter part of the last Ice Age. Many of those species continue to exist today, although 11 went extinct. The unique glacier-free valley supported such familiar species as wapiti (American elk), caribou, Dall sheep, and musk ox, but also 12-foot-high woolly mammoths, camels and saber-toothed cats. This refugium for tundra animals (and for the first human immigrants coming from Asia into central Alaska) was shielded from creeping two-mile-thick ice sheets by the mountains of the southern Alaska Range. (Distances are condensed in this idealized landscape.) Although the habitat covered hundreds of square miles, the ground was permanently frozen and the vegetation tough and sparse.

Silt deposits accumulated there during the late Pleistocene, and remained naturally refrigerated and undisturbed until the twentieth century, when they were flushed with high-pressure hoses to uncover gold deposits underneath. As a byproduct of these mining operations, many mummified carcasses turned up, in addition to bones and teeth and bits of flesh and fur of the late Ice Age animals that had lived here. Among the grazers were caribou, yak, saiga antelope (still roaming the Kazakhstan steppes today), large-horned bison, camel, mastodon, and wapiti, which attracted an array of formidable predators including the Eurasian lion (much larger than its African cousin), wolf, grizzly bear, fox, short-faced bear, and saber-toothed cat. In the far distance, for those with sharp eyes, a tiny image can be seen of several spear-wielding human hunters surrounding a giant ground sloth. When Matternes occasionally added such near-microscopic drawings, he never employed any magnifying lenses.

A group of hunters surrounds a giant ground sloth in the extreme background of the Fairbanks mural. When the preliminary painting was nearly finished, it contained no humans, but a new curator insisted that there was fragmentary evidence that ground sloths were present in Alaska at the same time humans had arrived there, so Matternes reluctantly obliged by creating this detailed miniature scene. However, it is so small relative to everything else in the painting that most viewers miss it altogether.

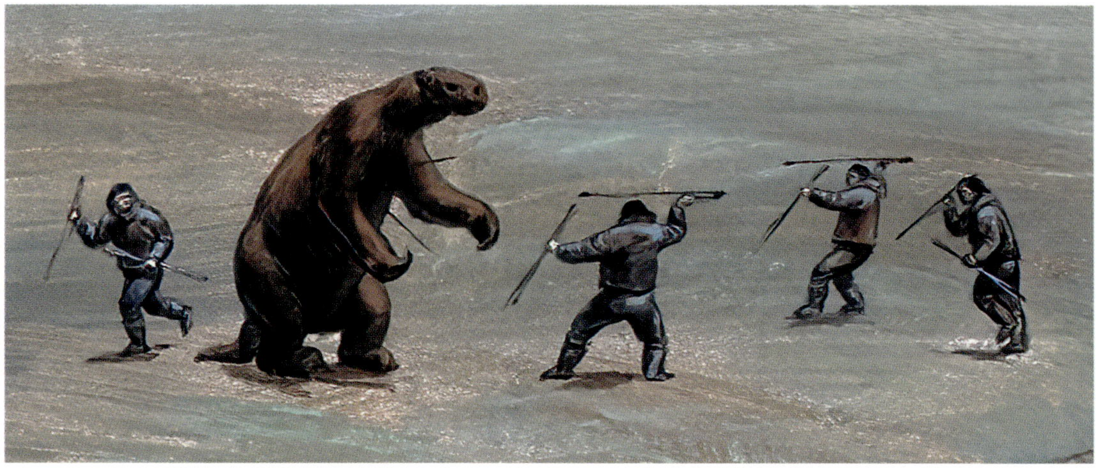

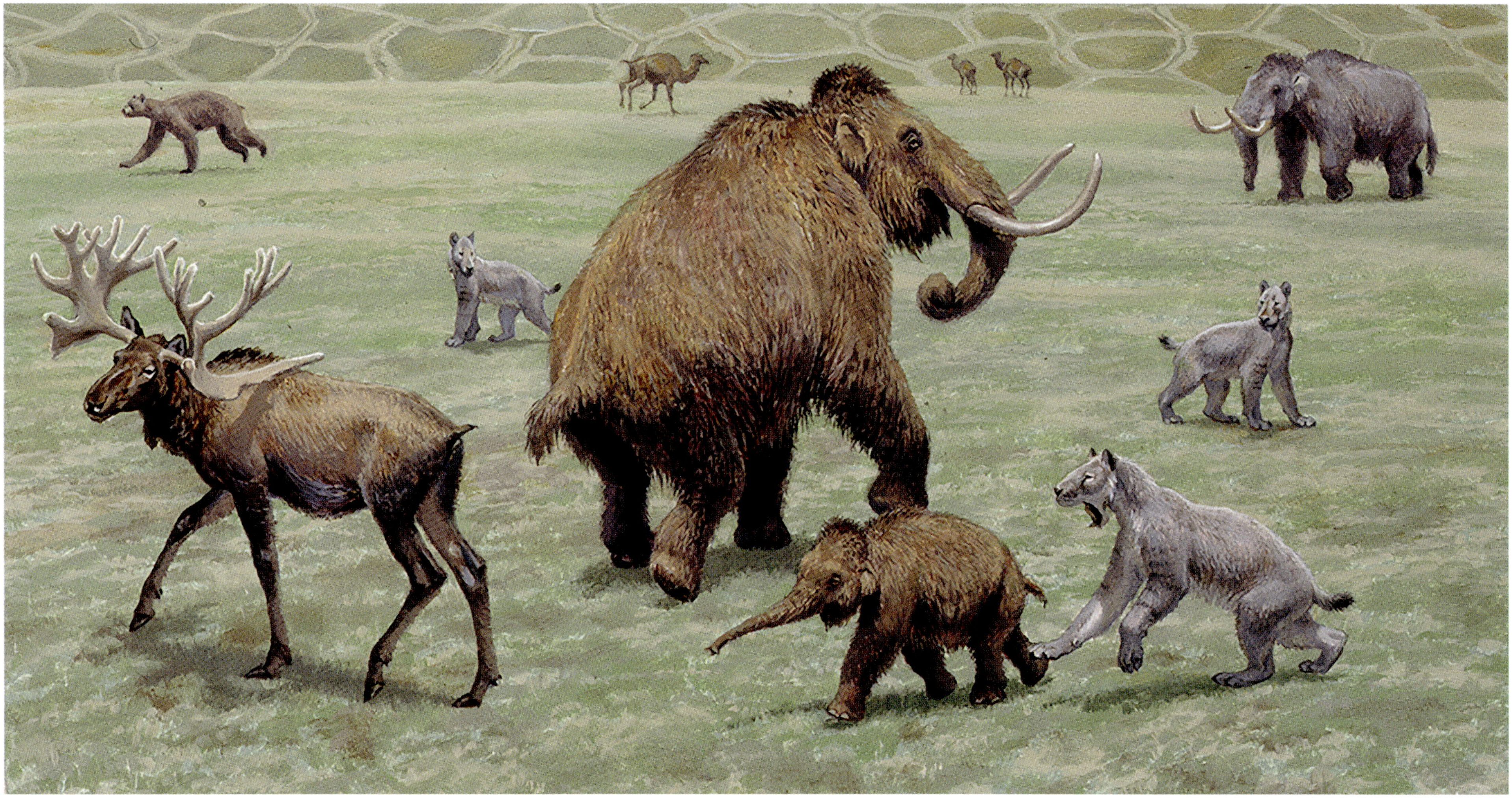

Detail from the Fairbanks mural shows a woolly mammoth mother with calf, which is being attacked by three saber-tooth cats. In the mid-distance on the right is a mastodon, while camels and a bear roam in the far distance. A large moose-deer, *Cervalces*, with its palmate antlers, walks away in left foreground.

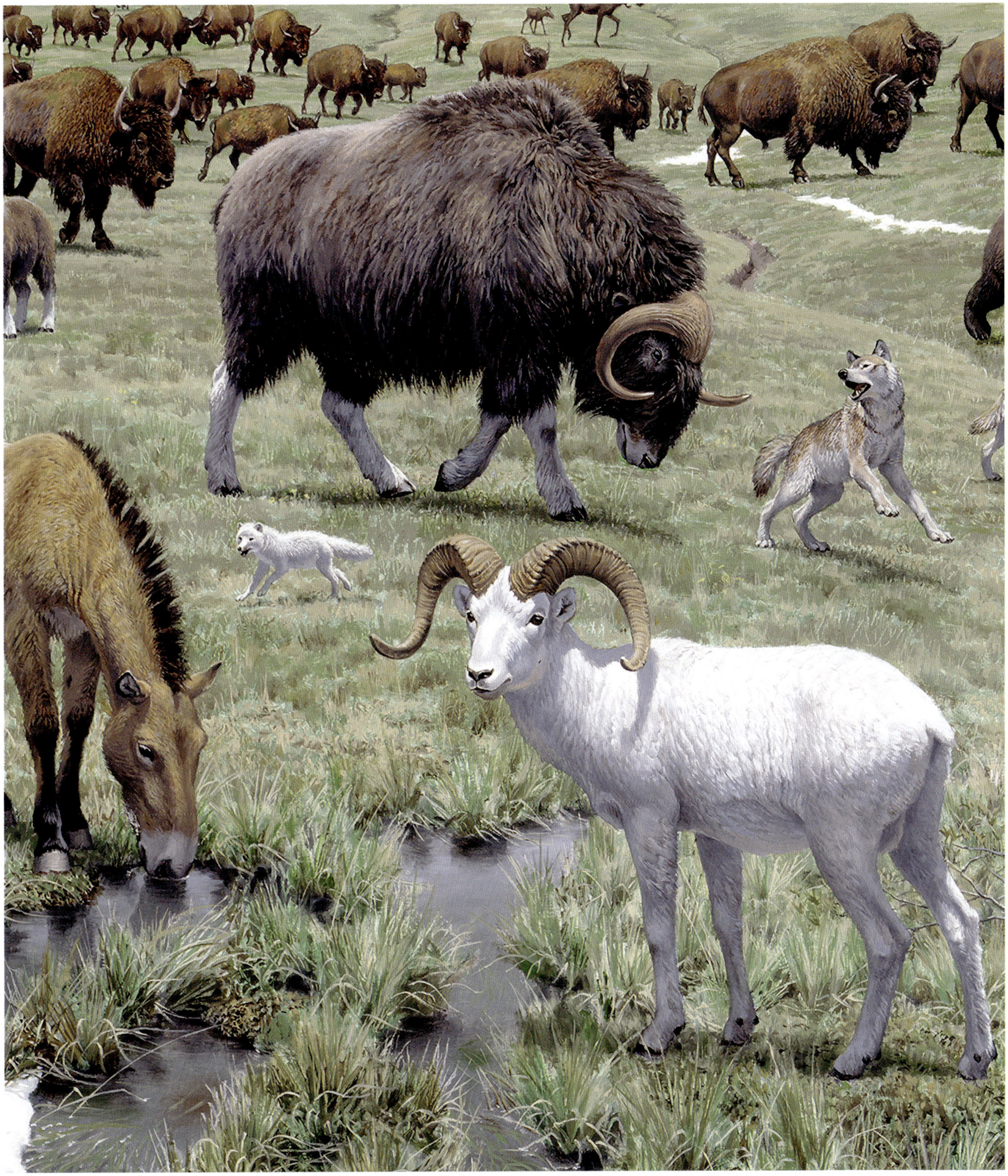

Grouping from the Fairbanks mural shows the giant musk ox *Ovibos* worrying a wolf. Behind them is a herd of bison, and in the foreground a Dall's sheep. Apparently, the sheep originally inhabited the plains with the others, but was eventually pushed up to higher elevations by human hunters and now is found only in the mountains.

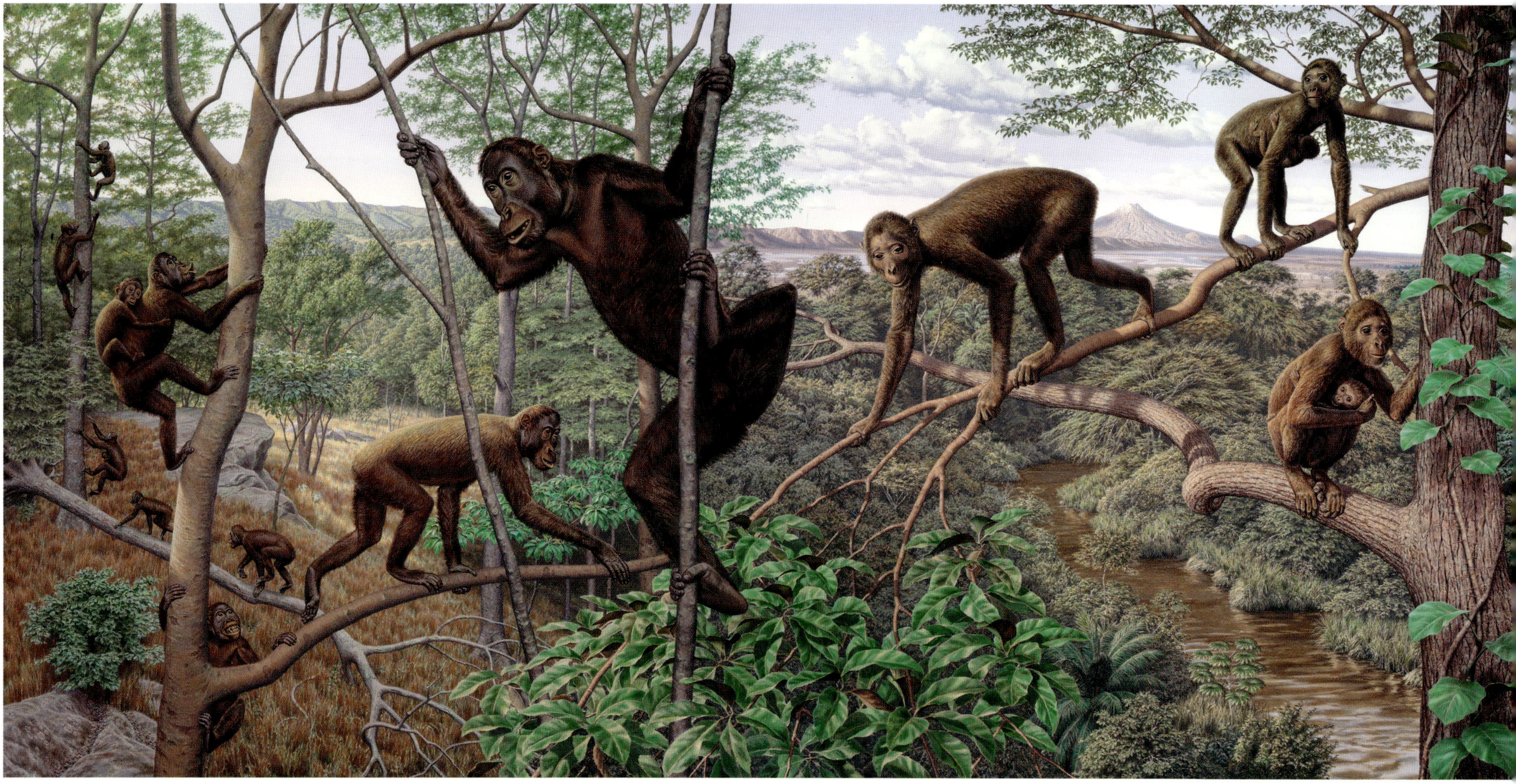

PRIMATES

Preliminary Color Sketch for AMNH Fossil Primates Mural

In 1992 Matternes was commissioned to paint a mural about 22 feet long for a new Hall of Human Biology and Evolution at New York's American Museum of Natural History. The challenge: a visual summary of the early history of the order Primates, which includes lemurs, monkeys, apes, and humans. The anthropology curator (Ian Tattersall) and designer (Willard Whitson) wanted him to create a continuous series of vignettes that would read from right to left, as the story of fossils brought to life. This masterwork, which has since been moved to the museum's Spitzer Hall of Human Origins, is not intended to represent a linear sequence; rather, it transports the viewer on a journey through time, pausing to capture snapshots of species as the order evolved and diversified over the more than 50 million years of the Age of Mammals. Beginning at the far right, two frisky squirrel-like *Plesiadapis* males confront one another in a dispute over a territorial boundary on the limb of a dawn redwood. Early primate relatives with gnawing front teeth, they lived in European and North American temperate forests during the Paleocene epoch, some 58–55 million years ago. Early ginkgo trees and conifers are shown as abundant in their forest habitat.

To the left of this scene is the vertical-clinging-and-leaping *Notharctus* that also lived in the extensive forests of Europe and North America, but about 10 million years later. Fossil skeletons of these lemur-like primates, best known from Wyoming, date from the middle Eocene epoch, around 45 million years ago. In addition to his usual meticulous restorations from the fossil evidence,

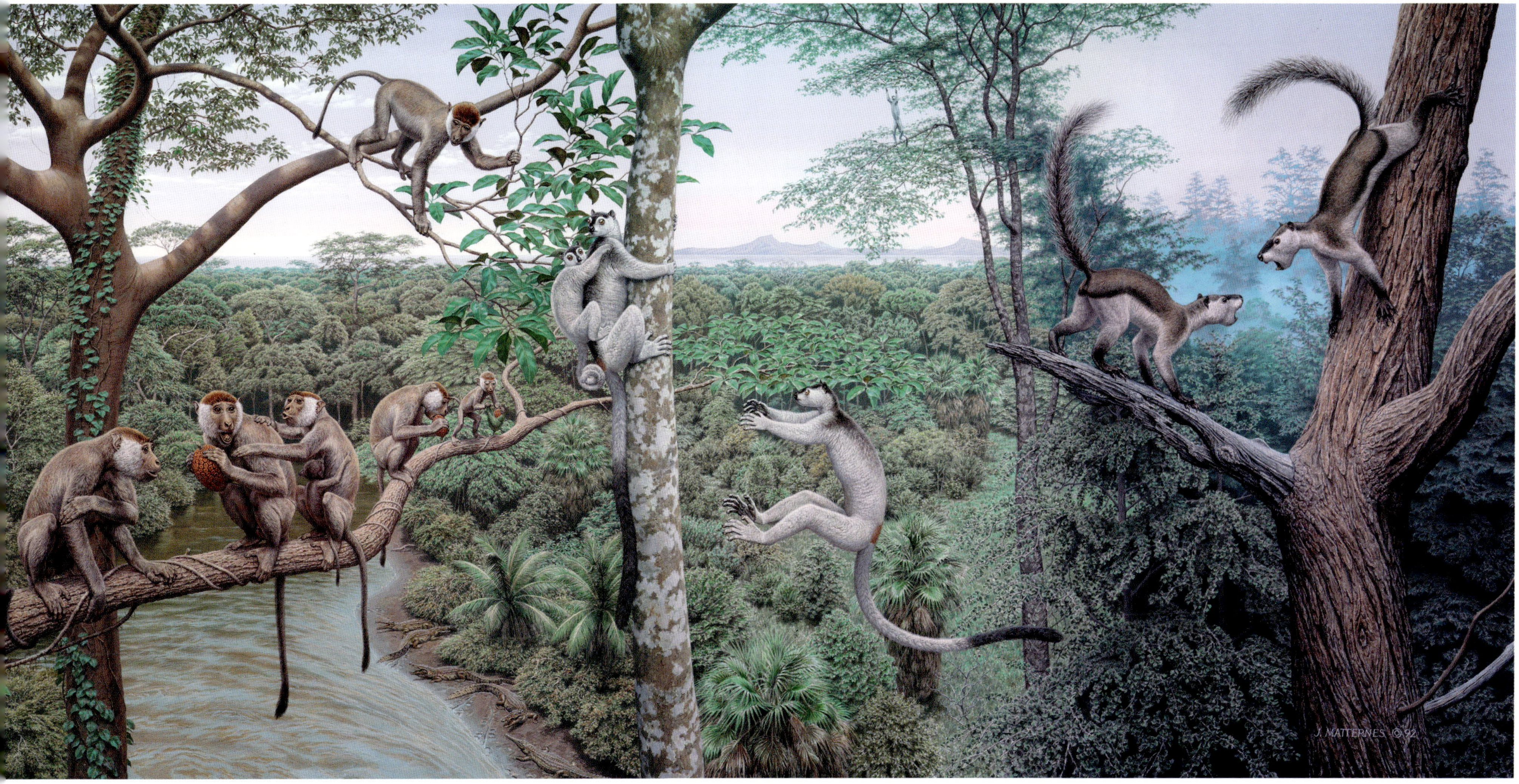

Matternes based their general appearance on living analogues, the indriid lemurs that are still found today on the island of Madagascar.

The central vignette features *Aegyptopithecus*, a 30-million-year-old early precursor of apes and monkeys that lived in northern Egyptian tropical coastal forests near the Tethys Sea (which later became the Mediterranean). Their lifestyle resembled those of arboreal monkeys; several of them are seen feeding on the large red fruit of an *Epipremnum* vine. Way below them, in a river that runs through the forest, a group of crocodilians is basking in the sun; they haven't changed much in tens of millions of years.

Continuing to the left, we encounter the 24-million-year-old *Proconsul*, an arboreal and tailless East African distant precursor of today's apes that had rather monkey-like limb proportions. In the background is a carbonatite volcano that spews a grayish-white ash. *Proconsul* fossil skulls and partial skeletons were first found in 1931 on Rusinga Island in East Africa's Lake Victoria, and initially were considered possible ancestral chimps. Its fanciful scientific moniker means "before Consul," after a famous chimpanzee who resided in London's Regent's Park Zoo at the beginning of the twentieth century.

On the far left is *Sivapithecus*, a hominoid known from India, Pakistan, and Turkey that lived between about 12 and 8 million years ago. Its head resembles that of an orangutan, but its body is more generalized. A troop of nine individuals is shown, implying a considerably more social animal than today's orang, which is fairly solitary. Matternes chose to show them hurriedly scrambling up trees to escape from an unseen predator.

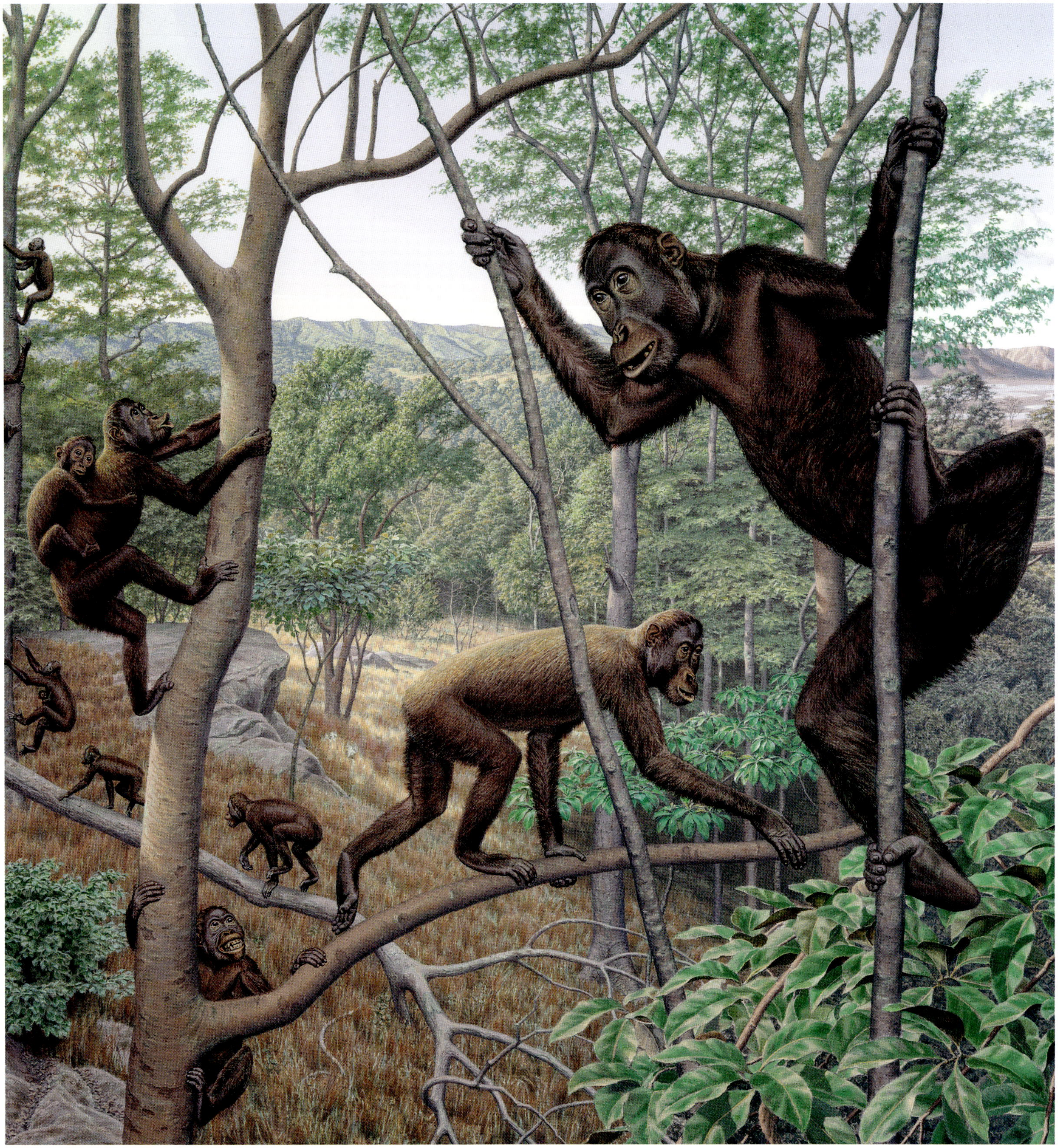

A group of *Sivapithecus* flee from danger by taking to the trees.

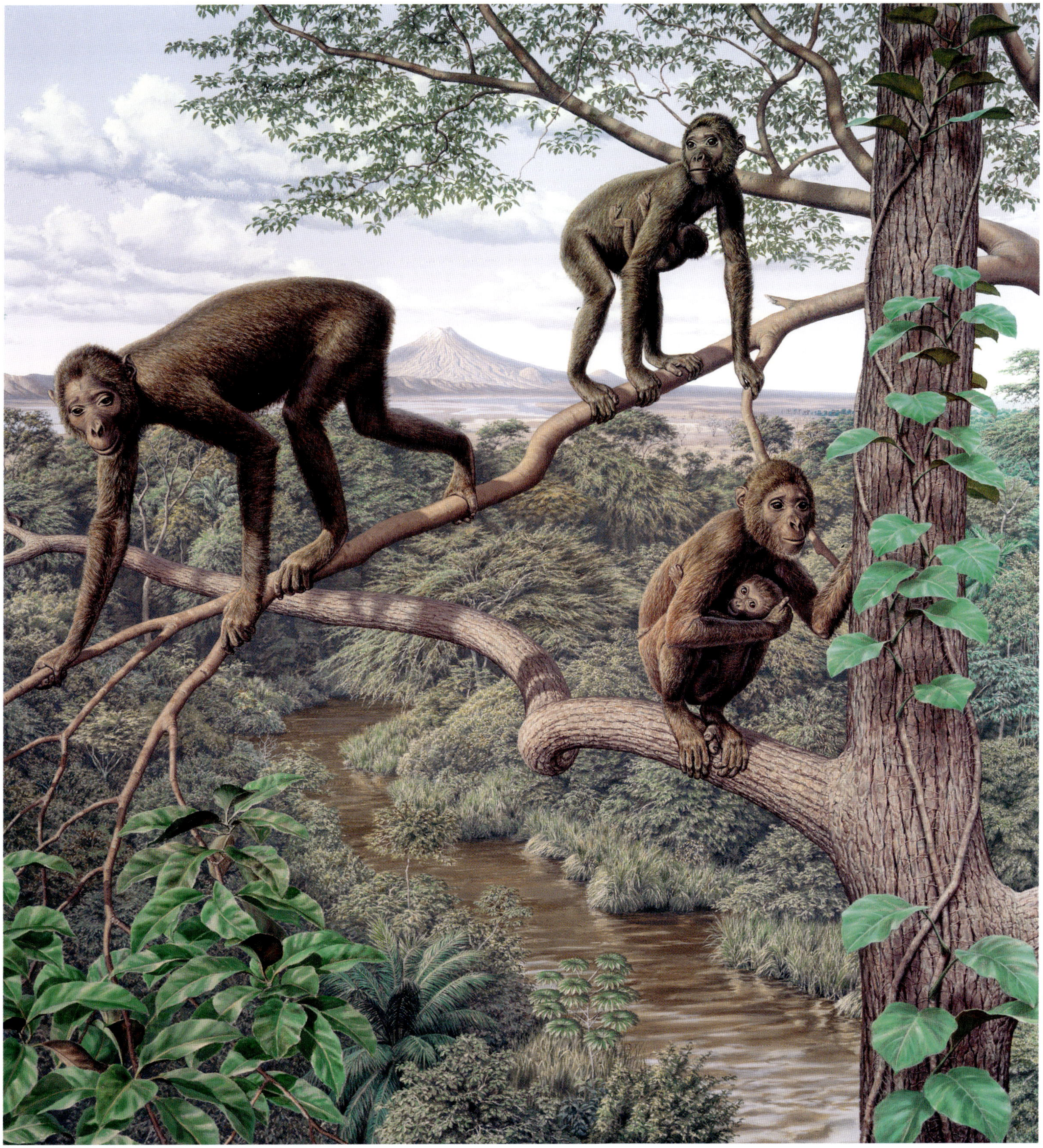

Proconsul, an early Miocene ape whose fossils were first discovered on Rusinga Island in Lake Victoria, East Africa.

SCALING THE FAMILY TREE

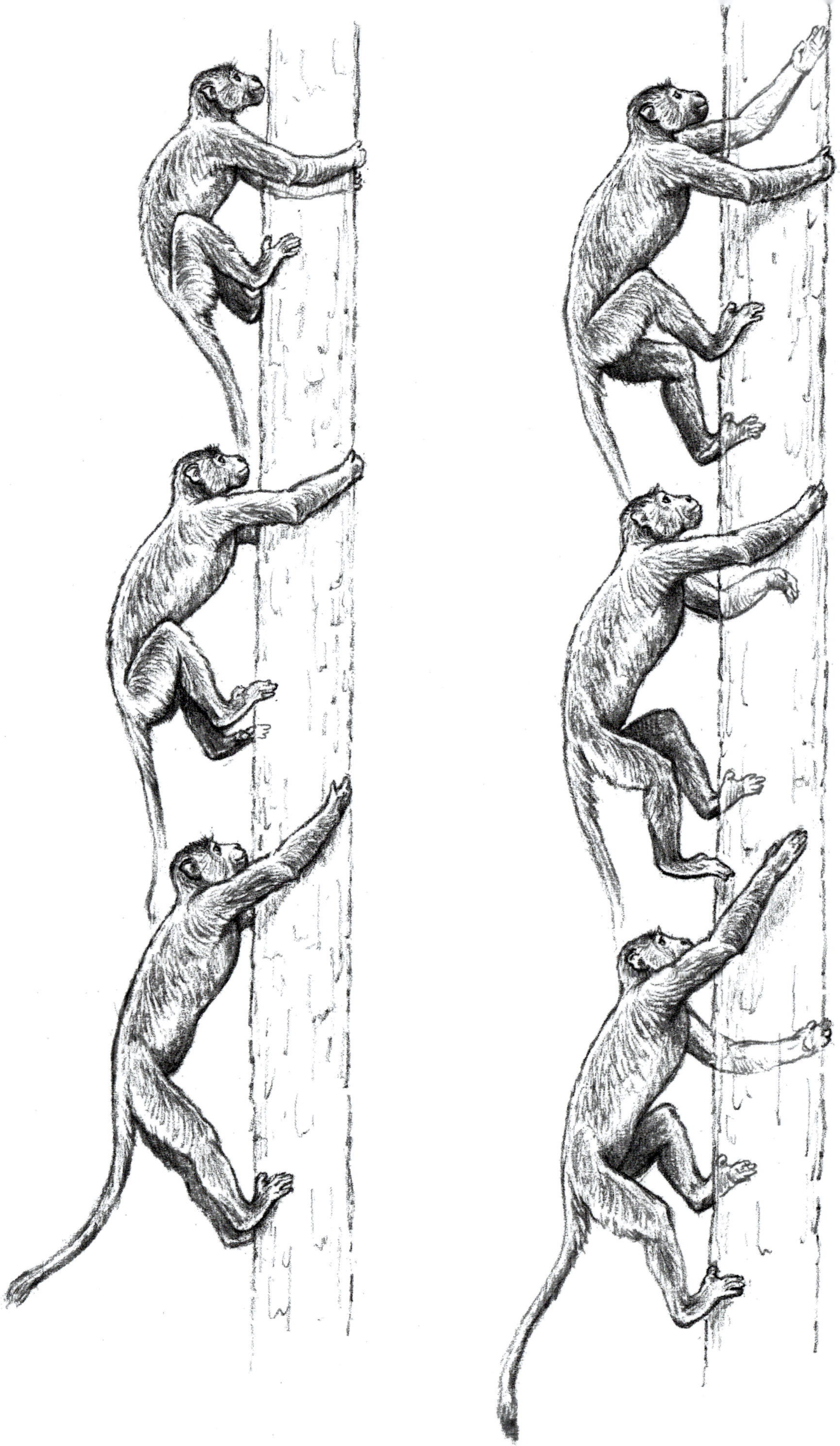

Sketches of *Aegyptopithecus*, an early higher primate with a monkey-like appearance that inhabited East Africa. One of Jay's techniques in sketching contemporary animals is to move them around on paper until he decides how best to show their positional behavior.

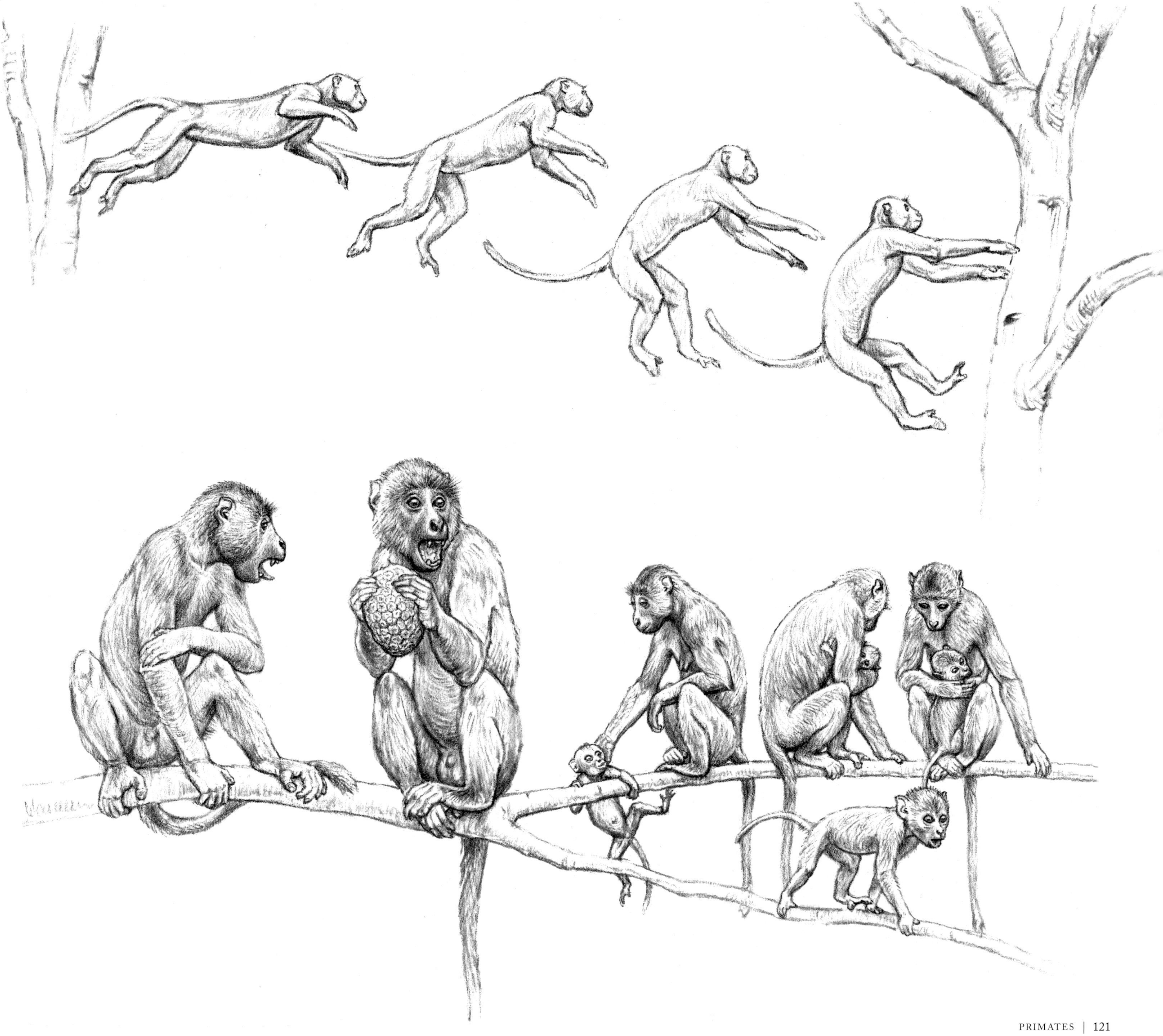

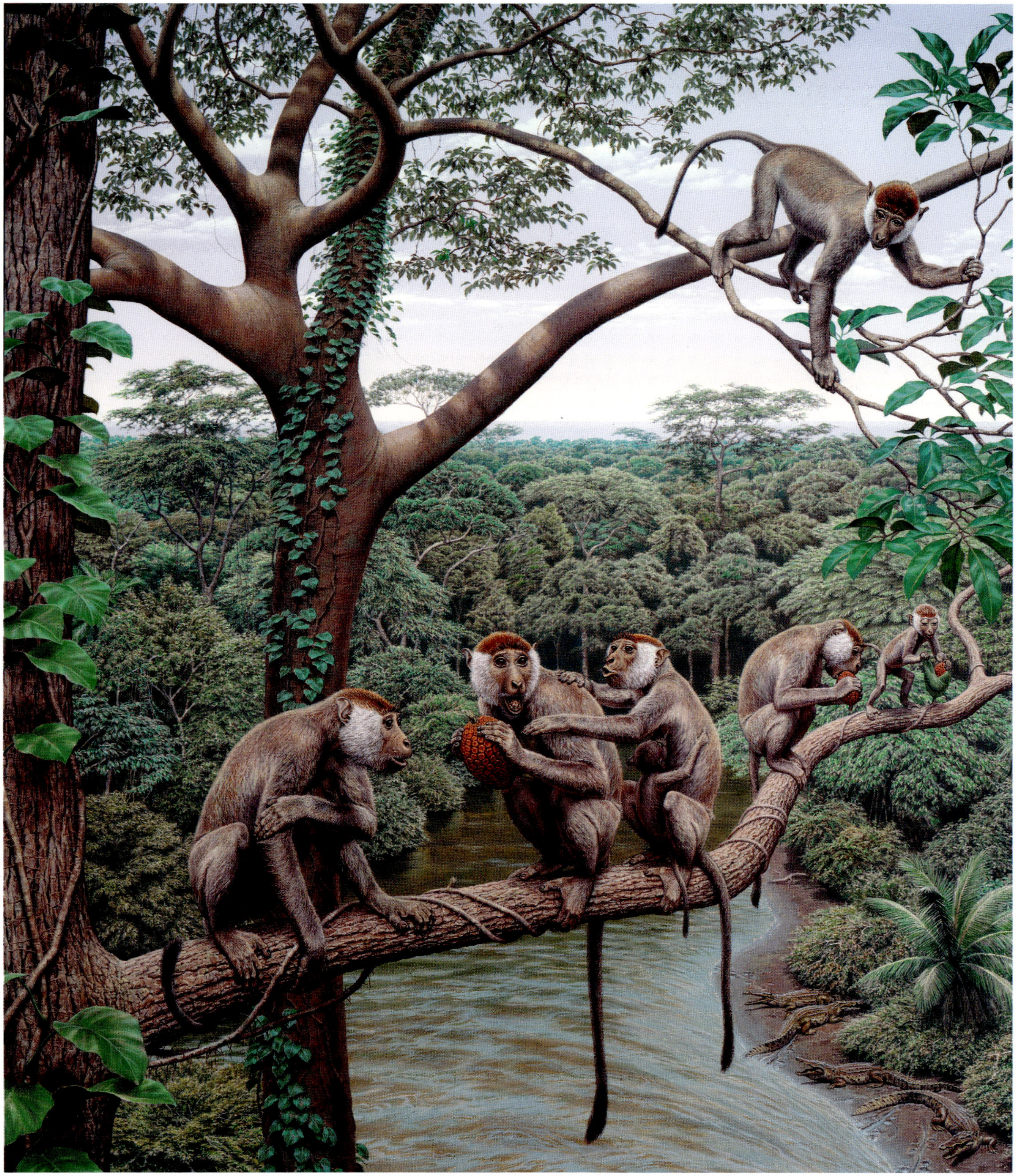

Aegyptopithecus congregate on a tree limb in the AMNH primate mural, totally unconcerned they are cavorting high above a crocodile-infested river. One individual is eating an *Epipremnum* fruit. Early Oligocene–Miocene of the Fayum, Egypt.

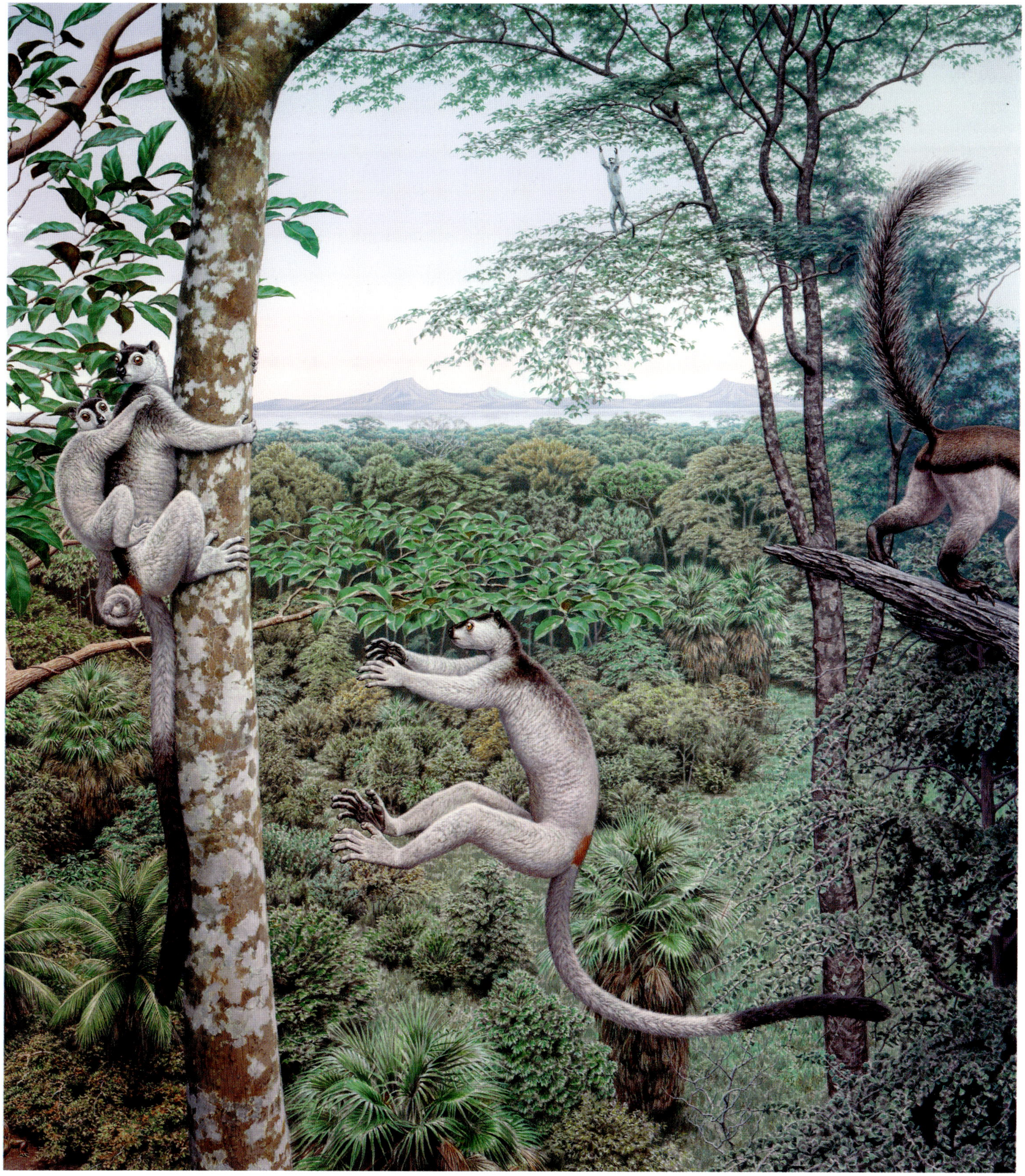

Smilodectes (*Notharctus*), a mid-Eocene lemuroid from the Bridger Formation of Wyoming, 50–46 mya, was an early leaper and tree climber.

LEMURS

Tattersall's Golden-Crowned Sifaka, and Brown Lemurs

Madagascar, the world's largest oceanic island, is home to a unique and charismatic fauna and flora. Best known among its endemic animals are the lemurs, which last shared an ancestor with our own lineage around 50 million years ago. Before human farmers reached Madagascar at around the time the Roman Empire collapsed, the island had given rise to diverse lemurs that ran from the size of a mouse to that of a gorilla. Today only smaller-bodied lemurs remain, the biggest of them belonging to the indriid family of vertical clingers and leapers. The diurnal indriid genus most widely distributed in Madagascar is *Propithecus*, commonly called sifakas; the most recently described species is the golden-crowned sifaka (*Propithecus tattersalli*), named for Ian Tattersall of the American Museum of Natural History, who was the first scientist to observe it in its remote northern forest. When Matternes was commissioned to do a portrait of this species (right), he paid attention not only to its distinctive physical features, but to its environment as well.

Jay also painted portraits of another species of lemur, opposite, that originated in Madagascar, but now also lives on Mayotte, a small island to the northwest. (The female is on the left, with her darker-faced male companion on the right.) As an arboreal quadruped, this lemur, *Eulemur fulvus*, is very different in its limb proportions from the long-legged sifaka. Poignantly, like other lemurs, it is completely dependent on its disappearing forest home.

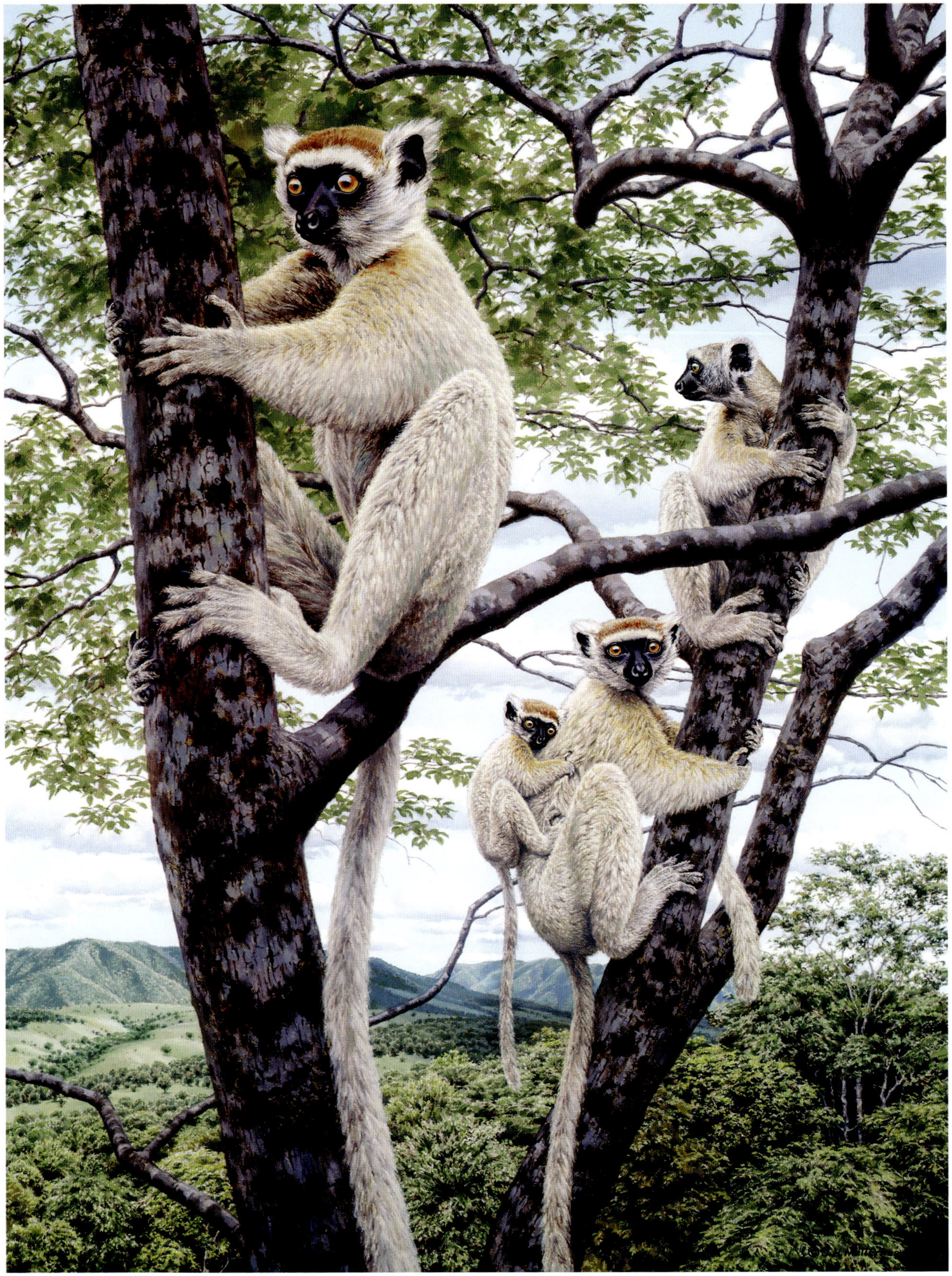

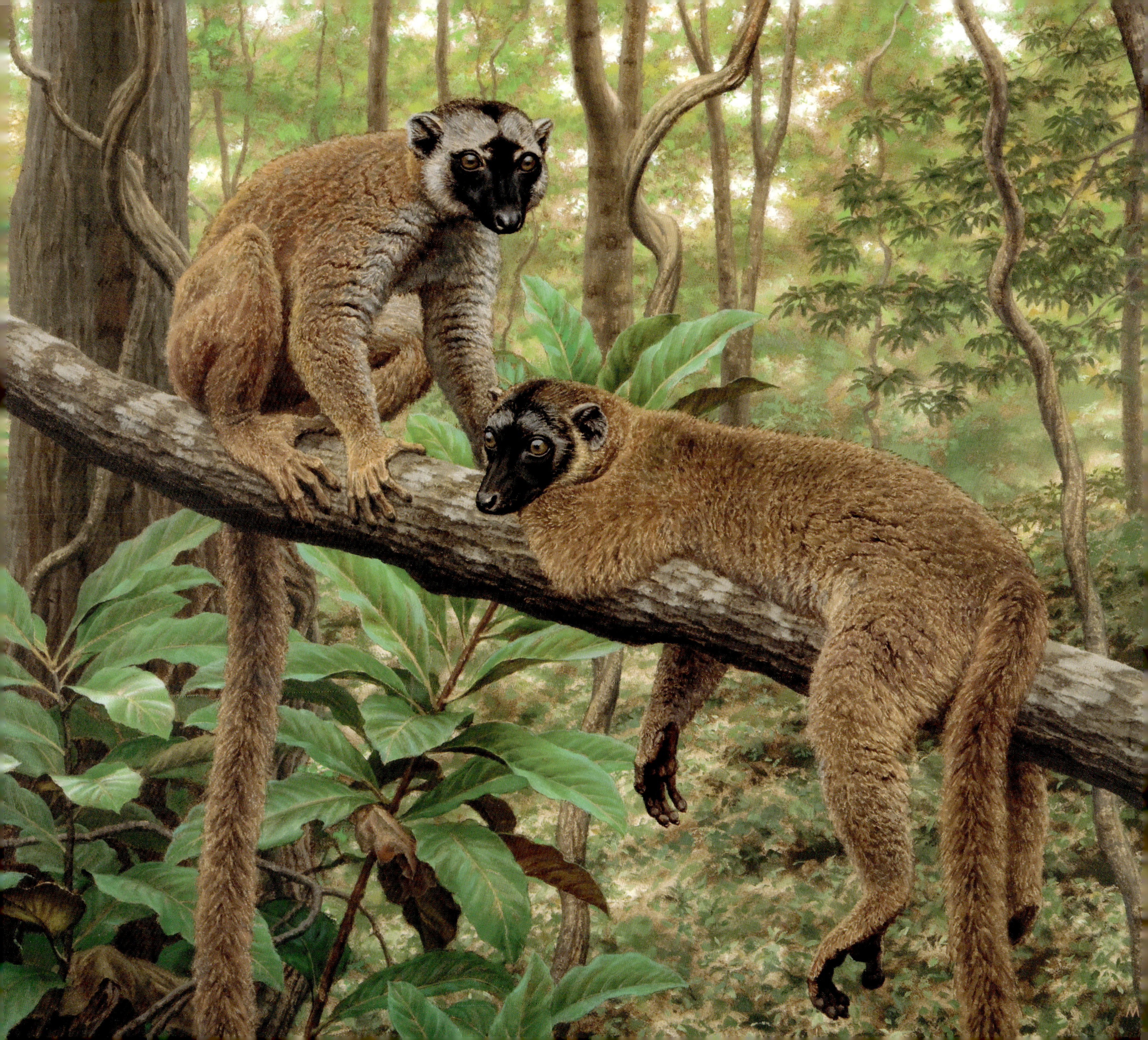

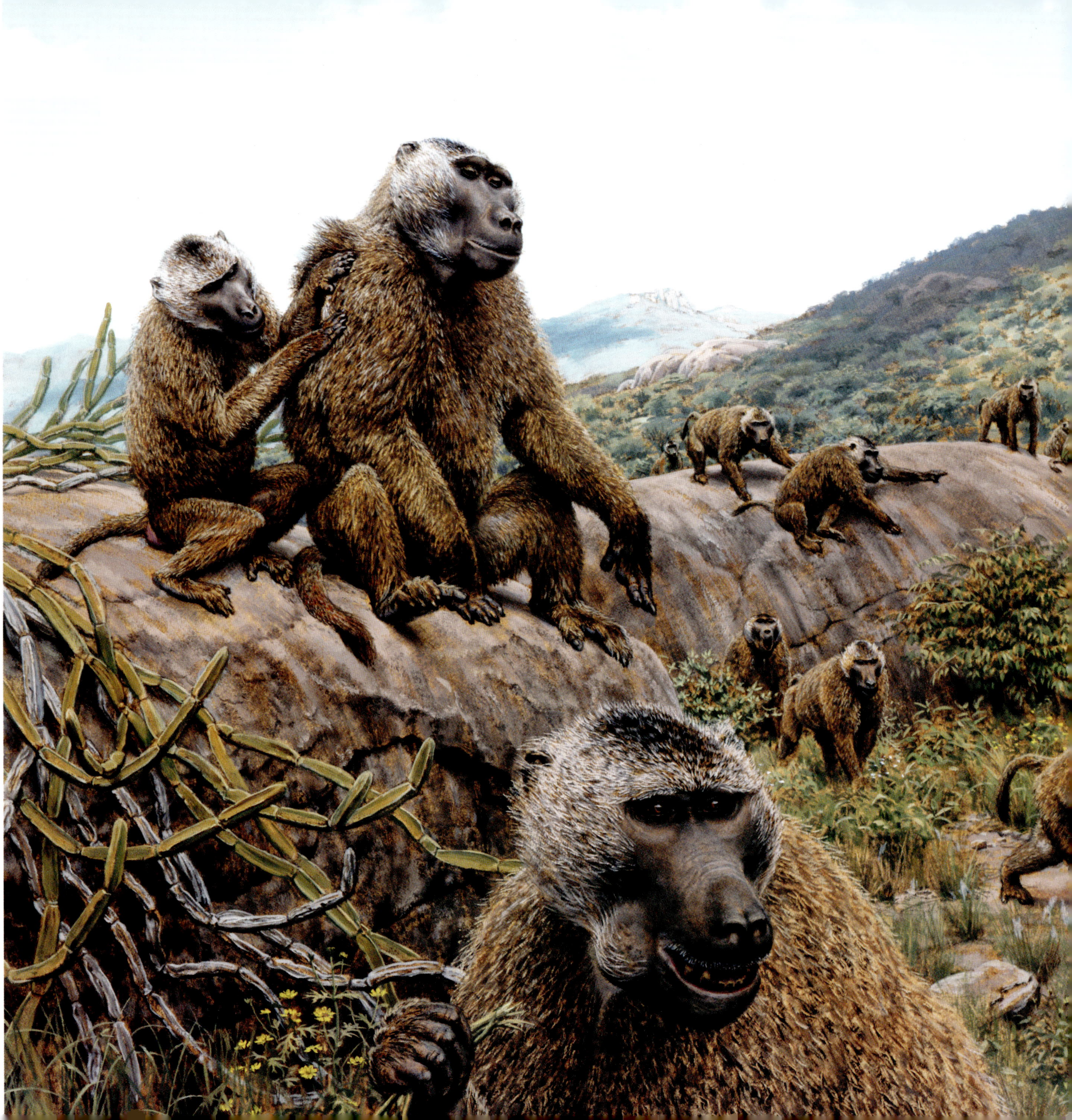

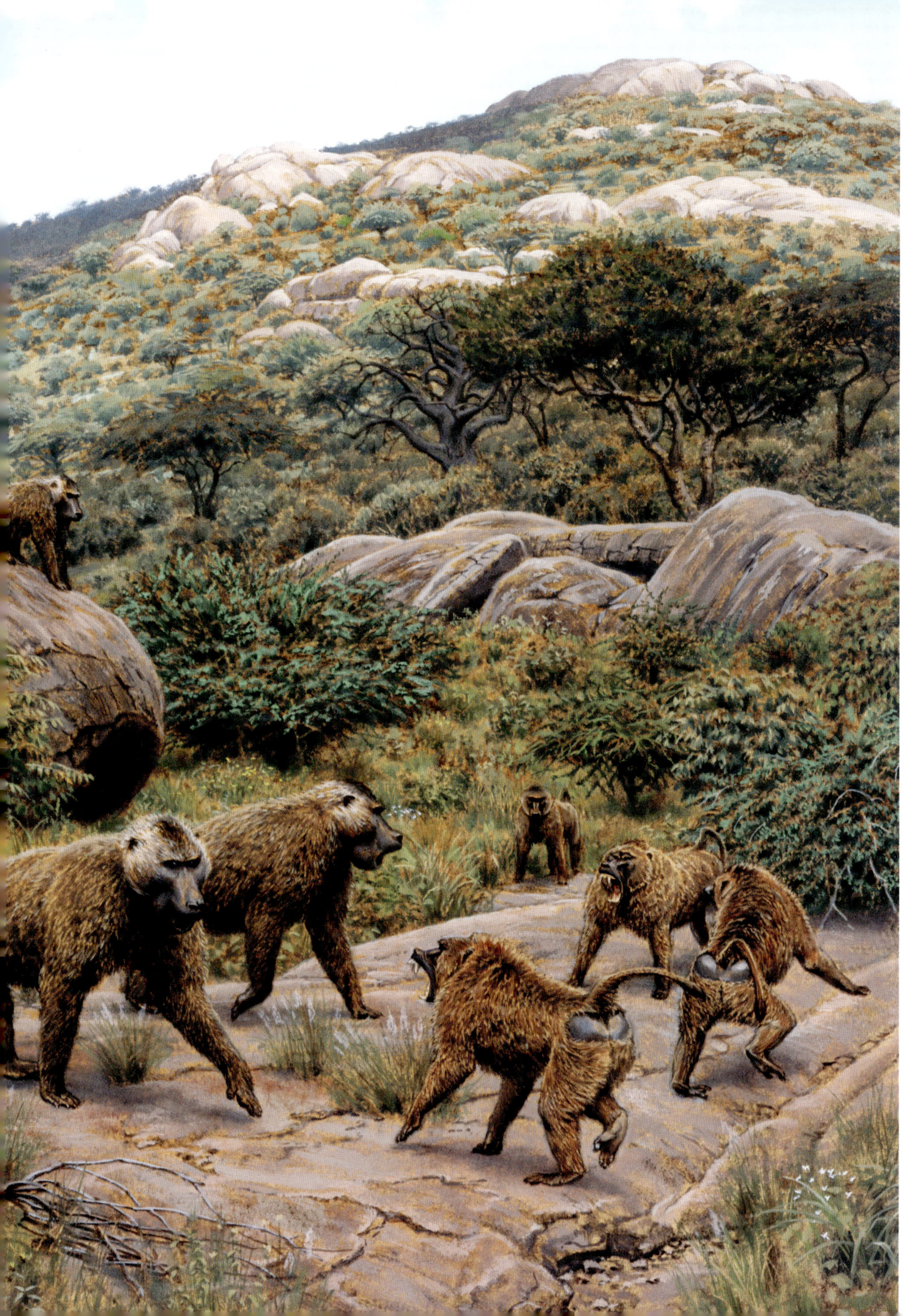

BABOONS

Giant Baboons

Famed paleoanthropologist Louis Leakey asked Matternes to illustrate an enormous baboon that lived 1.5 mya at Olduvai Gorge, Tanzania, for his 1969 book *Animals of East Africa*.

As Jay began measuring and drawing fragments of skull and limb bones at the Kenya National Museum in Nairobi, he was immediately impressed by the primate's huge dimensions. In life, this massive monkey would rival the size of a female gorilla; no living baboon species comes remotely close. Its relatively short rear limbs suggest that it was not a swift sprinter, but its formidable canine teeth would have stymied the most determined predator.

From its teeth, scientists concluded that it was more closely related to the *Theropithecus* group than to the so-called "dog-faced baboons" (*Papio*), the smaller species in the painting that is still common in Africa today.

The gelada baboon of the Ethiopian highlands is the only surviving representative of the *Theropithecus* group. However, during the Pleistocene there were several coexisting species, of which the giant was only one. Jay based their habitat on his visit to the hilly terrain of Tsavo National Park, Kenya.

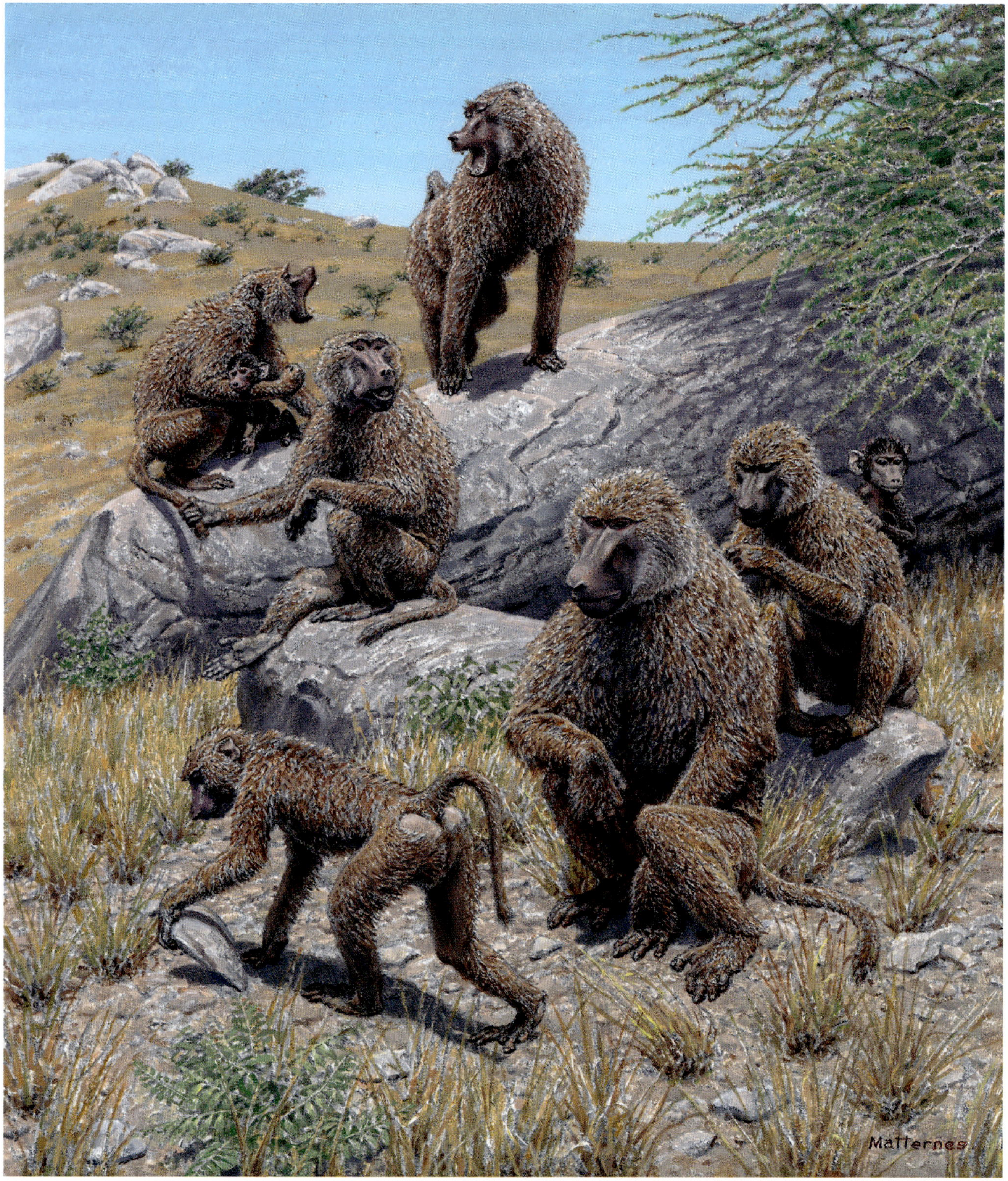

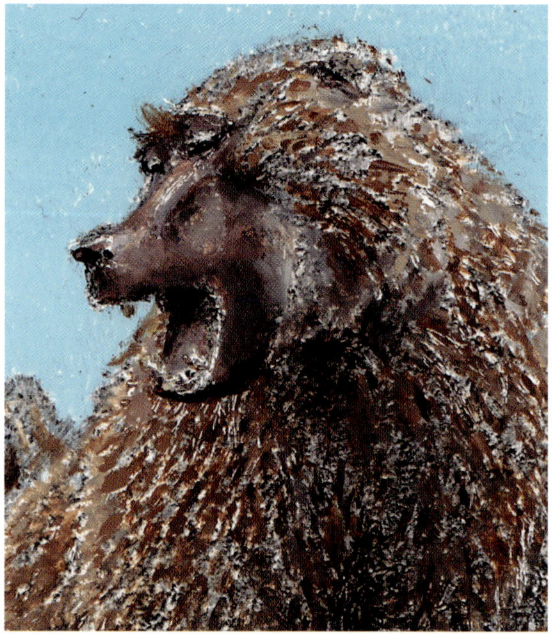

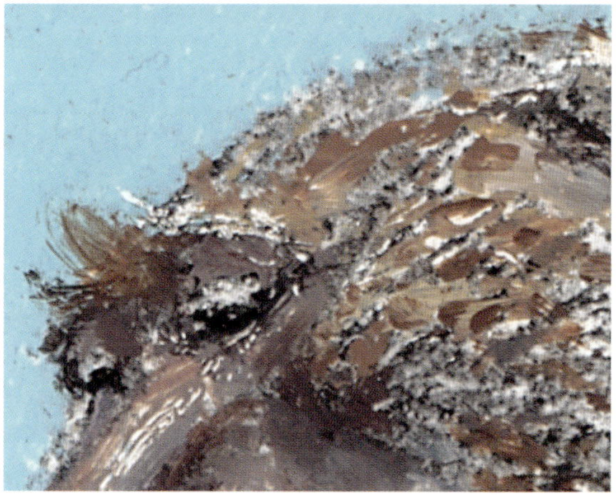

Olive or Anubis baboons (*Papio anubis*) during their morning foraging for scorpions, insects, small animals, and vegetation in their native East Africa. Extreme close-ups reveal one of Jay's techniques for rendering textured fur, by leaving blank white surfaces among the brushstrokes.

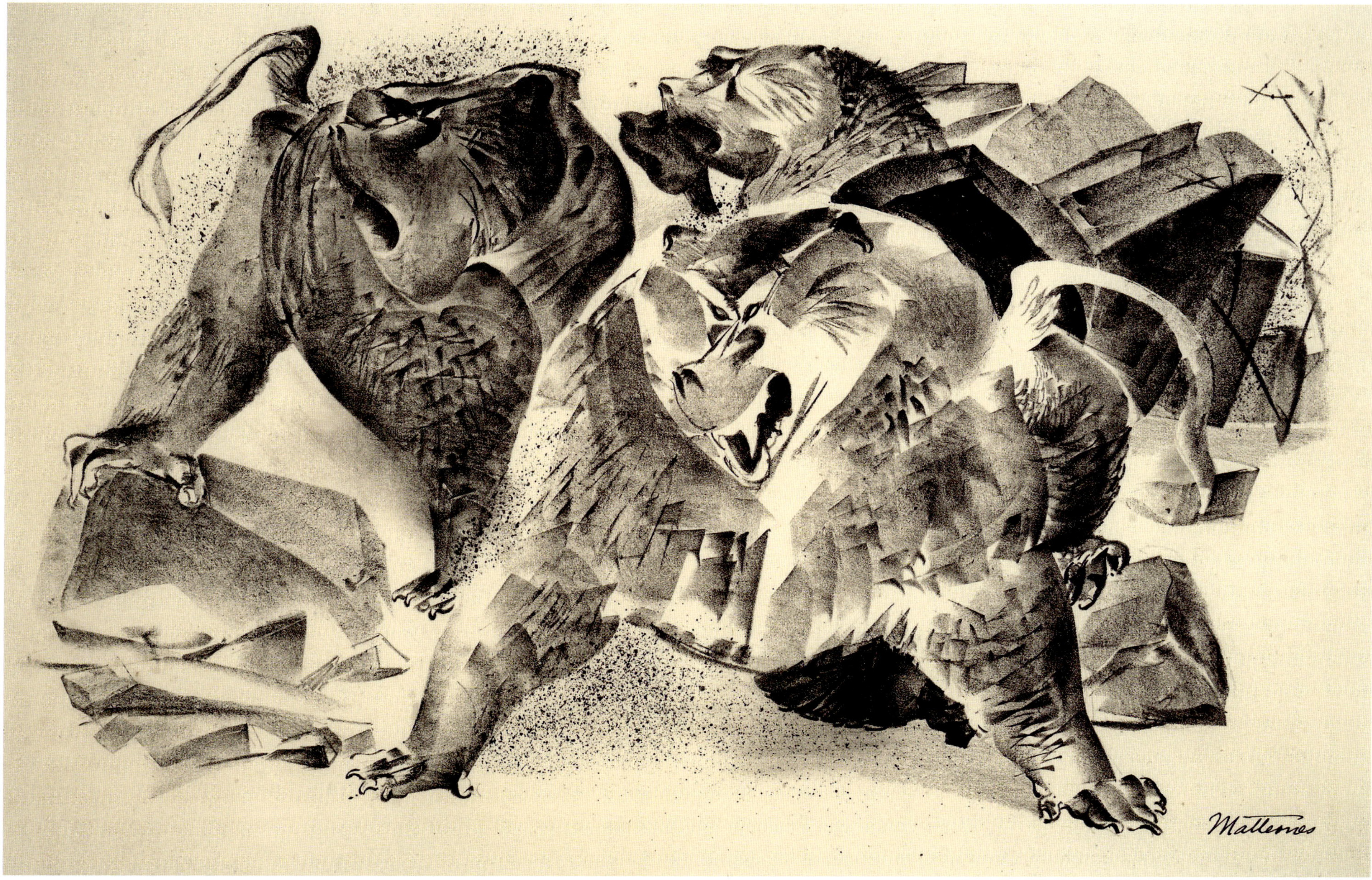

Matternes' lithographic print, *Art Deco Baboons*, was submitted as an art school exercise in 1954.

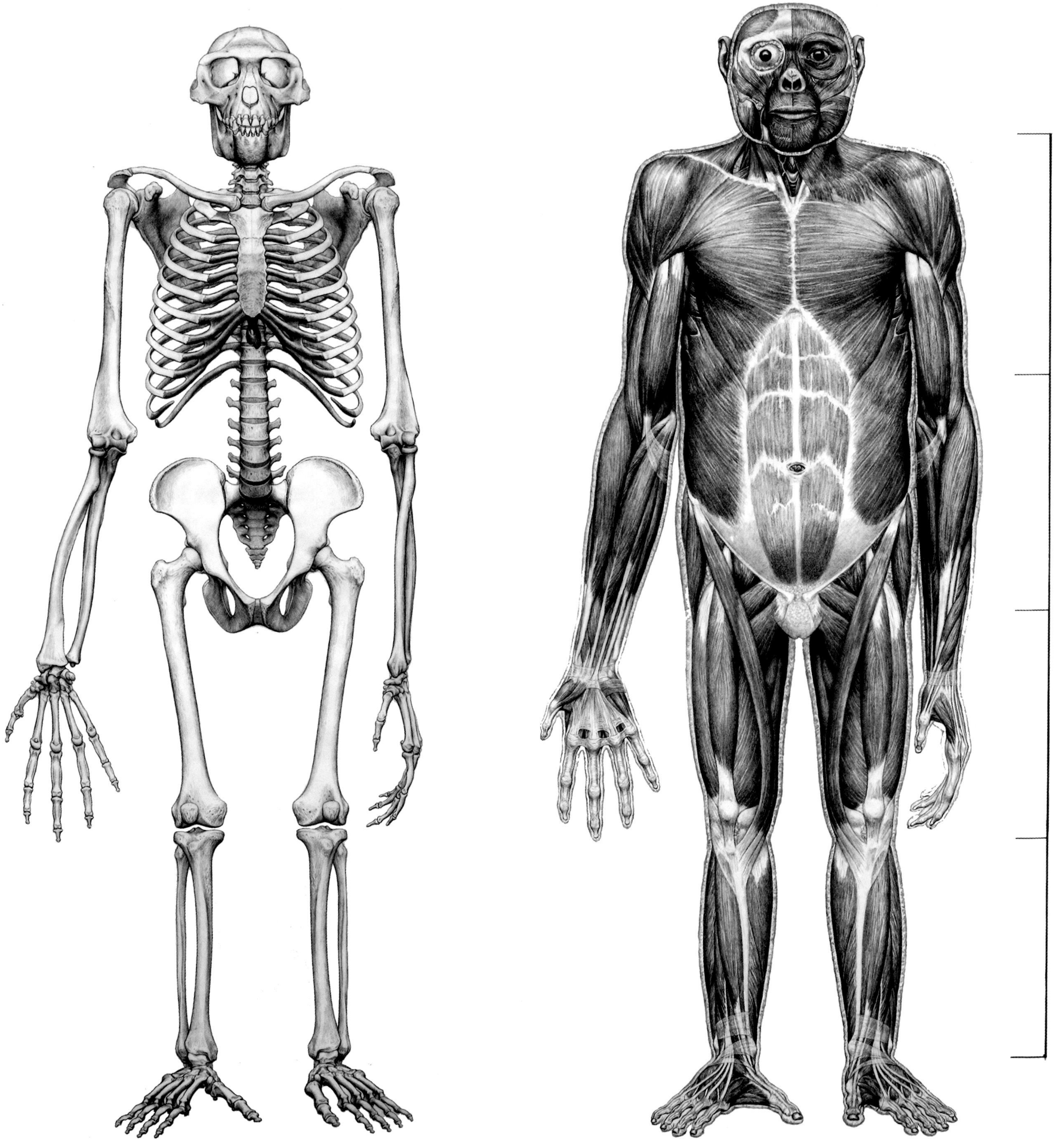

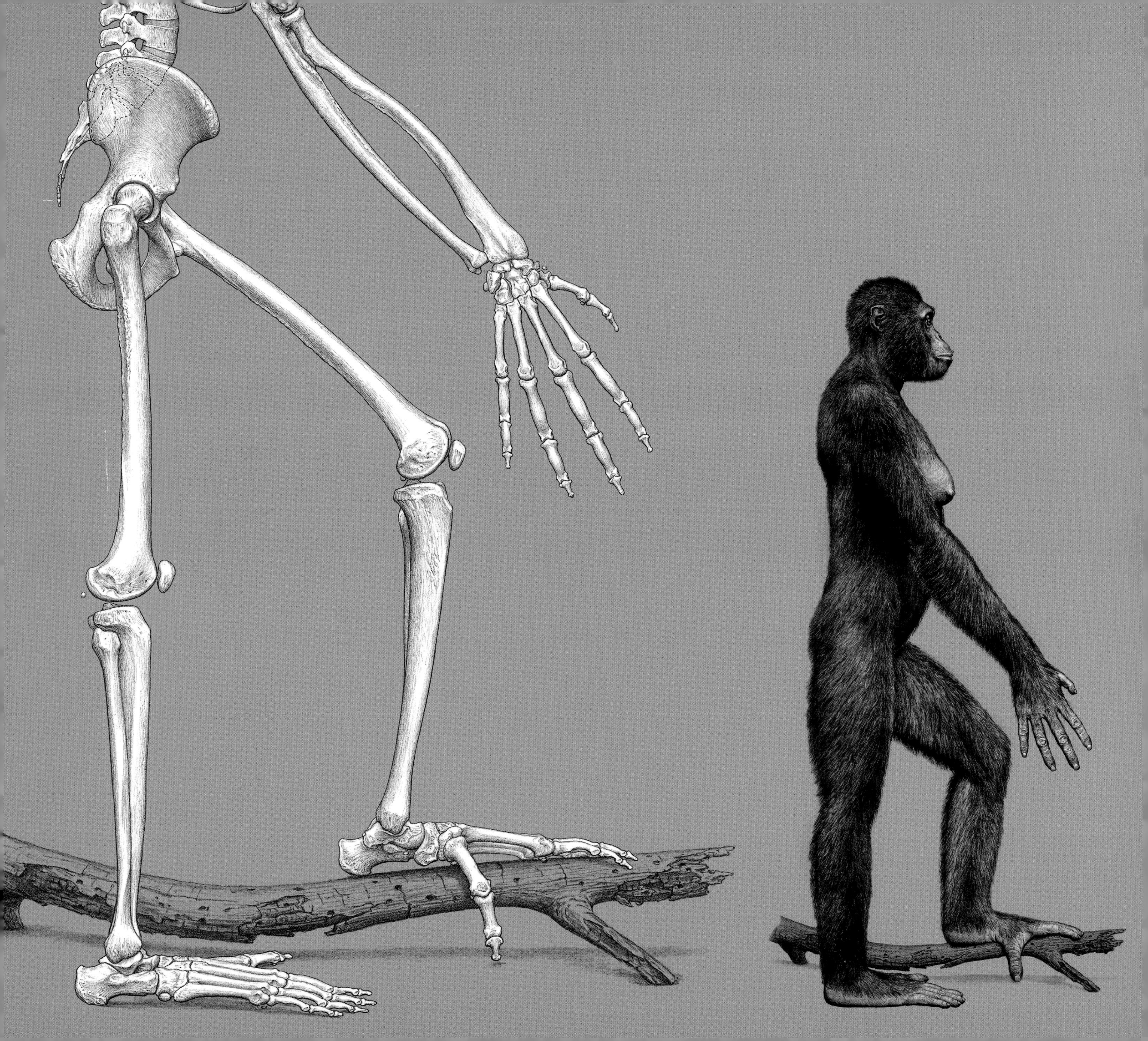

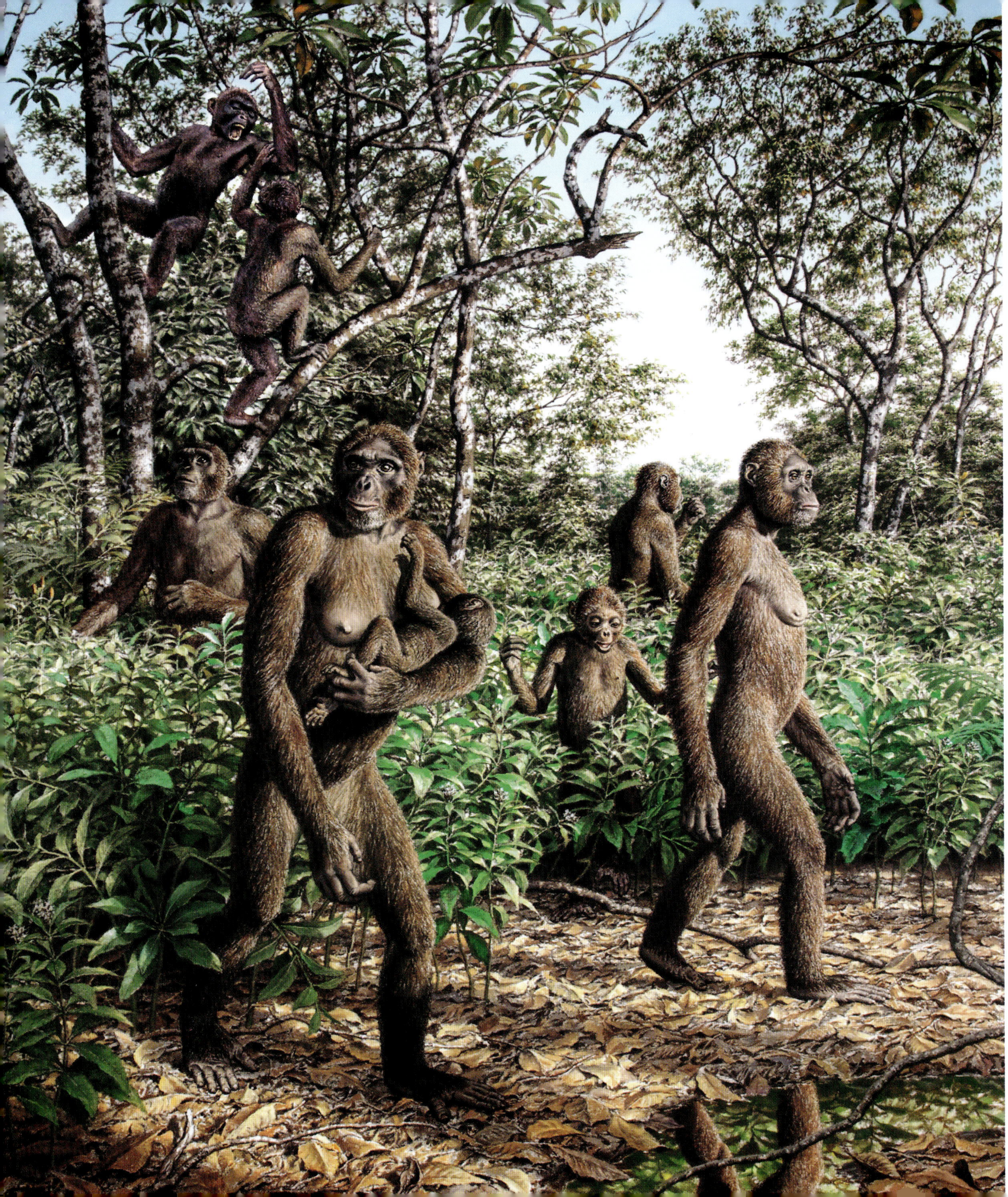

Ardipithecus was a unique primate, well adapted to walking upright on the ground as well as moving among the trees. They lived in tropical woodlands with low-growing trees and an open canopy that fostered dense undergrowth, like similar areas in Ethiopia today.

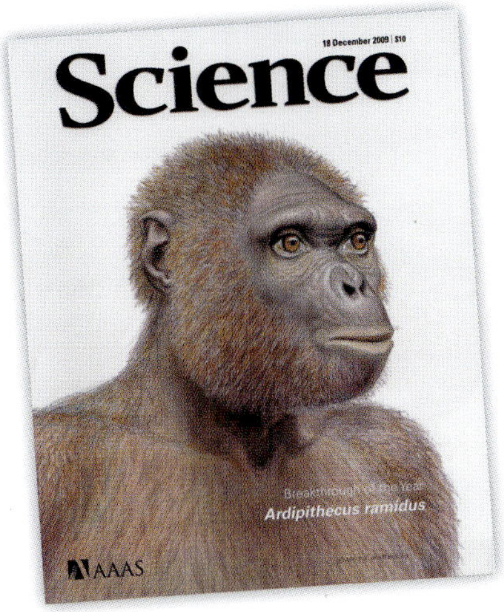

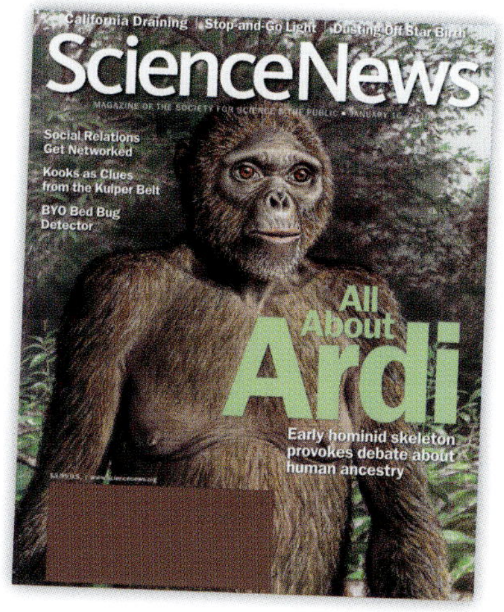

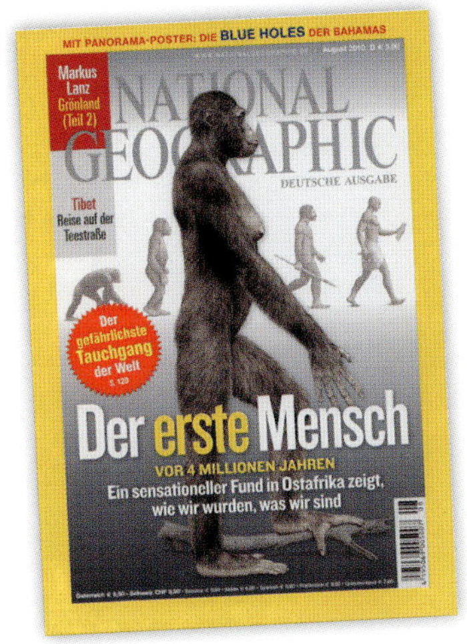

Ardi made an international iimpact, with Jay's restoration garnering the covers of science magazines worldwide when the discovery was finally revealed in *Science* magazine on October 2, 2009.

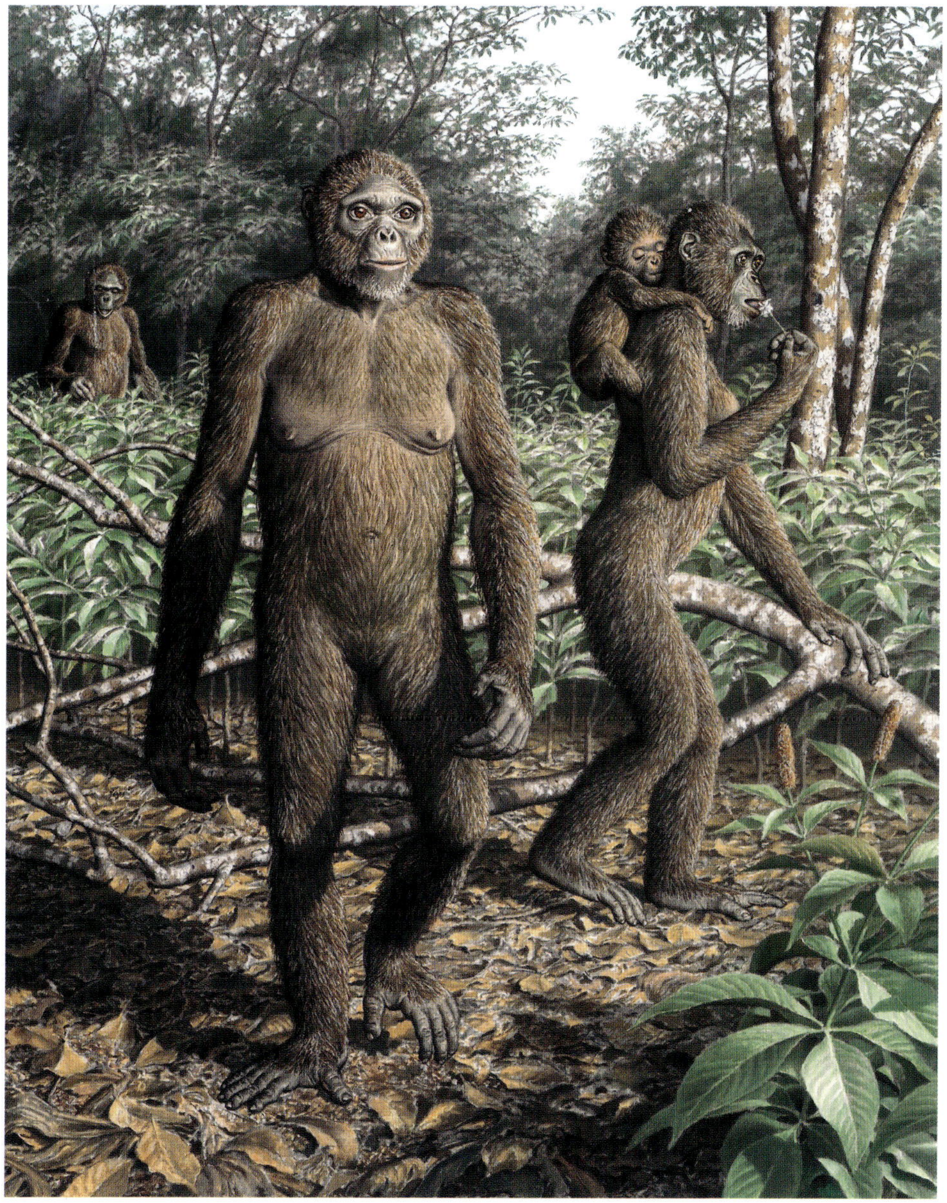

of "science," perhaps unintentionally minimizing the enormous contribution of his informed and experienced artistic sensibility. Sometimes the press insultingly credited his drawings to an anonymous "staff artist."

Subsequently, Jay completed two paintings of an Ardi family strolling through dense woodland as some of their youngsters play in the trees. The adults are nursing infants or gathering fallen fruits from the forest floor. Modern chimpanzees occasionally go after colobus monkeys and other small mammals, and it is possible that Ardi might once in a while have hunted small animals for food, but Jay would not represent such a scene without solid scientific evidence for it.

Jay's drawings and paintings of Ardi, both the skeleton and the fleshed-out restorations, have become the standard images by which the creature is known in thousands of books, newspapers, television documentaries, and science journals. His cover portrait for *Science* of an *Ardipithecus* female, along with the full-body and other reconstructions inside the magazine, flashed around the world. *Time* magazine placed the discovery and analysis of *Ardipithecus* as number one on its list of the year's top ten scientific achievements, and its spokesmen were heaped with numerous additional accolades.

For Jay, though, the eleven years he spent working on and off on his Ardi reconstruction represented purely a labor of love; there was nothing in the scientific budget to pay an artist, and nobody sought to reimburse him for his unique and expert contribution. None of that mattered to Matternes. He knew this was an important fossil, and, as he has remarked, "I wanted to be the one to bring it back to life."

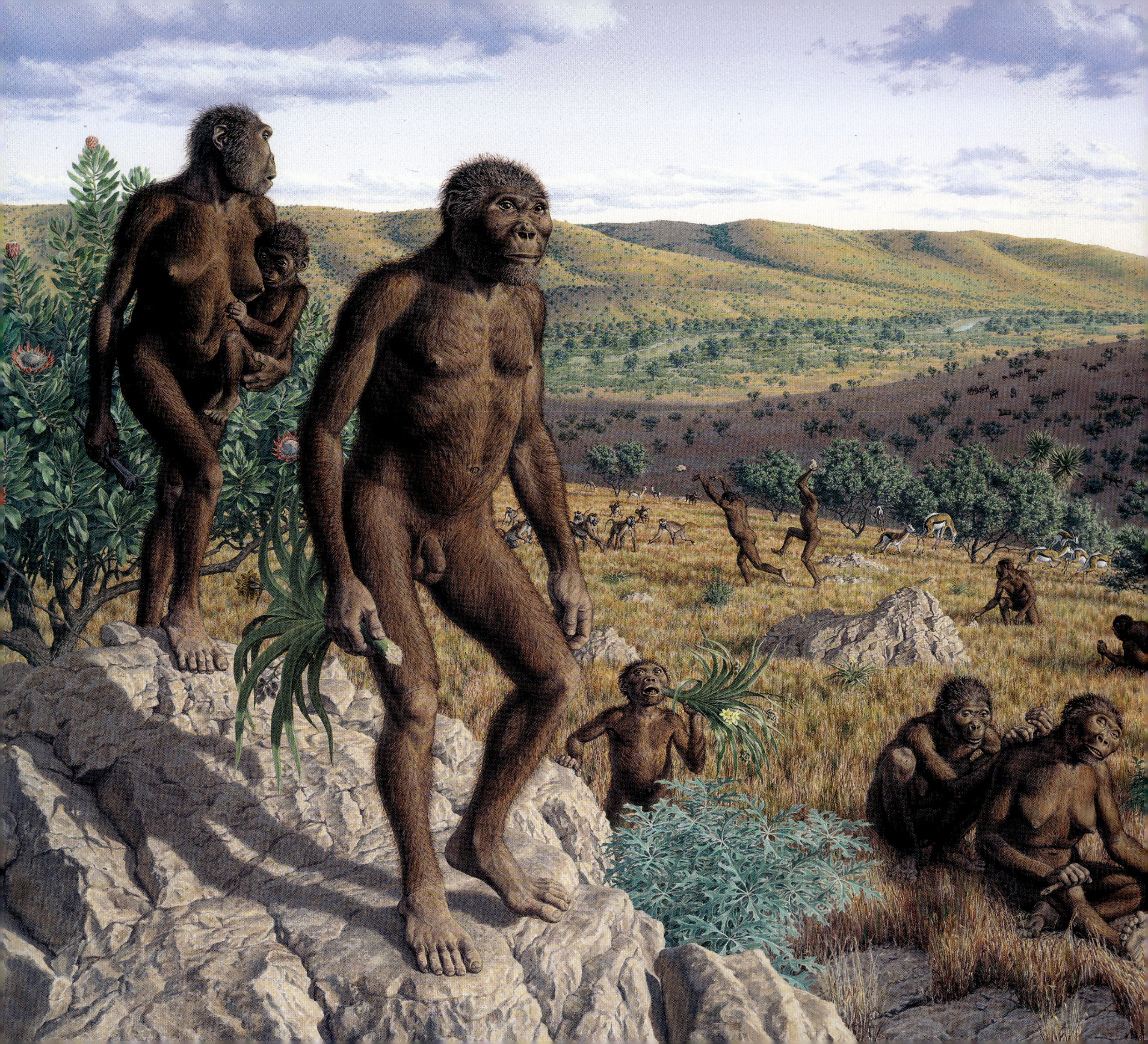

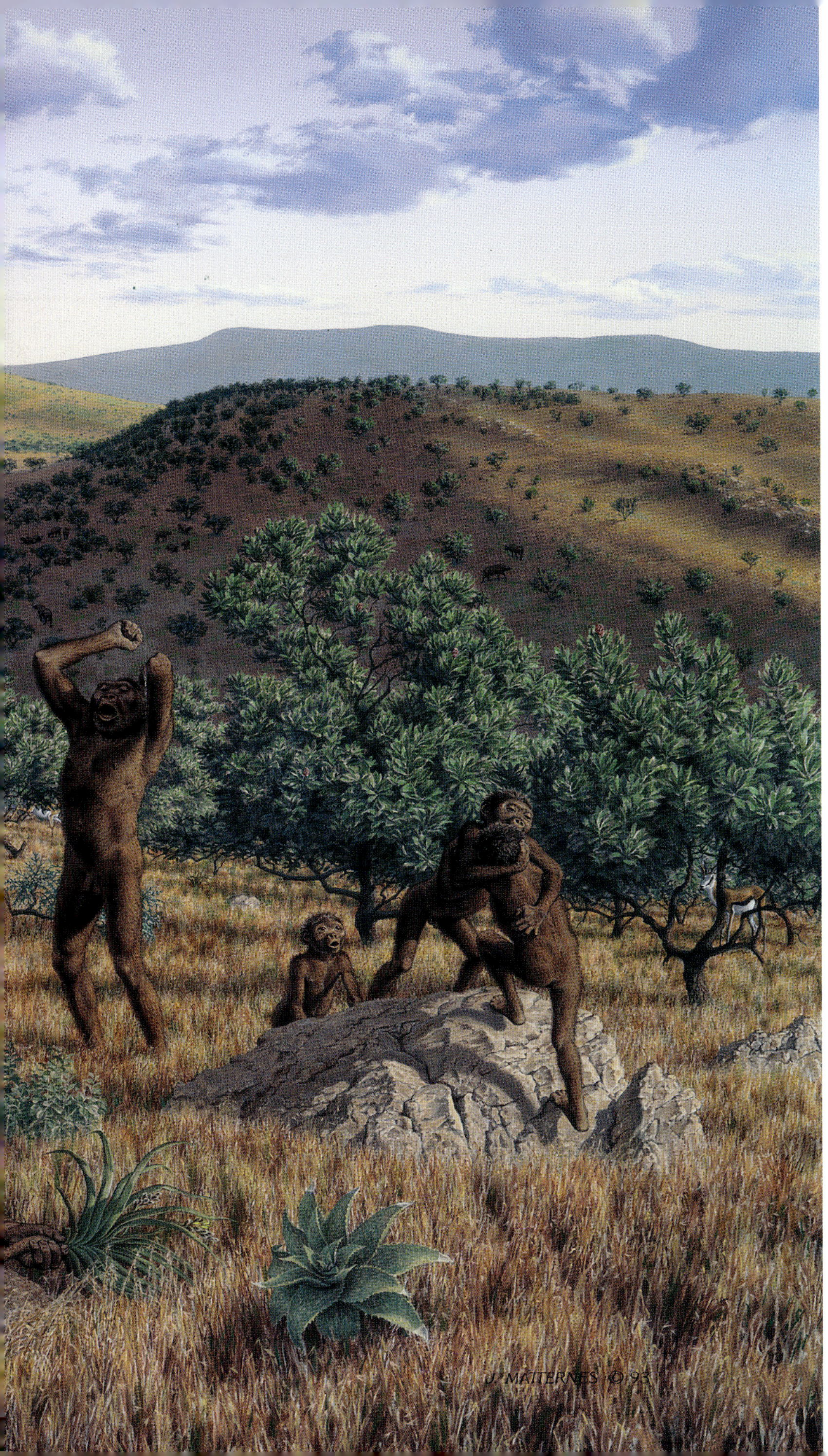

PARANTHROPUS
(ROBUST AUSTRALOPITHS)

Paranthropus Habitation Group

Jay's large painting shows a *Paranthropus* group foraging on South Africa's high veldt about 1.75 million years ago in a habitat dominated by short, shrubby *Protea* trees. An adult female and her infant accompany a male, who is holding a favorite edible plant, *Hypoxis rigidula*. Beneath him, a youngster chomps a stem base, while in the background young adults hurl rocks at an unwelcome baboon troop. Two females concentrate on intense social grooming, while a male calls attention to himself with a stretching display, two juveniles engage in rough-and-tumble play, and an infant looks on. *Hypoxis* is still common around Swartkrans Cave, where many specimens of *P. robustus* have been discovered.

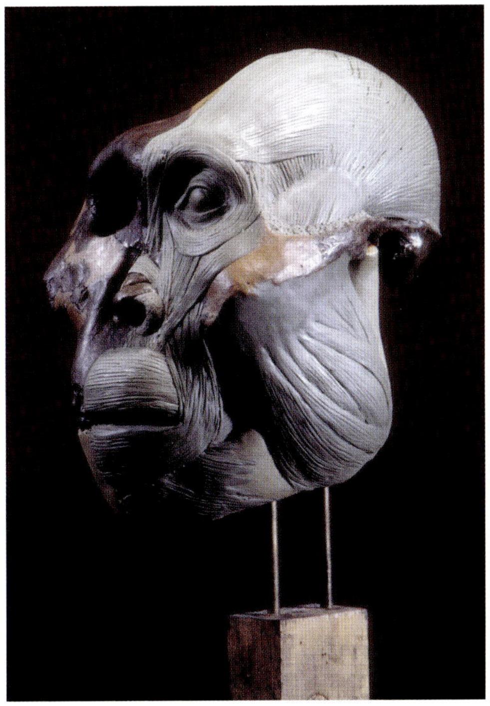

Partial life-size sculpture by Matternes (1960) of the famous OH 5 *Paranthropus boisei* skull from Olduvai Gorge, found by Mary Leakey in 1959. Originally nicknamed "Nutcracker Man" for its enormous molars, the specimen was dubbed *Zinjanthropus boisei* in a 1959 scientific paper. Louis Leakey frequently referred to the fossil affectionately as "Dear Boy."

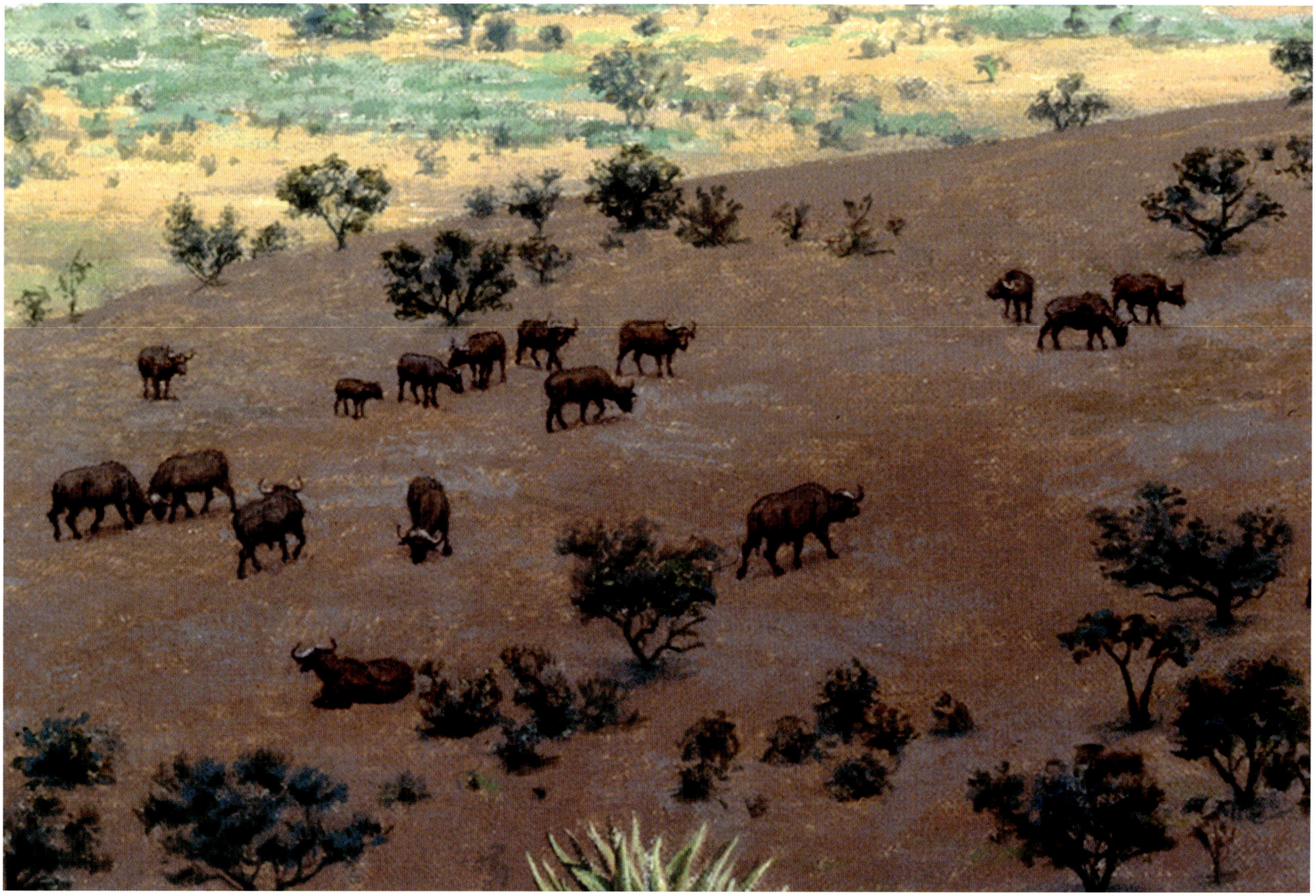

In the background of the *Paranthropus* re-creation (previous page), this herd of Cape buffalo is reproduced as a series of indecipherable specks. Greatly enlarging them reveals that Matternes painted them as fully formed miniatures, in such typical bovine attitudes as cows tending their calves and two males sparring.

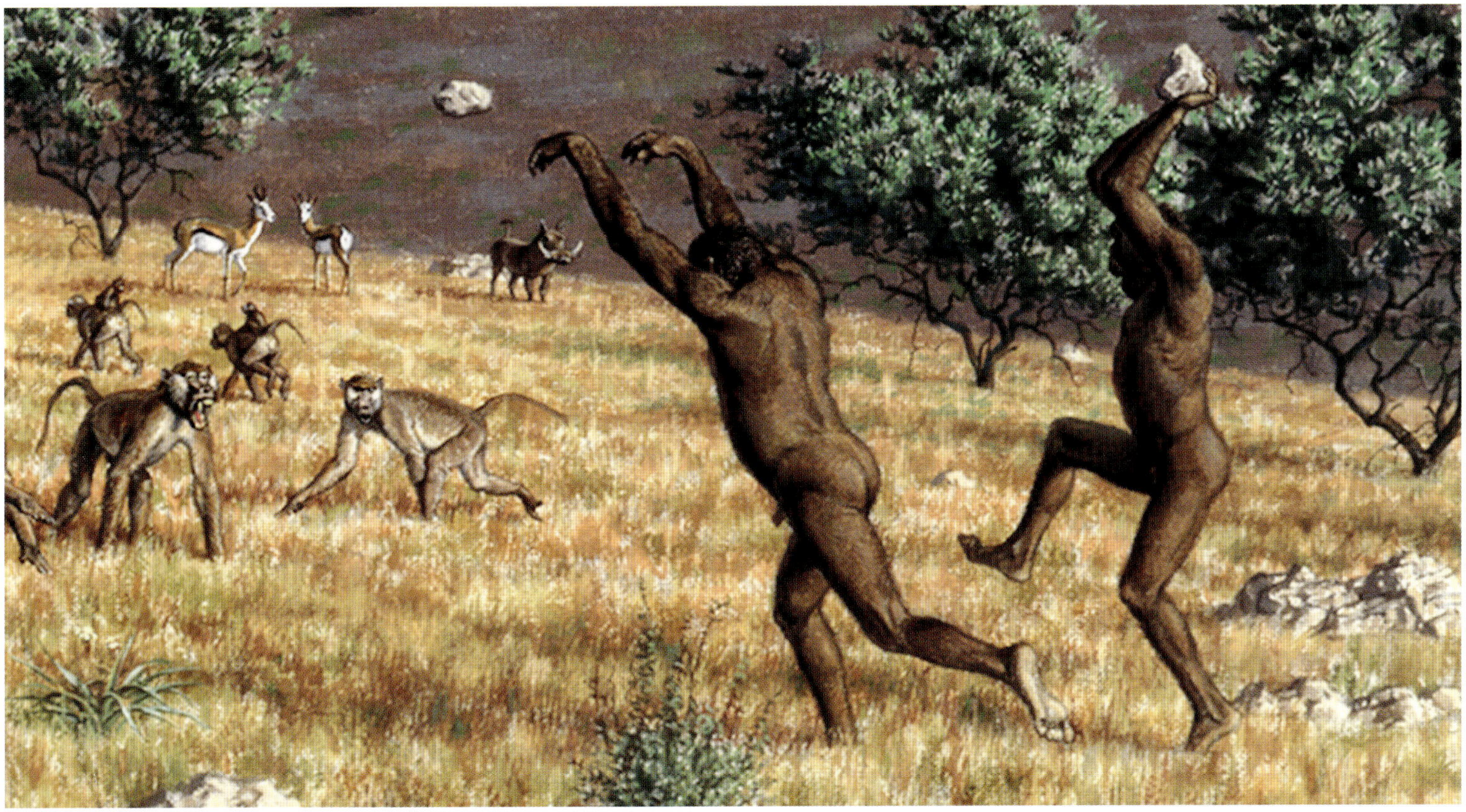

Left: Throwing stony projectiles, *Paranthropus* males discourage a troop of yellow baboons (*Papio cynocephalus*) from infringing on their foraging territory. In the background are a warthog and pair of springbok antelope. Below, left: Adult females engage in mutual grooming during a break in their foraging. Below, right: A toddler munches the shallow roots of a *Hypoxis*. When paleontologist Francis Thackeray visited the site in the 1990s to photograph foliage and trees for Jay, he encountered a boomslang (*Dispholidus*), one of the world's deadliest snakes, staring into his lens from a few inches away—"a hazard of existence in Africa common to both *Paranthropus* and myself!"

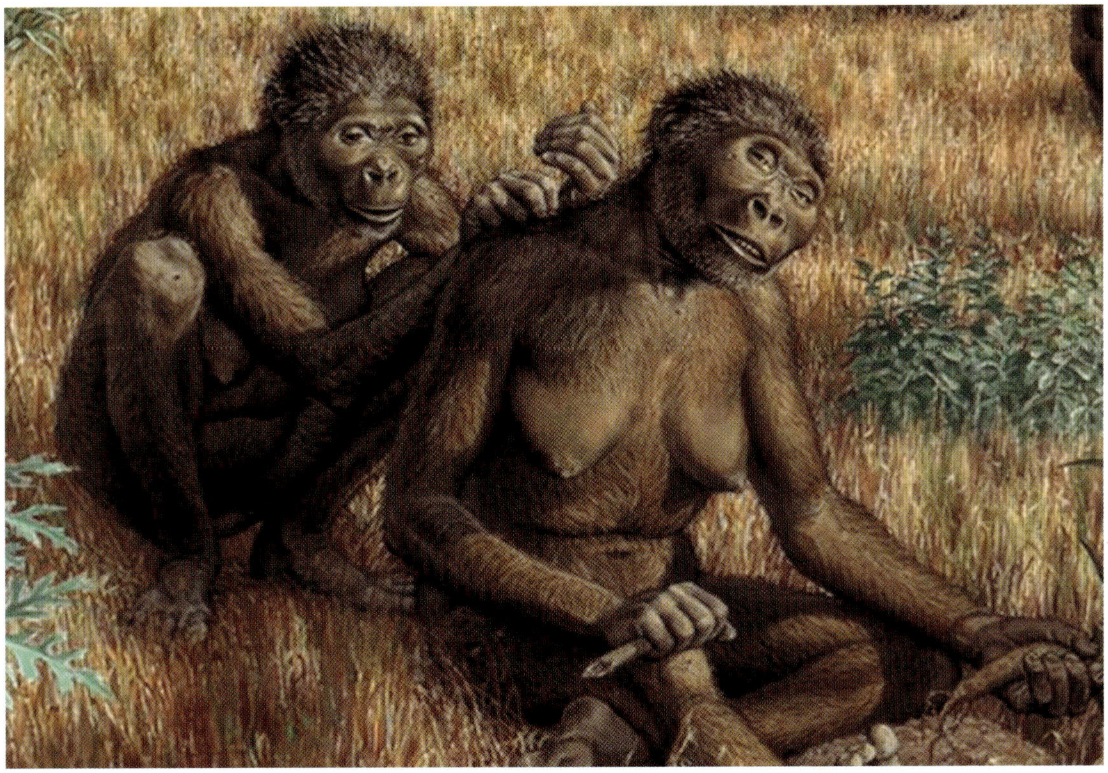

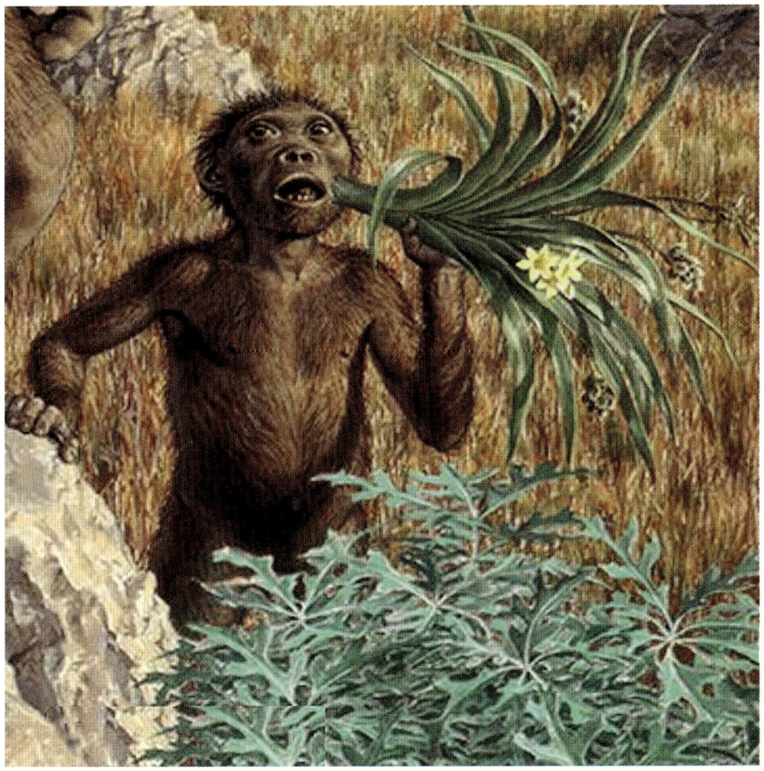

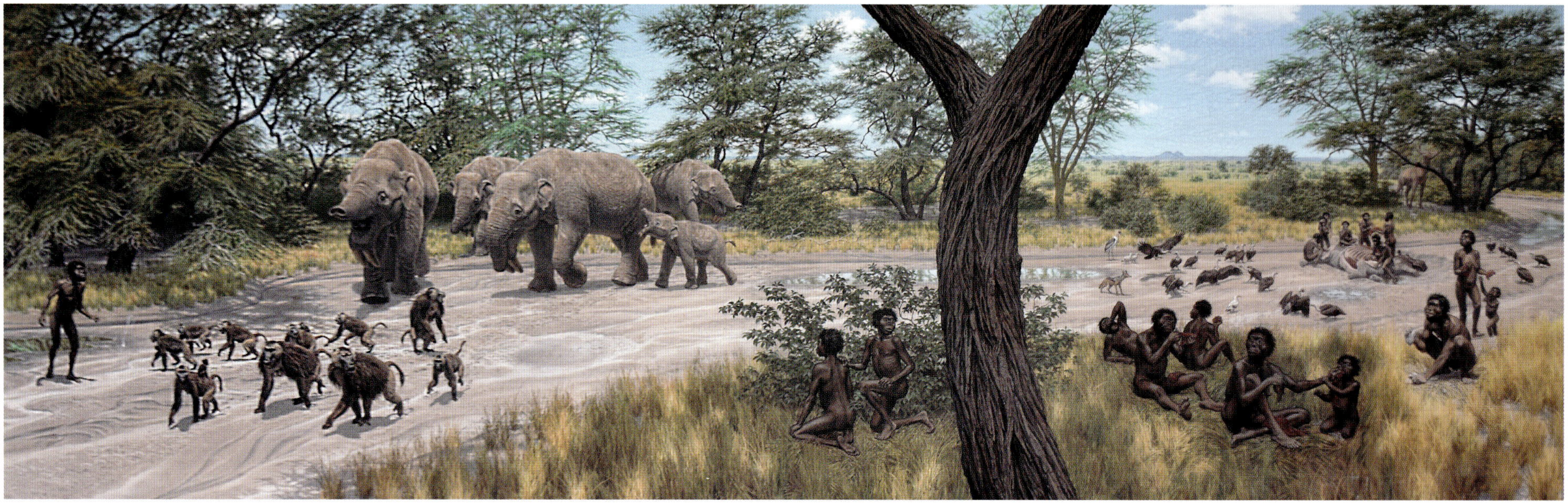

HIPPO FEAST AT ILERET

An impromptu feast is taking place, almost three million years ago, in a dried-up stream bed in what is now the Turkana Basin in Kenya. These hominins, depicted by Matternes as happily gorging on raw meat, are of indeterminate species but might be *Homo* (or *Kenyanthropus*) *rudolfensis*, a fossil skull of which, with its singular deep, flat face, had been recovered from the area a few years previously. This species was contemporaneous in East Africa with *Homo ergaster*, *H. habilis*, and perhaps also *Paranthropus boisei*.

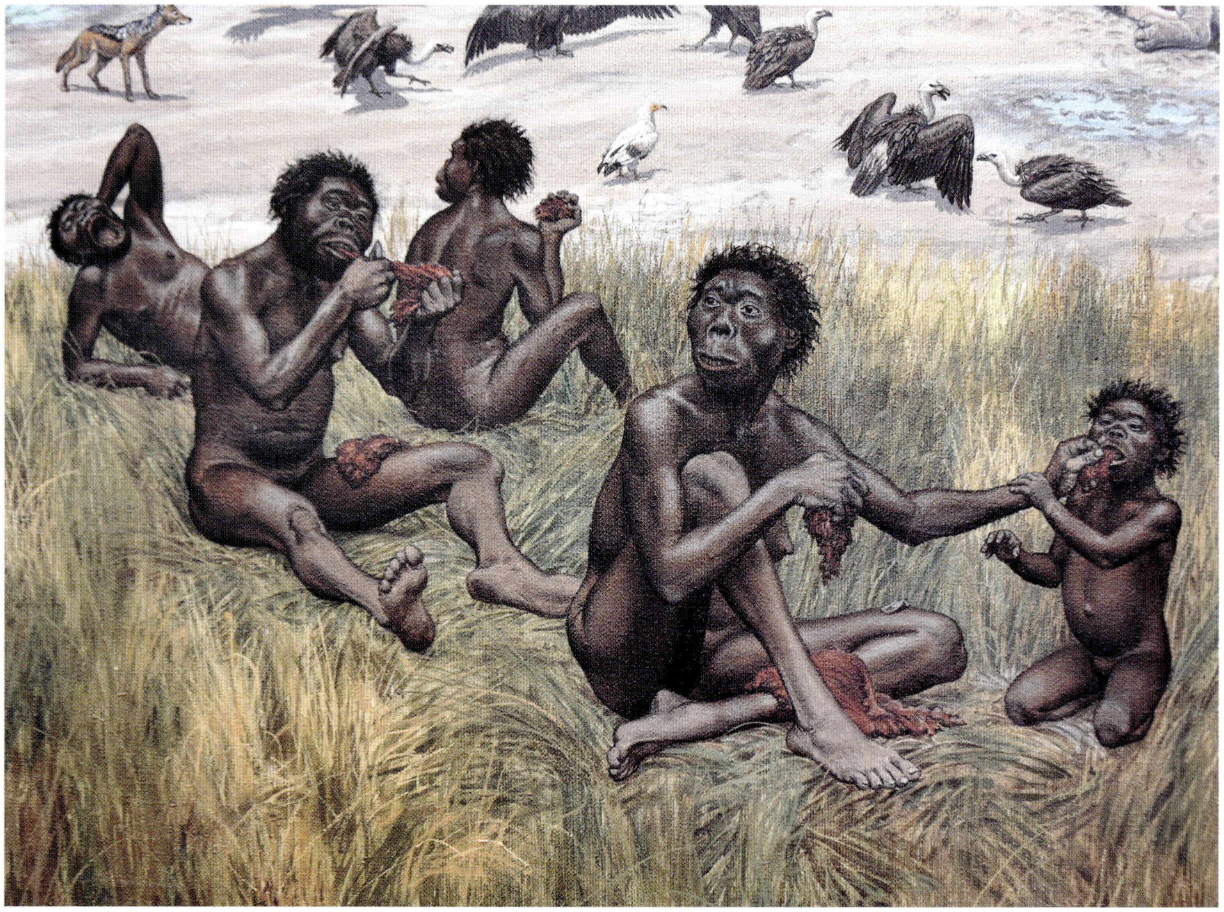

The 1974 painting illustrates discoveries by Richard Leakey and his team 20 miles north of Koobi Fora, Kenya. At this location, known as Ileret, some extraordinary features included the skeletal remains of a complete hippo, riddled with cut marks, together with many of the sharp stone tools used in butchering the carcasses.

If the scene appears cinematic, that is no accident. Matternes was commissioned by the British film company Survival Anglia to restore this "hippo butchering site" for its television film *Bones of Contention* (1976). Producers commissioned him to design a horizon-

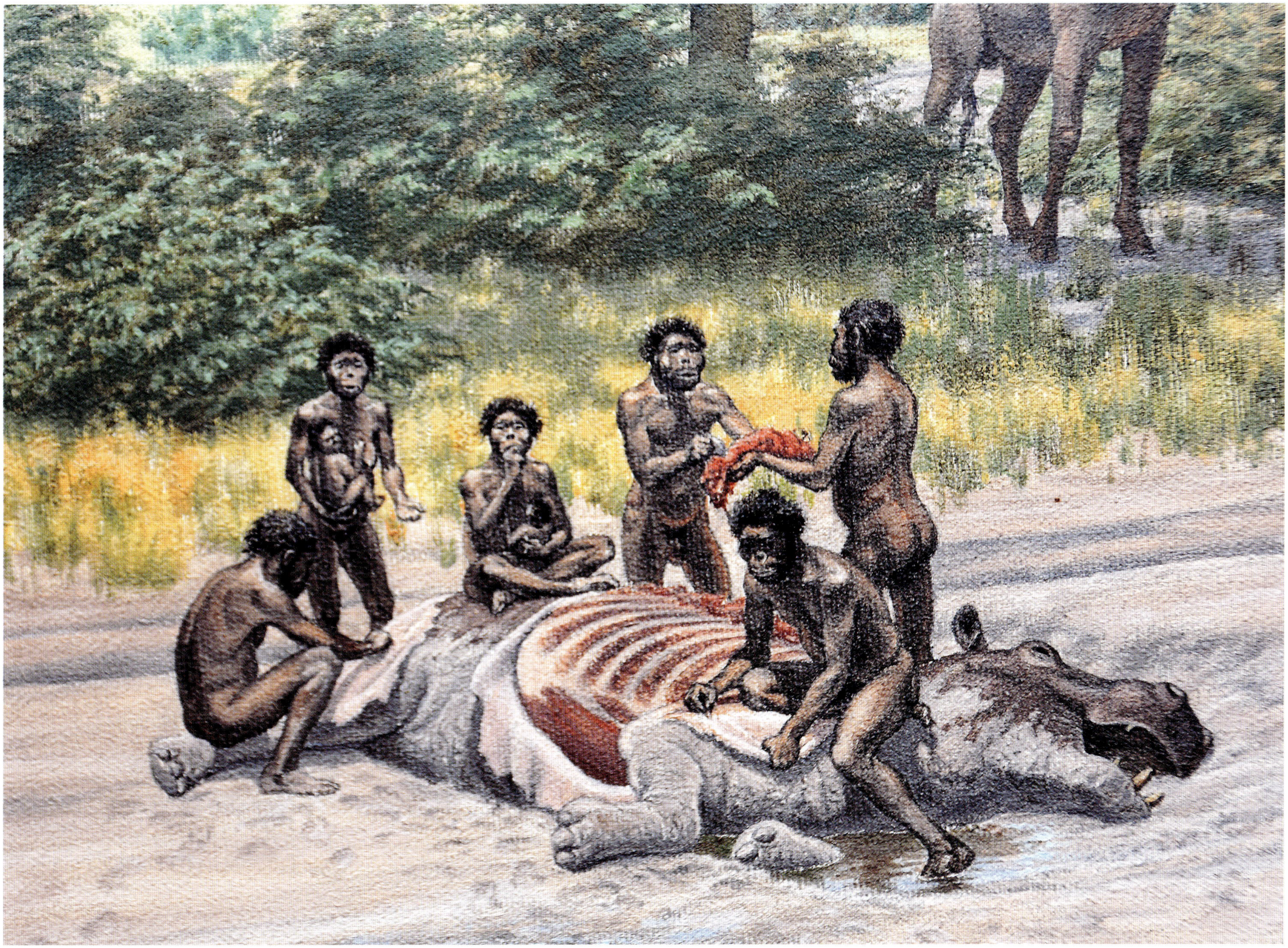

Hominins butcher a hippopotamus carcass with stone tools at Ileret, Turkana Basin, Kenya, more than 2.9 million years ago.

Right: As the hominin family shares the meat, a collection of hopeful scavengers arrives, including a maribou stork, jackals, and several kinds of vultures. Below: A troop of *Theropithecus* baboons surveys the scene, but does not stop.

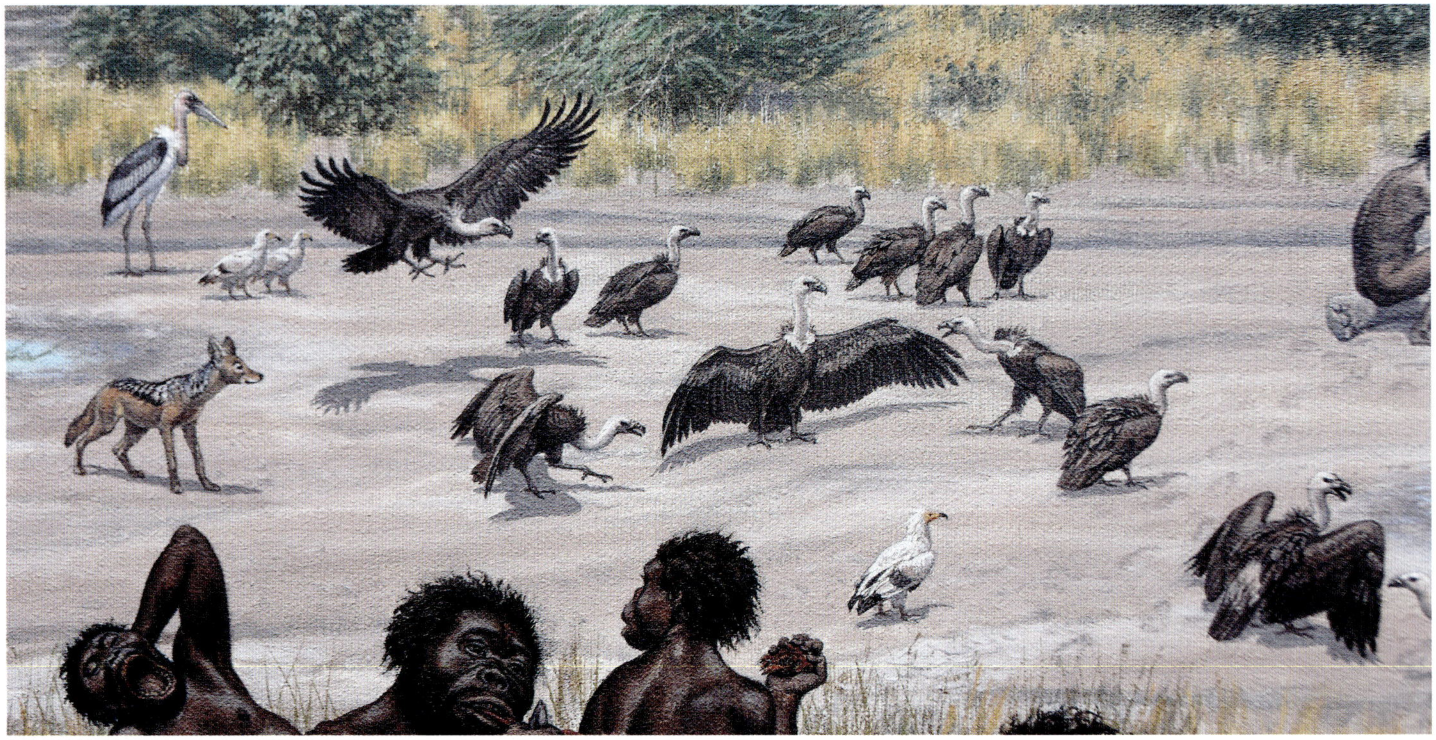

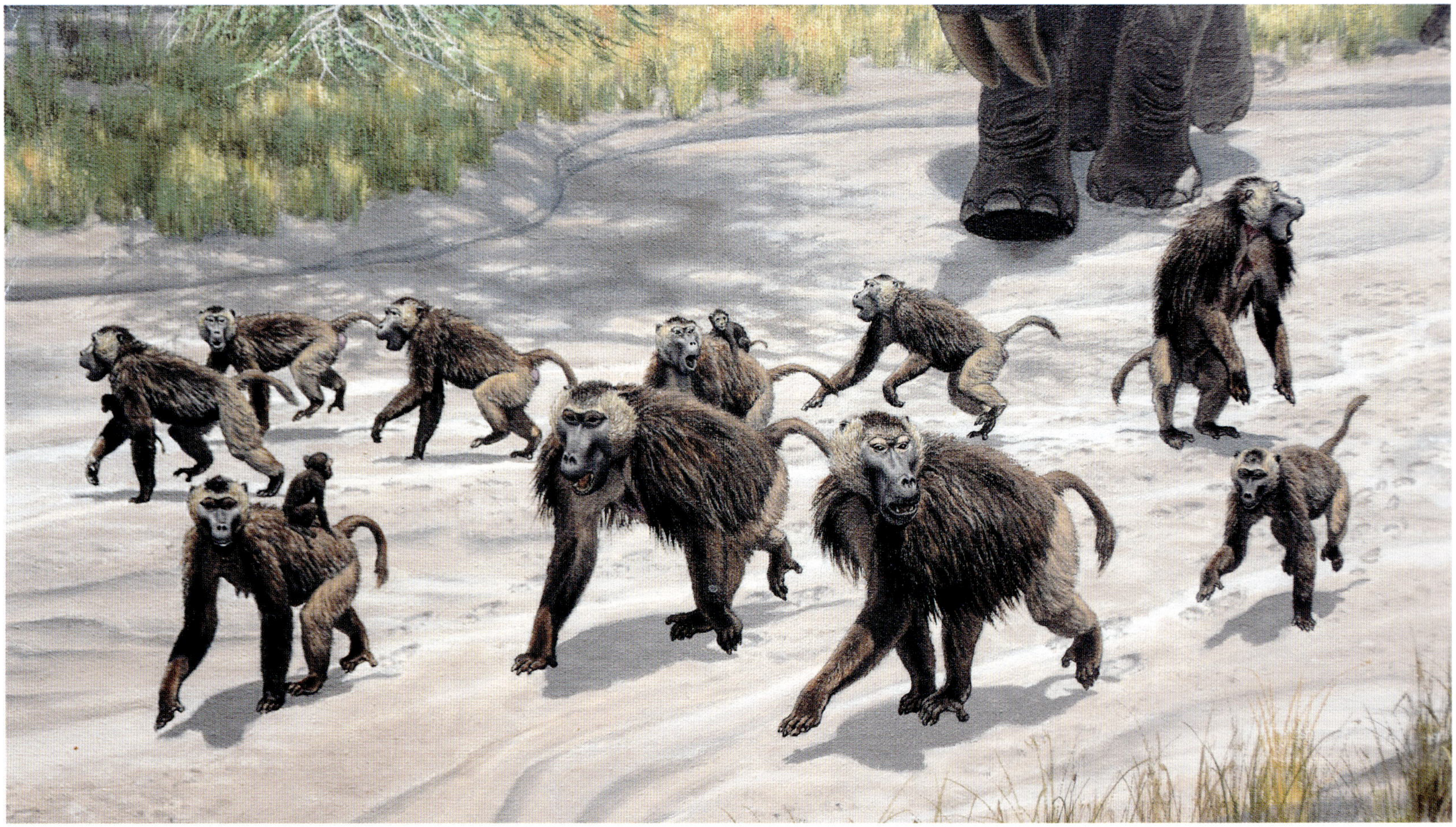

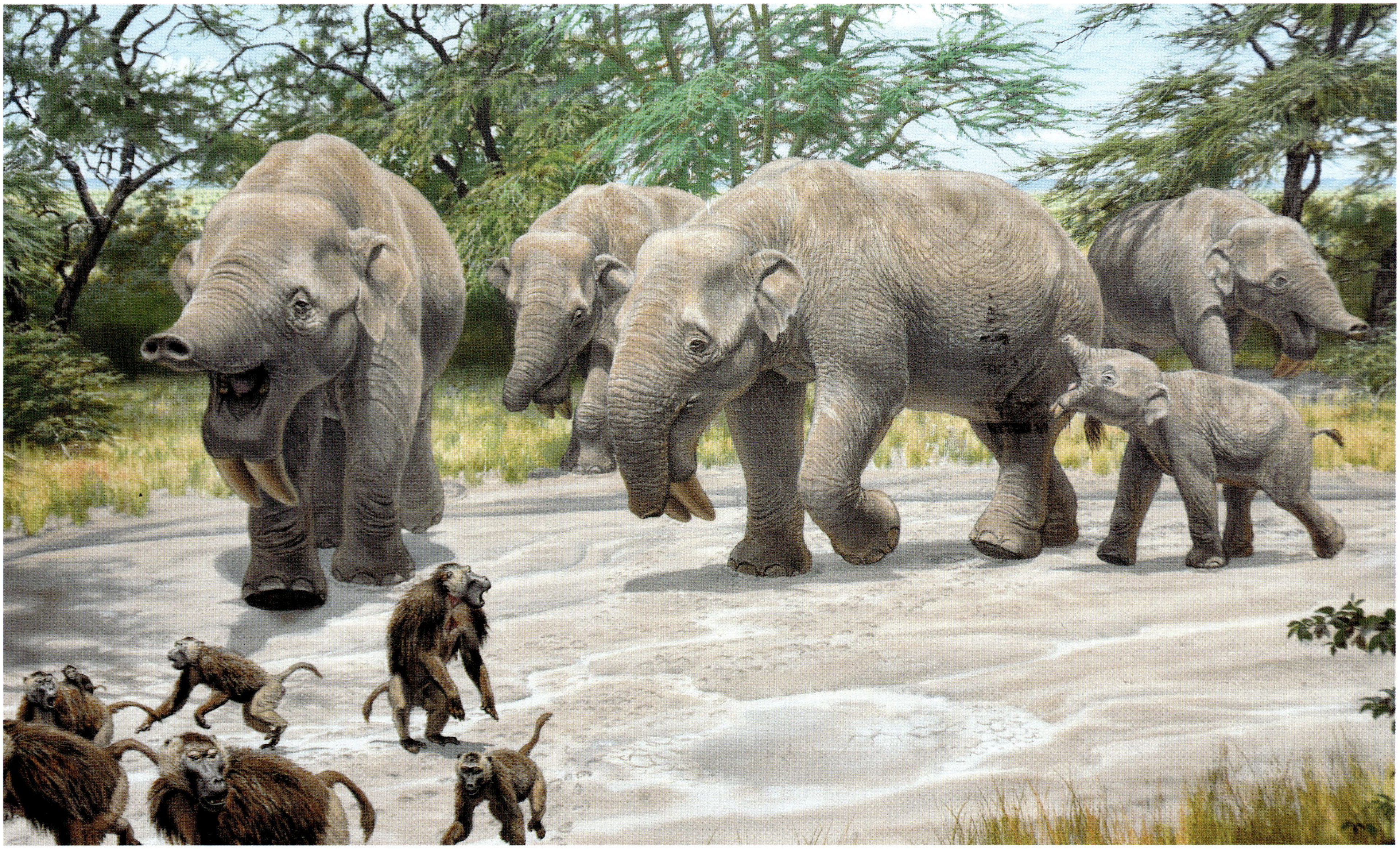

A herd of early elephants, *Deinotherium*, walk along the dry streambed, possibly in search of water.

tally elongated painting, so that the camera could slowly pan it as a narrator described the scene. After the documentary was aired, the filmmakers donated the painting to the Kenya National Museum in Nairobi, where the director, Frederick Manthi, had it photographed especially for this book.

In the distance, six males cut up a hippo near a dried-out stream bed and distribute meat to their extended family. Ten individuals are receiving pieces of the carcass, including three eating, two feeding their children, and one fabricating more stone tools for the work at hand. Seven of the al fresco diners are simultaneously watching a herd of elephant-like *Deinotherium* walking along the nearby streambed, perhaps searching for water. They are preceded by a troop of large *Theropithecus* baboons that meander along paying scant attention to the pachyderms or the humans. Small scavengers, including vultures, jackals, and storks, approach the butchers hoping to make off with scraps. Larger predators, such as hyenas or lions, have not yet arrived; when they do, the picnickers will disperse, taking away whatever they can carry.

More recently discovered footprints exposed in the 1.5-million-year-old sedimentary layers at Ileret are the oldest evidence of modern human-like foot anatomy. Their size is consistent with estimates for *Homo ergaster/erectus*, and they are morphologically distinct from the famous 3.75-million-year-old footprints at Laetoli, Tanzania. The Ileret prints are direct evidence that, by 1.5 million years ago, hominins had evolved an essentially modern human foot and style of bipedal locomotion.

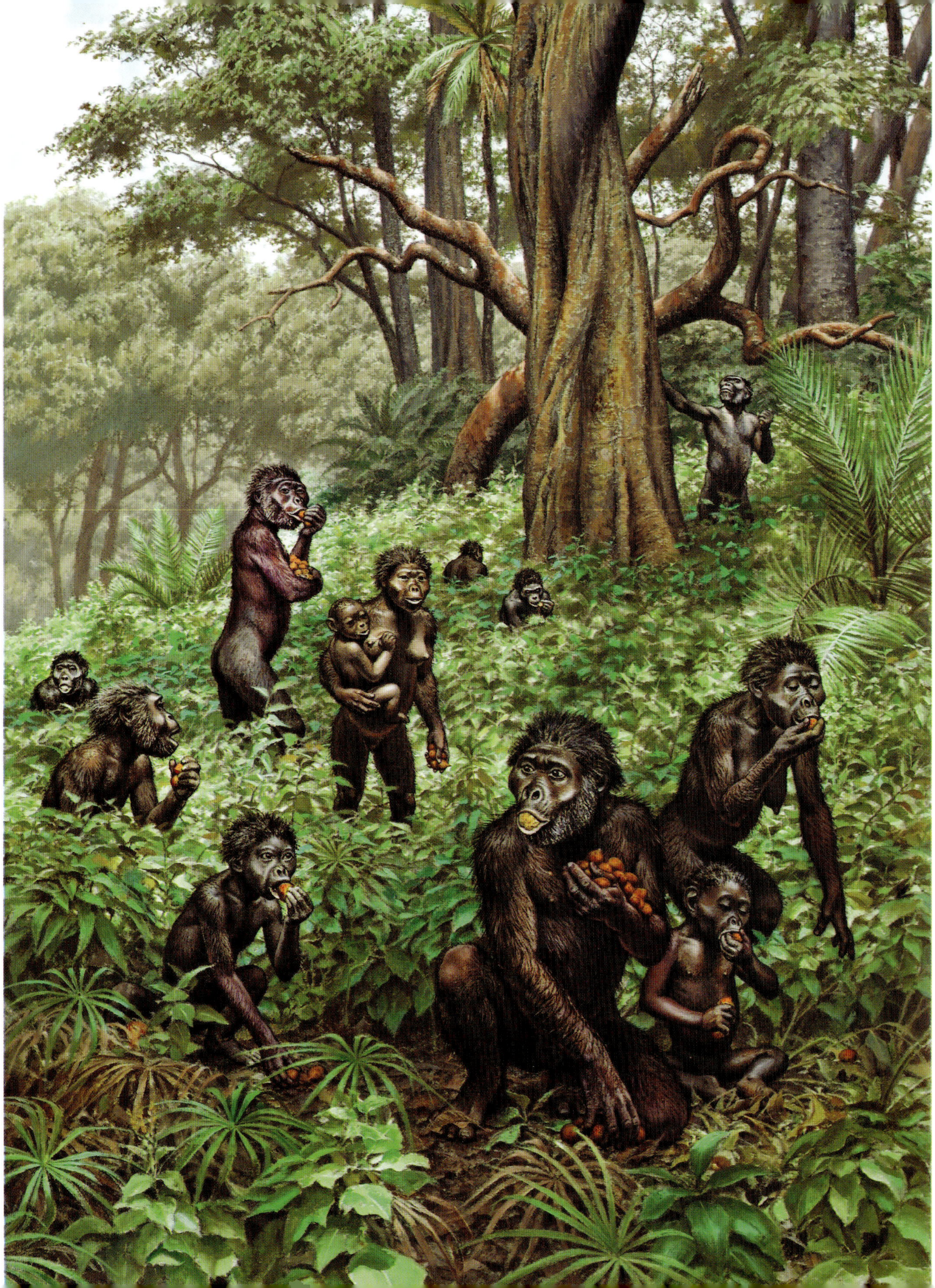

Family Feast

This view of an australopith family gorging on wild figs is set 3.5 million years ago in upland woodlands like those still bordering Lake Manyara in present-day Tanzania. Matternes has twice visited the area, near Olduvai Gorge, which still teems with wildlife, to photograph forest landscapes and vegetation as references for *National Geographic* paintings.

The hominins are eating ripe figs that have fallen from surrounding trees. Some sit to enjoy their meal and others continue searching the forest floor, while a male in the foreground stuffs his mouth with all it can hold.

Matternes is known for carefully delineating such details as hooves and toes, but at the time of this painting experts still debated the correct restoration of this hominin's fragmentary fossil feet. Were they fully straightened for upright walking, or did they retain a strong, flexible big toe for grasping branches? One anthropologist urged the artist to leave the question open, so Matternes obliged by obscuring the male's controversial tootsies with (what else?) a handy fig leaf!

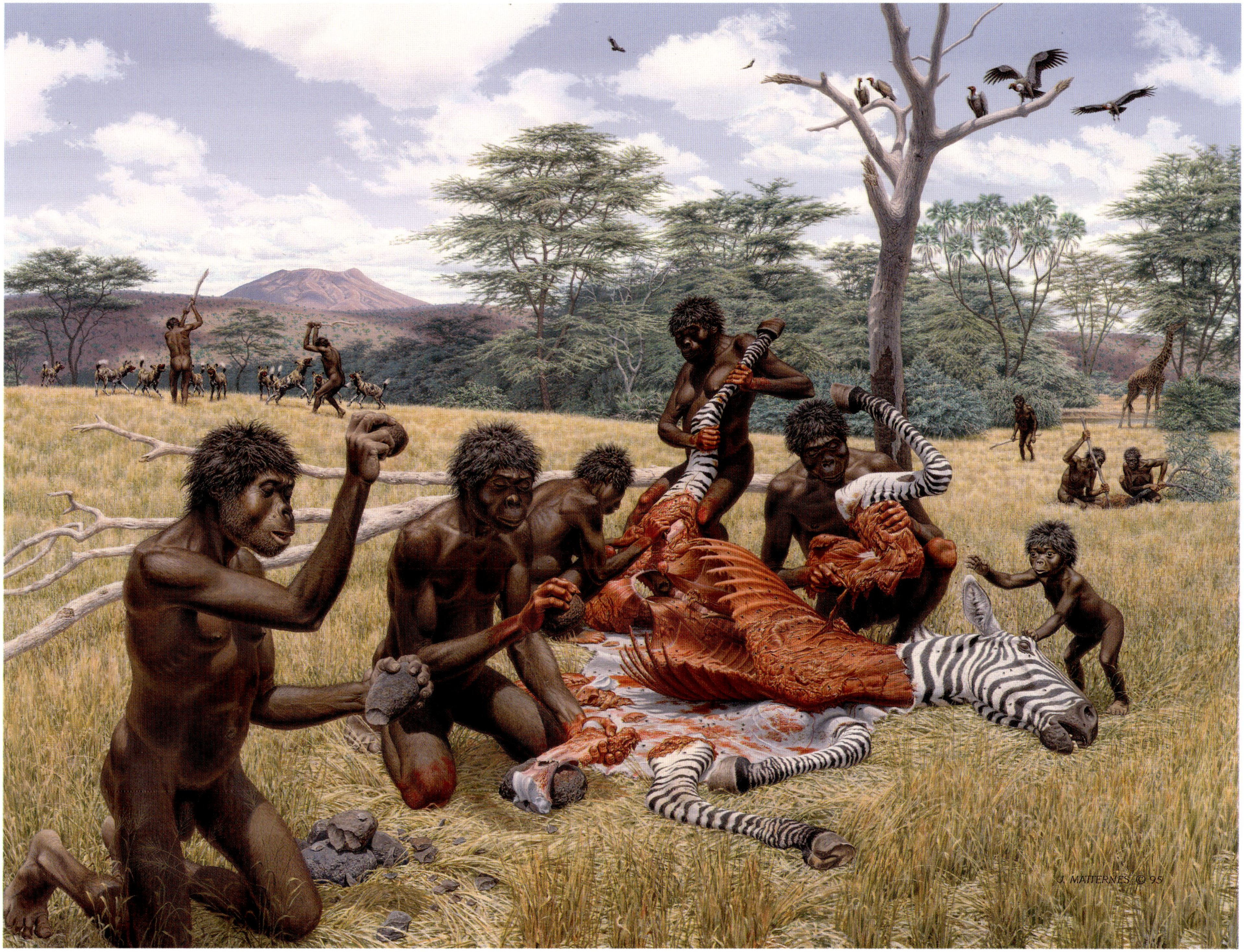

Dismantling a Zebra

Early species of our own genus, such as *Homo habilis* or *H. ergaster,* probably butchered animal carcasses, like this zebra, as a group effort. In the distance, several hunters attempt to drive a pack of wild dogs away from the kill, which the hominins may have stolen from them.

HOMO ERECTUS AND PRAIRIE FIRE

This painting from 1982 reflects the view, then still popular, that the famous Peking Man (*Homo erectus*) had mastered the art of using fire. The evidence was the blackened appearance of its 770,000-year-old fossil bones, but subsequent tests established that the darkening was caused not by fire, but rather by the deposition of manganese from cave fill. (Later, Jay also learned that the smoke produced by burning prairie grasses would have been white, not black.)

The earliest intimations that hominins had deliberately burned fires in hearths come from South Africa about a million years ago, well before Peking Man's time, but fire does not become a routine feature at archaeological sites until about 500,000–400,000 years ago. Matternes vividly imagines the drama and excitement of the hominins' early experiments with using fire—a major event in prehistory.

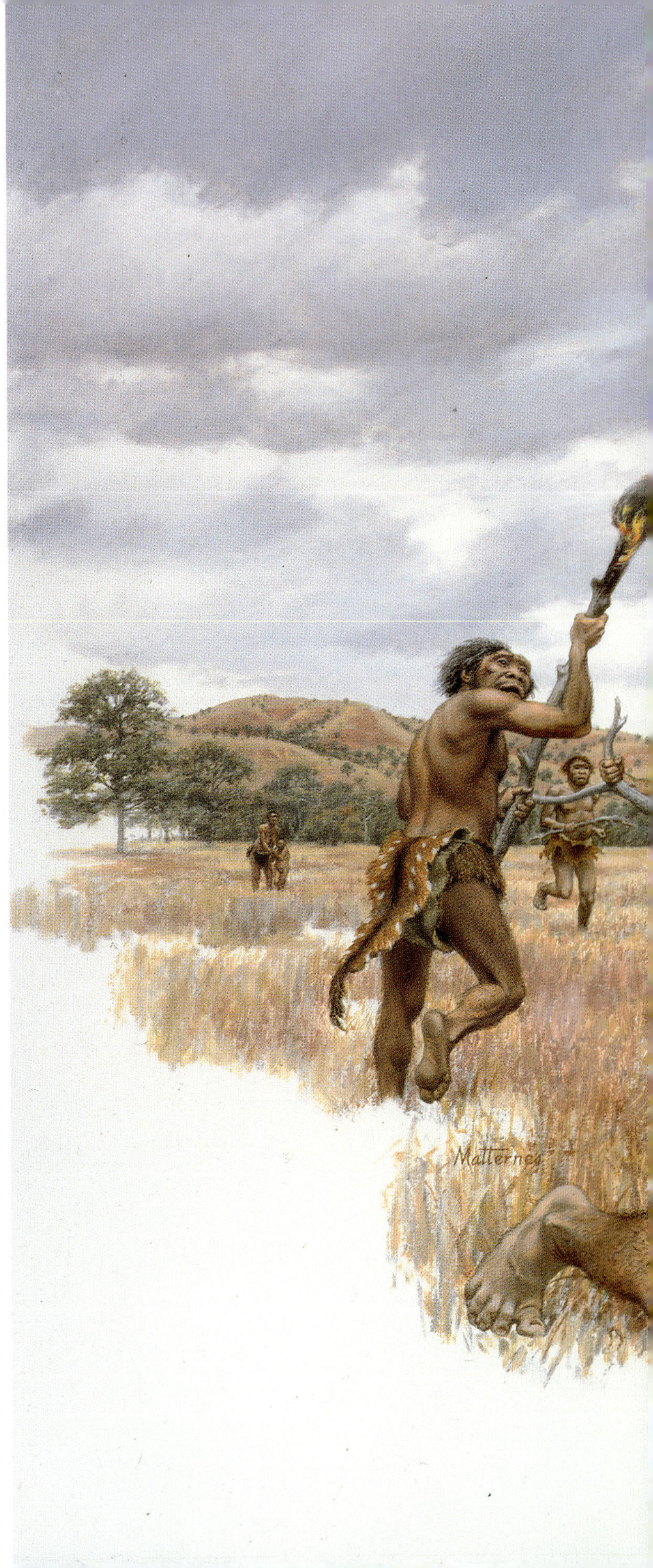

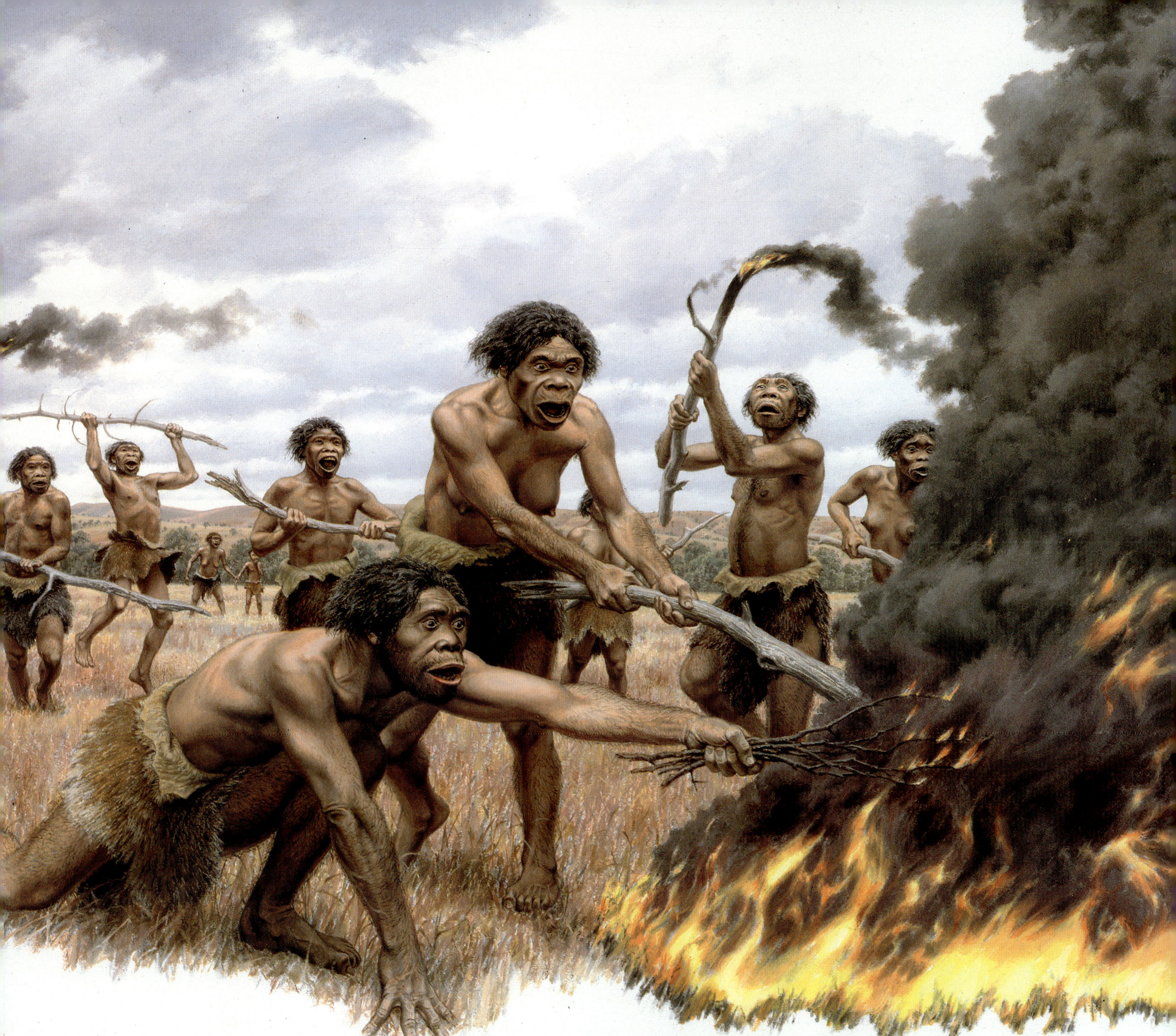

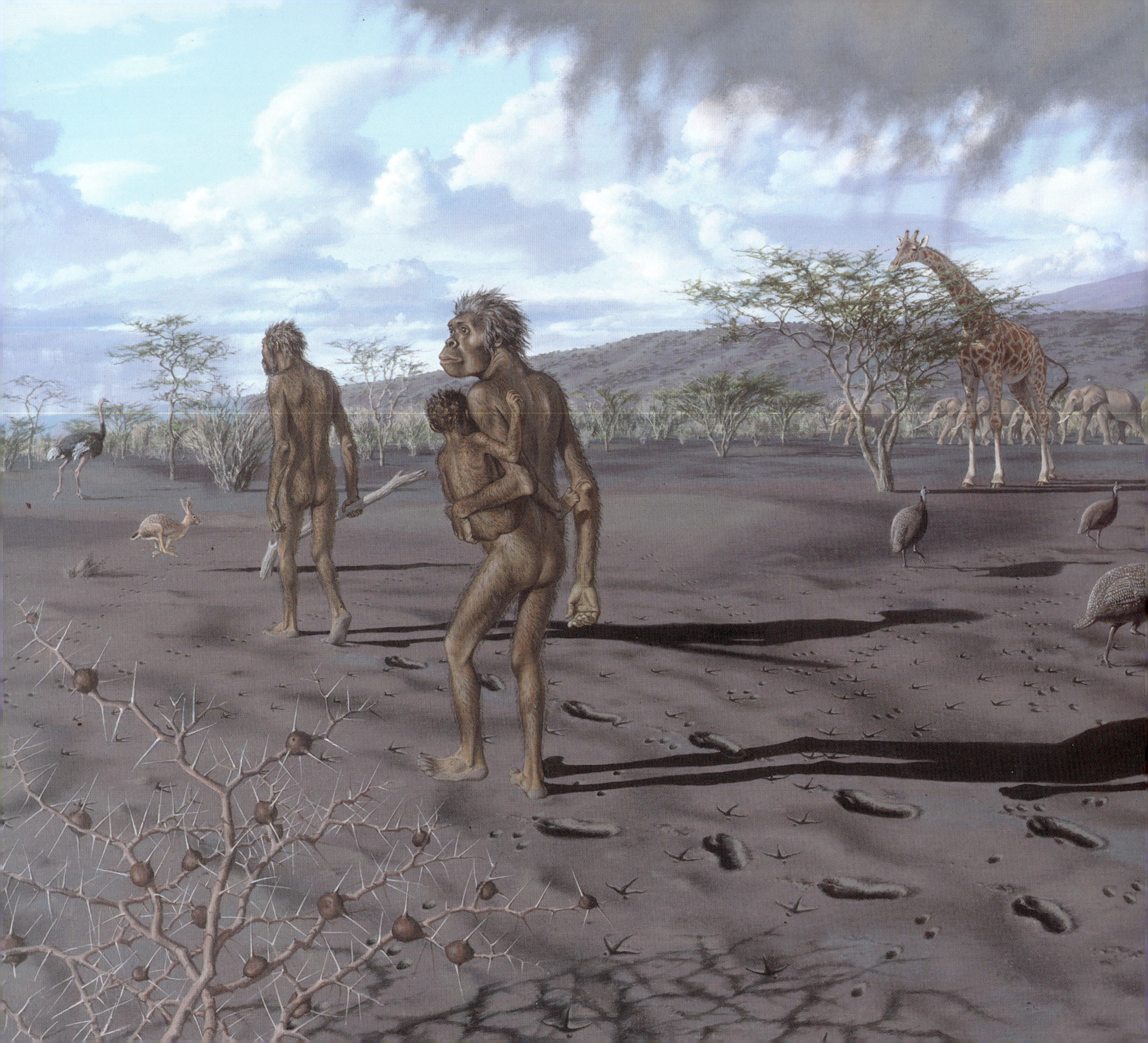

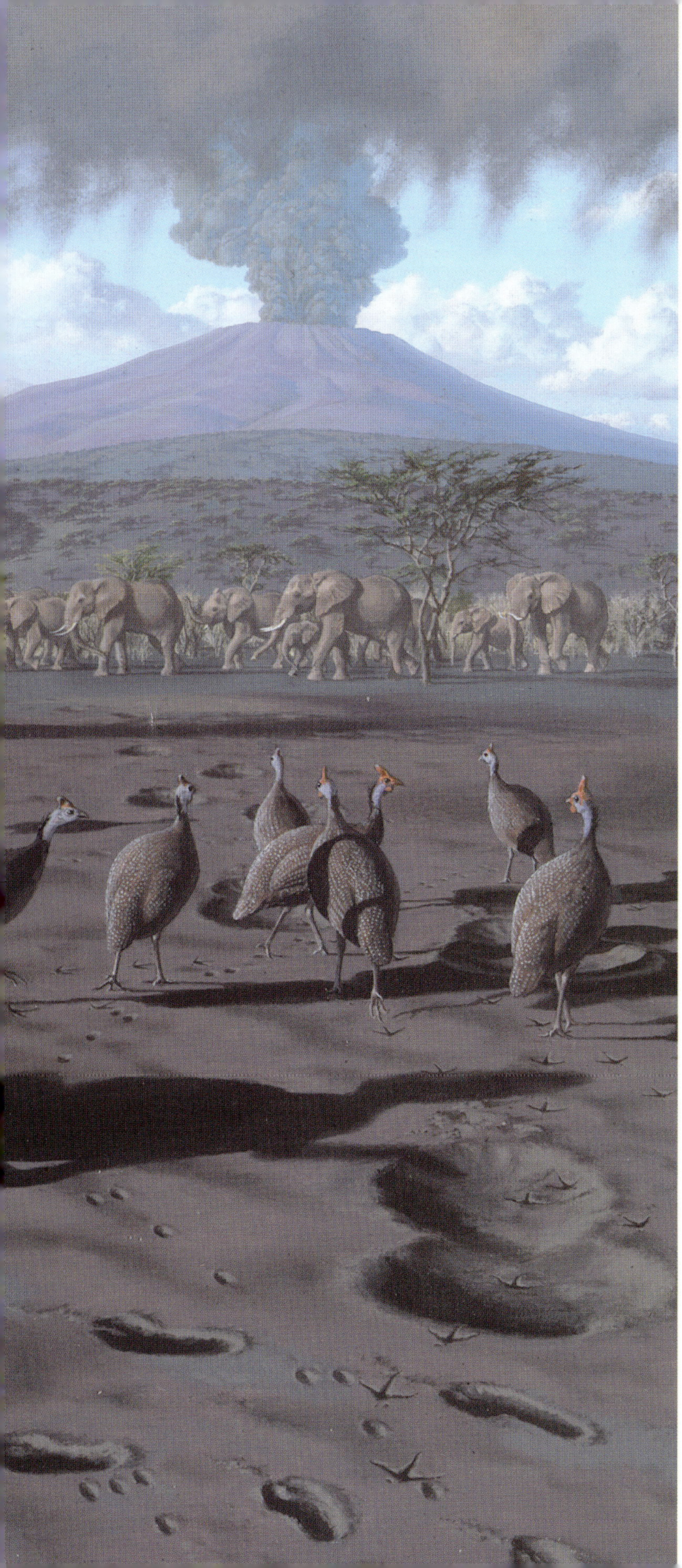

LAETOLI FOOTPRINTS

The 1970s produced extraordinary discoveries in paleoanthropology. None was more astonishing than the series of hardened footprints of a very remote human relative that was found by Mary Leakey's group at Laetoli, a site not far from Olduvai Gorge. Sites in this area had already yielded fossils of *Australopithecus afarensis*. Most physical traces of early humans consist of petrified bones and teeth, from which the behaviors of their possessors have to be inferred. But in this case, it was behavior itself that was fossilized.

Some 3.5 million years ago, a tall volcano, now deeply eroded as a low hump on the horizon, spewed out dense showers of volcanic ash that settled in thin layers on the surrounding landscape. One such layer was fresh on the ground when a rainstorm gave the dry ash the slurry consistency of wet plaster of Paris. Numerous animals, ranging in size from guinea fowl to elephants, wandered across this muddy terrain, leaving distinct trails. Soon the tropical sun dried out those footprints, hardening them. More ash fell later, burying and preserving the prints, until erosion exposed them millions of years later, to the delight of paleoanthropologists. In one place, two parallel hominin footprint trails are preserved, running arrow-straight for almost thirty yards, clearly made by upright bipeds who did not use their hands to steady themselves. The feet that made them were relatively modern in form, with a great toe in line with the others, and hints of both transverse and longitudinal arches. One trail was made by a small individual, who left clear impressions. The other was larger, and less distinct.

In 1979 the National Geographic Society commissioned Matternes to paint this reconstructed scene for the press announcement of the find; he was conducted to the site by Mary Leakey herself. One early suggestion, followed by the artist, was that the smaller individual had been carrying a child. Although interpretation of the prints is still debated, no one has succeeded better in re-creating the foreboding atmosphere that must have prevailed at Laetoli as the volcano continued to spew ash during that frozen moment in hominin history.

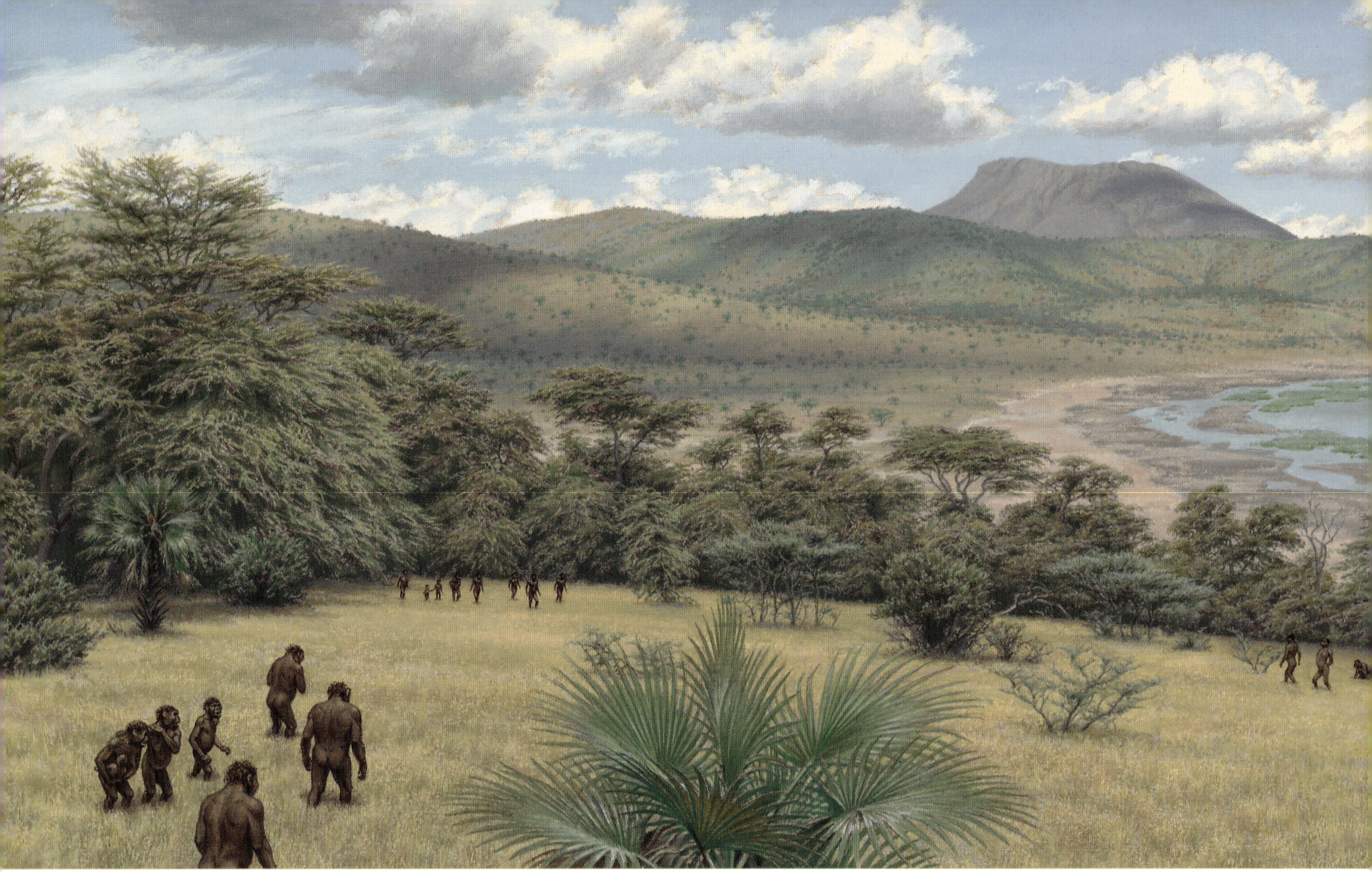

ONCE WE WERE NOT ALONE

Today we human beings are the only hominins left on Earth, but for most of the history of our subfamily several different kinds of hominins have lived on the planet at the same time—sometimes even sharing the same landscape. In this dramatic 1999 panorama of Lake Turkana in the early Pleistocene, painted for *Scientific American*, Matternes shows four different species of hominins coexisting near the lake.

When these contemporaneous related species roamed the area around two million years ago, Lake Turkana, formerly known as Lake Rudolf, was much more extensive than it is today—more like one of America's Great Lakes. Jay surveyed the present lake's dimensions with Richard Leakey's team in the 1970s both by flying over it, and by hiking along its perilous crocodile-populated shore.

Leakey and his geologists envisioned a grassy savanna lakeshore with a sandy beach and many shallow lagoons fringed with reeds. Broad, braided

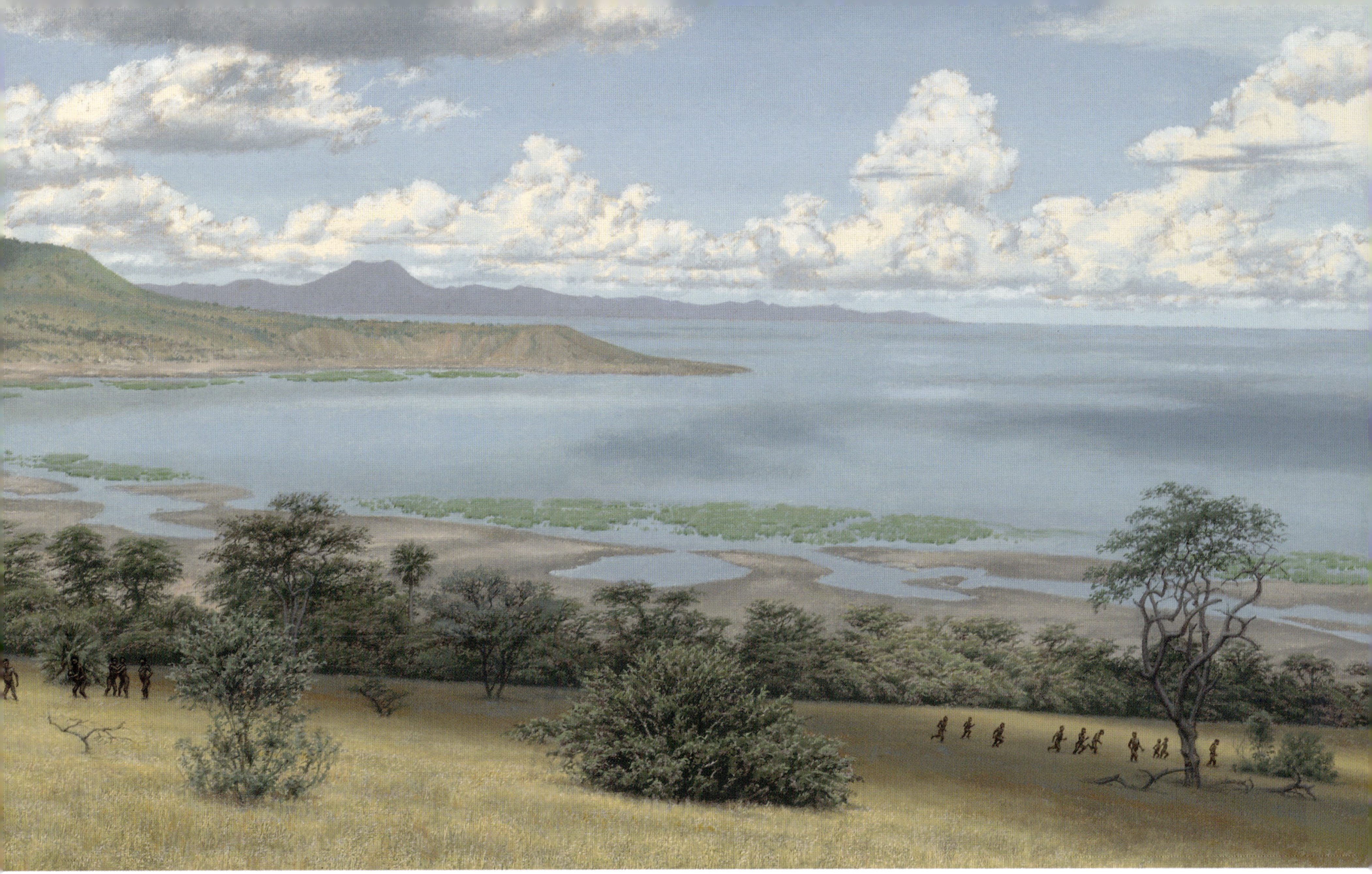

streams in strips of gallery forests flowed through silty flats into the ancient lake, and many volcanic cones punctuated the horizon. Jay used an elevated vantage point to present an inclusive view of the lake, along with its associated lagoons and reed beds.

The hominin groups are intended only to indicate variety; Matternes didn't attempt to depict precise appearances or possible interactions. The four forms include the robust australopith species *Paranthropus boisei*, two species of *Homo* (*H. habilis* and *H. ergaster*), and the enigmatic *Homo* (or *Kenyanthropus*) *rudolfensis*. About two million years ago they all managed to coexist in the very same local environment, yet today the entire planet does not support any hominin besides *Homo sapiens*. Anthropologist Ian Tattersall, author of the *Scientific American* article, comments that "there is a profound lesson here for us."

MARATHON MAN: EVOLUTION OF BIPEDAL RUNNERS

Next to an expanded brain and spoken language, both more recent acquisitions, bipedality has long been considered humankind's most crucial adaptation. But recent anatomical and physiological studies have added a new wrinkle: early hominins may have been excellent long-distance runners, or at least trotters. Hominins walked on two legs for several million years before they begin to develop larger brains, rapidly perfecting a bipedal adaptation that permitted efficient running. Compared to other apes, we have much springier leg and foot tendons, shoulders that rotate independently of the neck, hundreds of sweat glands for cooling the head and body, and numerous shock-absorbing joint structures and ligaments in the knees, spine, and ankles.

Speed itself seems to have been less crucial for survival than the combination of reasonable speed with exceptional endurance. American Indians and traditional African hunters are known to have routinely exhausted fleet-footed deer and antelope by running them down on foot for many hours, or

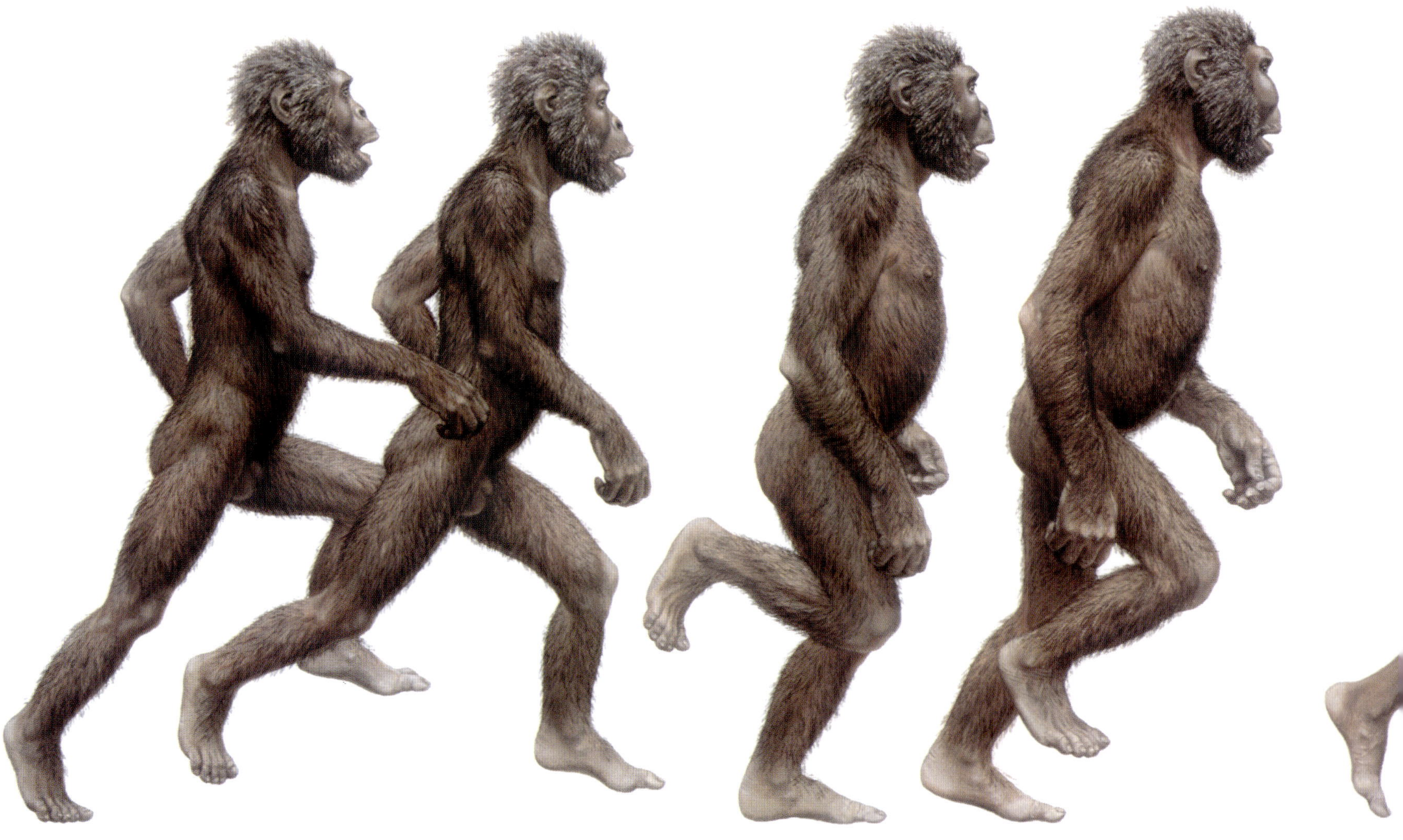

even days, because their prey needed to rest periodically to pant heat away, while they didn't.

New York City hosts an annual marathon; in the 2019 event, 53,627 people ran 26.2 miles in from 2 to 8 hours. Only a few years ago, it would have seemed inconceivable that so many city dwellers could run such distances while incurring very few injuries; but evidently, the stamina necessary for endurance running is built into the human condition.

In this illustration for a 1985 article on human evolution for *National Geographic* magazine, Matternes begins with the lightly built australopiths, which are followed by the heavier *Paranthropus*, then by *Homo erectus* and varieties of *Homo sapiens* that include the Neanderthals. A well-worn visual cliché depicts them as a lineup of walkers; the truth may be that for millions of years we have been literally trotting for our lives.

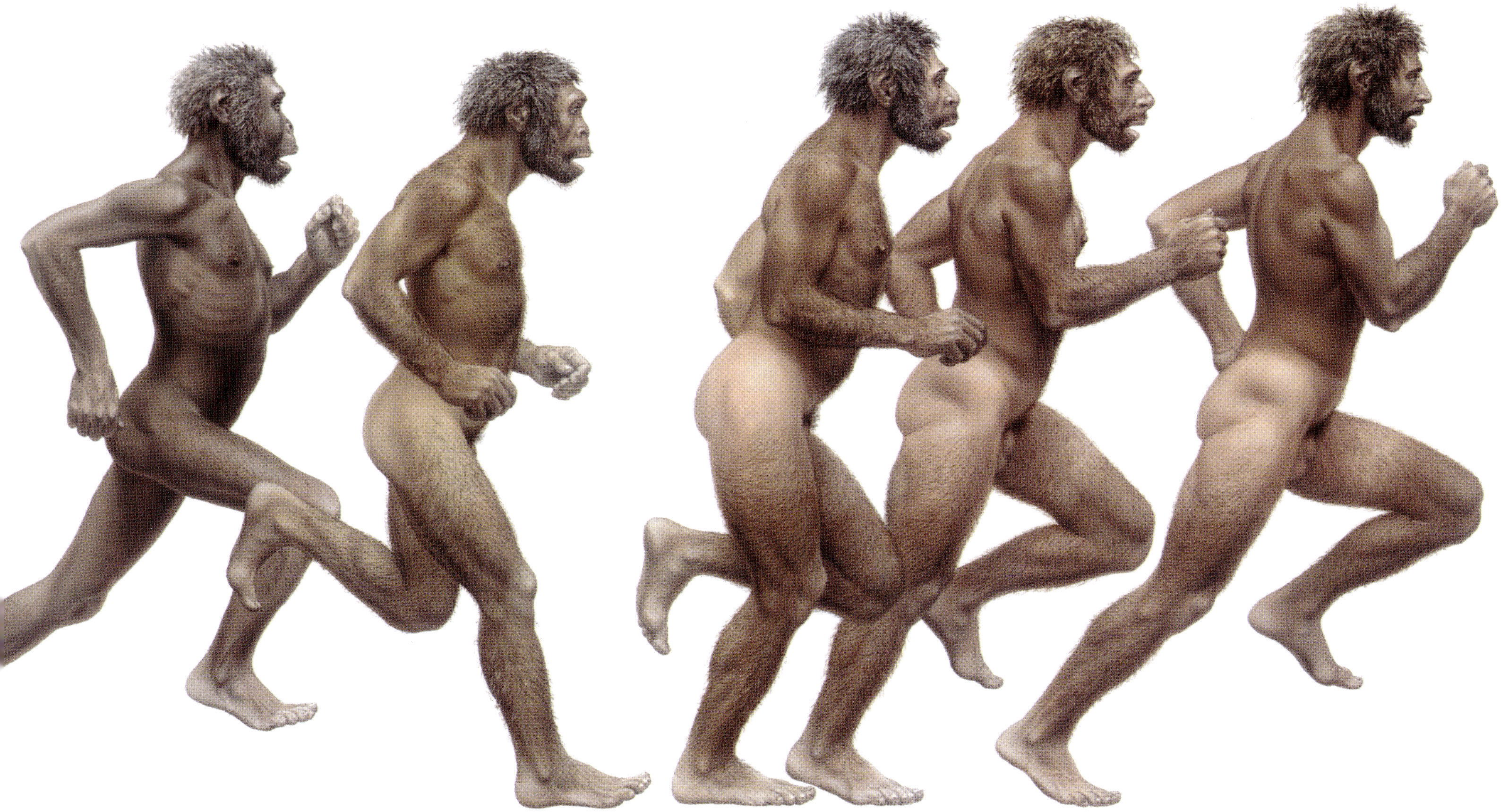

Australopiths Steal a Hyena's Kill

Early hominins drag a hartebeest head uphill to their rock shelter, about three million years ago. Matternes' charcoal drawing for the popular Time-Life book *Early Man* (1965) depicts audacious australopiths keeping furious hyenas at bay while stealing their dinner. Although fabricated tools had not been found with these hominins, the South African anatomist and pioneering paleoanthropologist Raymond Dart suggested in the 1950s that they might have utilized animal teeth, bones, and horns as cutting, boring, and scraping tools—the so-called "osteodontokeratic culture." That "culture" was subsequently abandoned as fanciful, but it remains probable that some australopiths were adept scavengers of predators' kills.

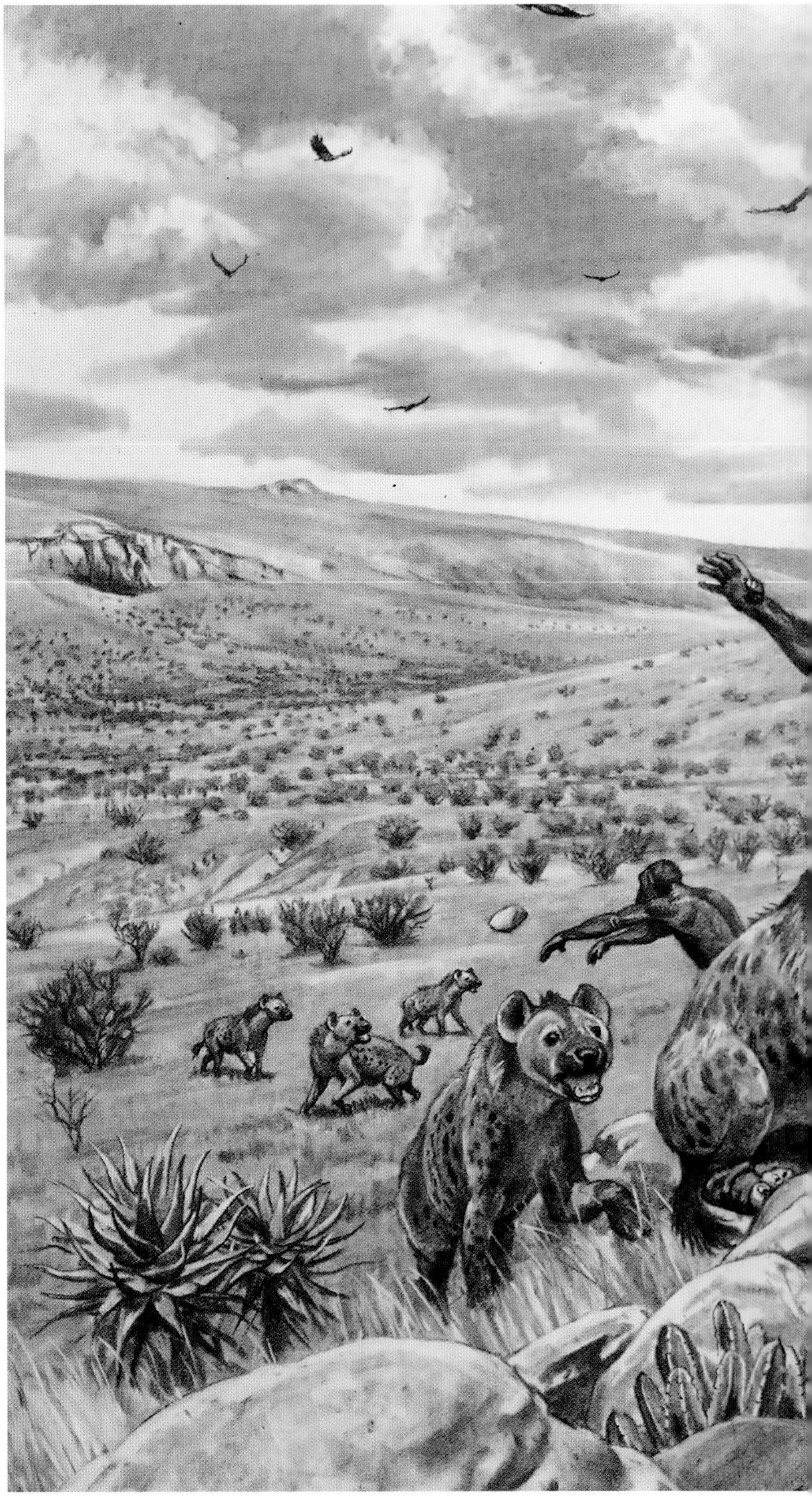

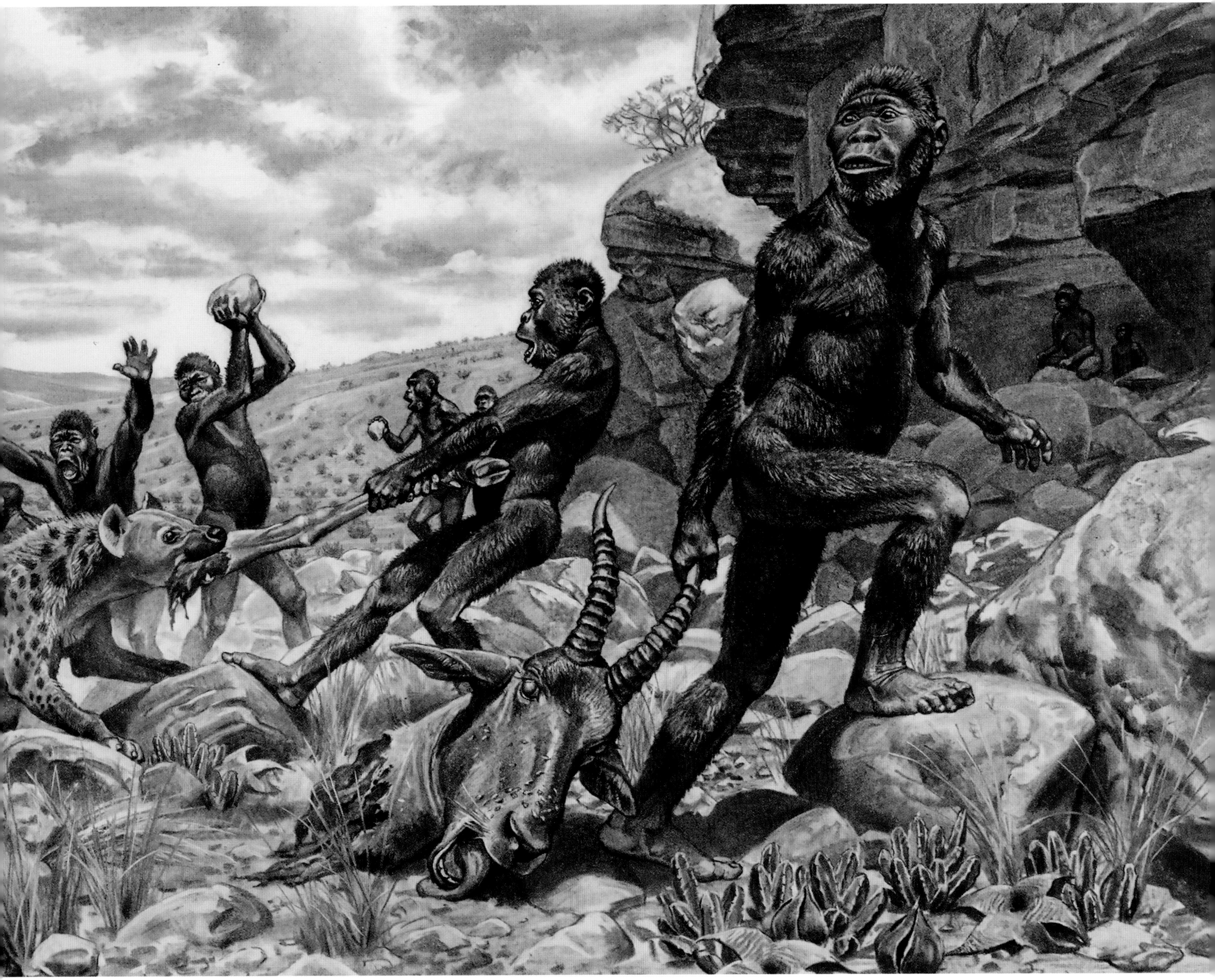

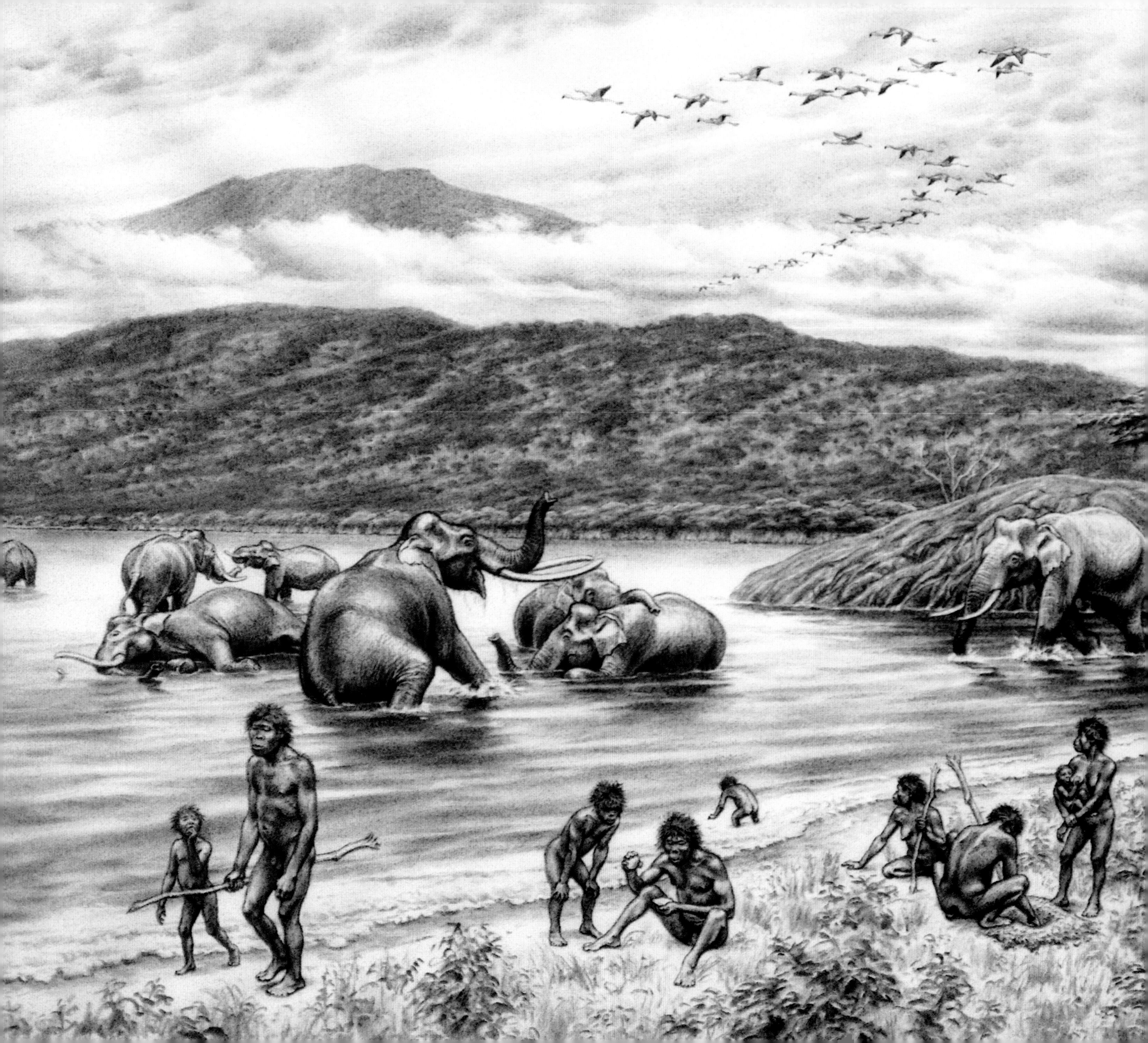

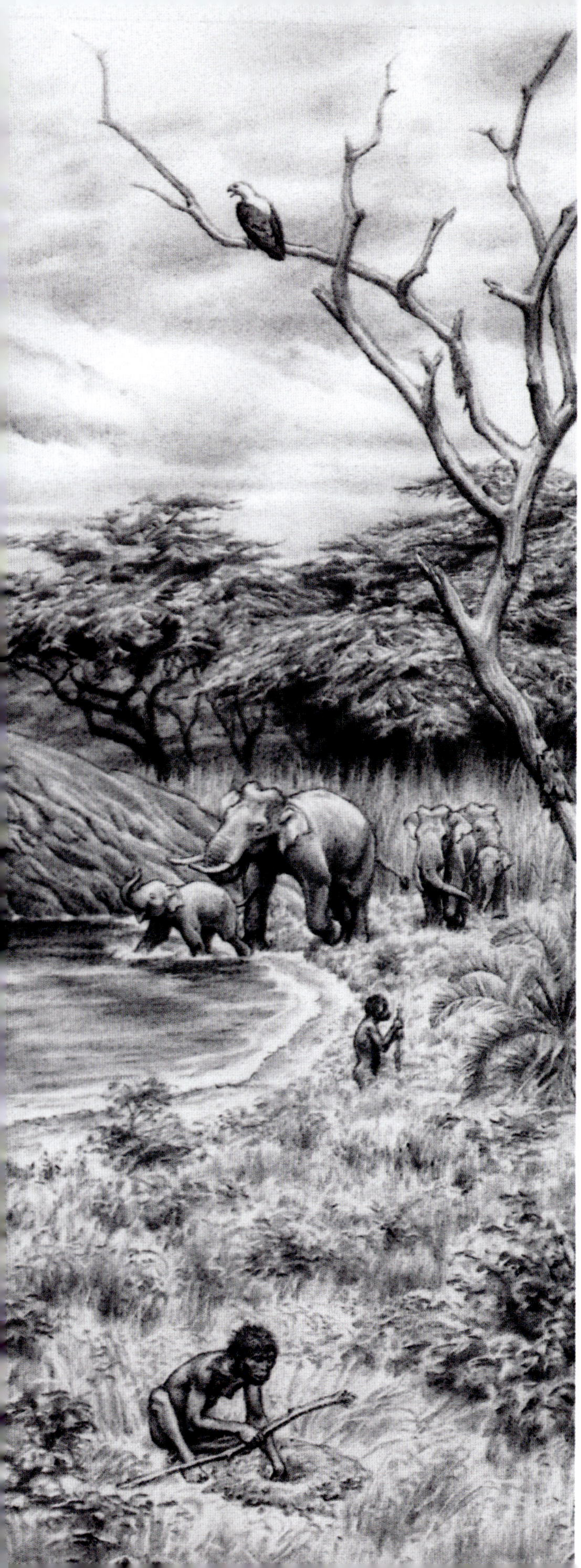

OLDUVAI LAKESHORE 1.8 MILLION YEARS AGO

Protected from predators by the proximity of the extinct proboscidians *Elephas recki*, a group of *Homo habilis* digs for tubers and roots in the ground cover on the shores of an ancient lake at a time when Olduvai was much wetter than its present arid landscape. One of the hominins fashions a crude stone tool, while a juvenile watches and learns. The frothy edge of the alkaline lake is bordered with elephant grass and acacia trees. In the right foreground is a dome of solidified lava that had flowed upward through a crack in the overlying rock layers, and could have served as a platform for pelicans and cormorants. An African fish eagle perches on a drowned snag of acacia, while a small flock of flamingos passes overhead. In the background rises the volcano Lemagrut, five thousand feet higher than it is today.

H. habilis vs. *Paranthropus*: An Iconic Confrontation

In this illustration for the Time-Life book *Early Man* (1965), the combative *Homo habilis* individuals in the foreground are holding crude flaked stone tools. While their sharp cutting edges suggest that these implements were used primarily to prepare food, the editors imagined that they might also have doubled as weapons to cut or slash. At the time the illustration was made, *Paranthropus* was not yet known to have fabricated stone tools, but during a battle members of either species might have simply picked up and thrown anything at hand.

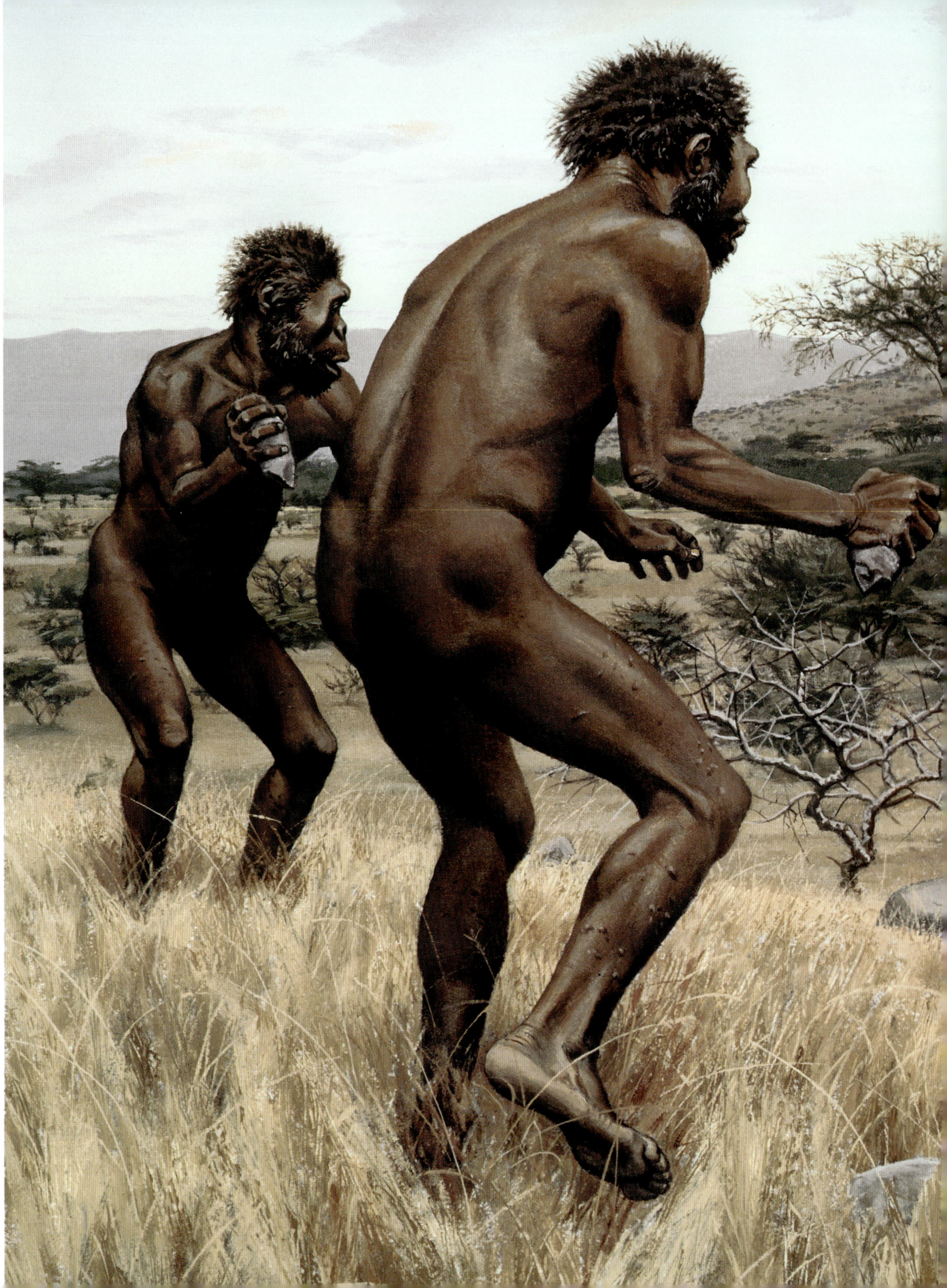

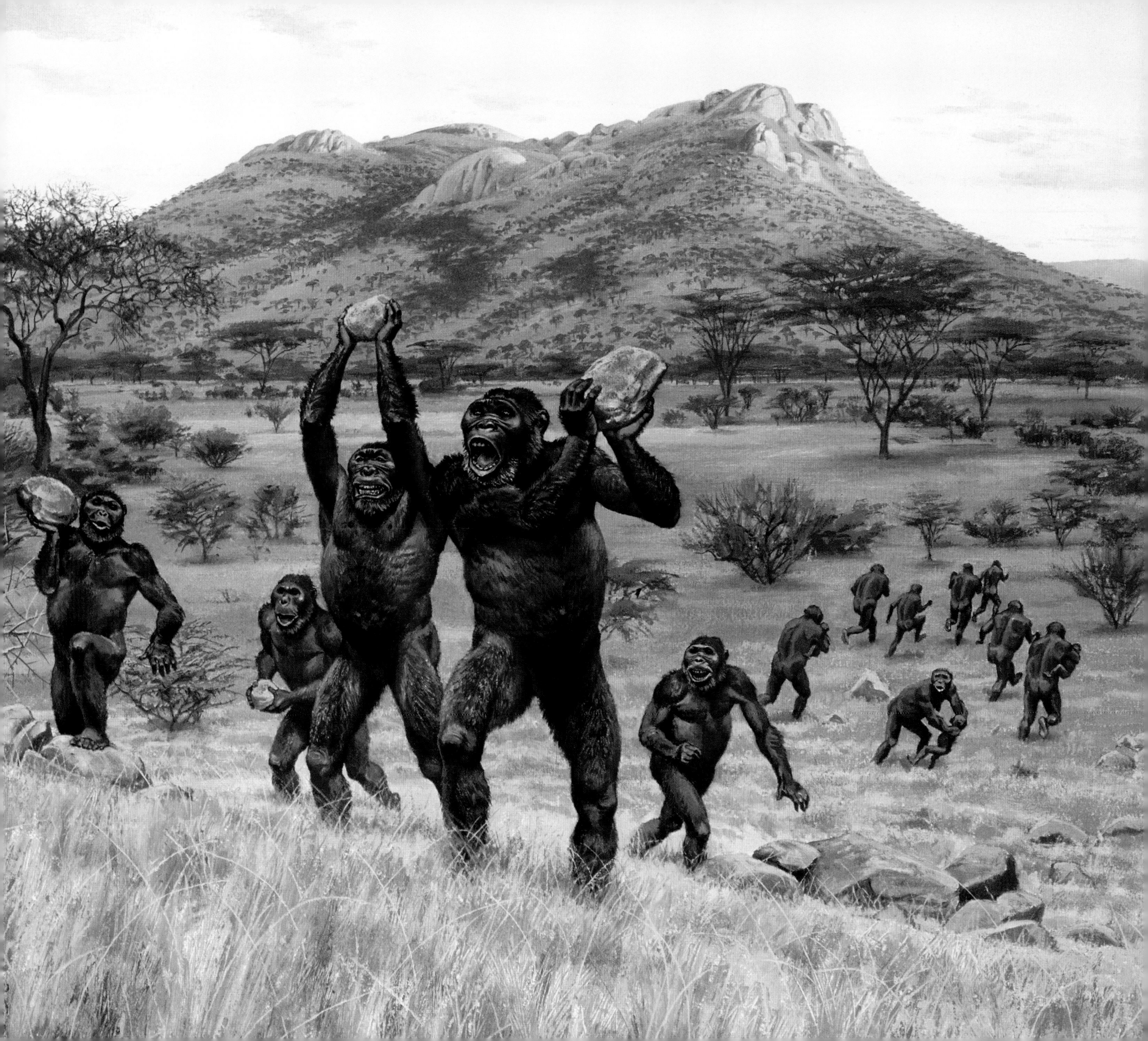

"A FAMILY ALBUM"

Four million years ago members of the hominin group we informally call the "austrolopiths" ranged over the African continent, from South Africa to Ethiopia and Chad. Some species were more lightly built than others, but most were about 3 to 4 feet tall, with small brains and large faces as reflected in Jay's portrait of *Australopithecus africanus*. Features of the upper body indicate they were still agile tree climbers, even as the structure of their pelves and legs affirms that they moved upright on two legs when on the ground. They ranged widely in diverse environments from forest to bushland.

Our ancient relative *Homo habilis* ("handy man") was so named in 1964 because it was likely the maker of the 1.8-million-year-old stone tools found with its fragmentary fossil remains at Olduvai Gorge, Tanzania. The new species was controversial at the time, and the rather miscellaneous assortment of 2.5- to 1.5-million-year-old African fossils since assigned to them remains problematic, although Jay has neatly contrived to catch its basic qualities. The earliest hominin that can confidently be assigned to our genus *Homo* is *H. ergaster*, first known from 2-million-year-old fossils in South Africa.

Named for a skeleton accidentally discovered in Germany's Neander Valley in 1856, *Homo neanderthalensis* was the first fossil human species to be described. This large brained but low-skulled hominin was a clever stoneworker, a sophisticated user of its environment, and a very close relative of *H. sapiens*, but it did not process information in exactly the way we do. Although genomic studies suggest some interbreeding between the two species when *H. sapiens* exiting Africa encountered Neanderthals in their European and western Asian homeland around 45,000 years ago, the two apparently found themselves in competition of some kind, resulting in the Neanderthals' rapid disappearance. Jay's portrait captures the strange familiarity of this mysterious relative.

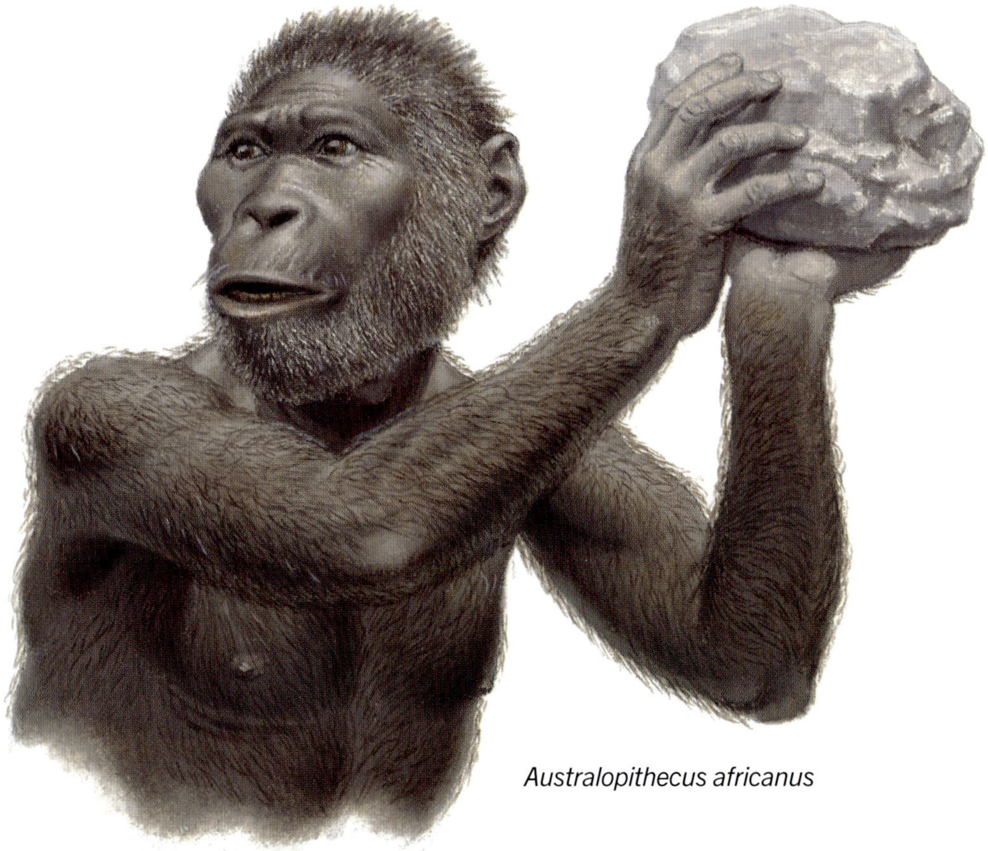

Australopithecus africanus

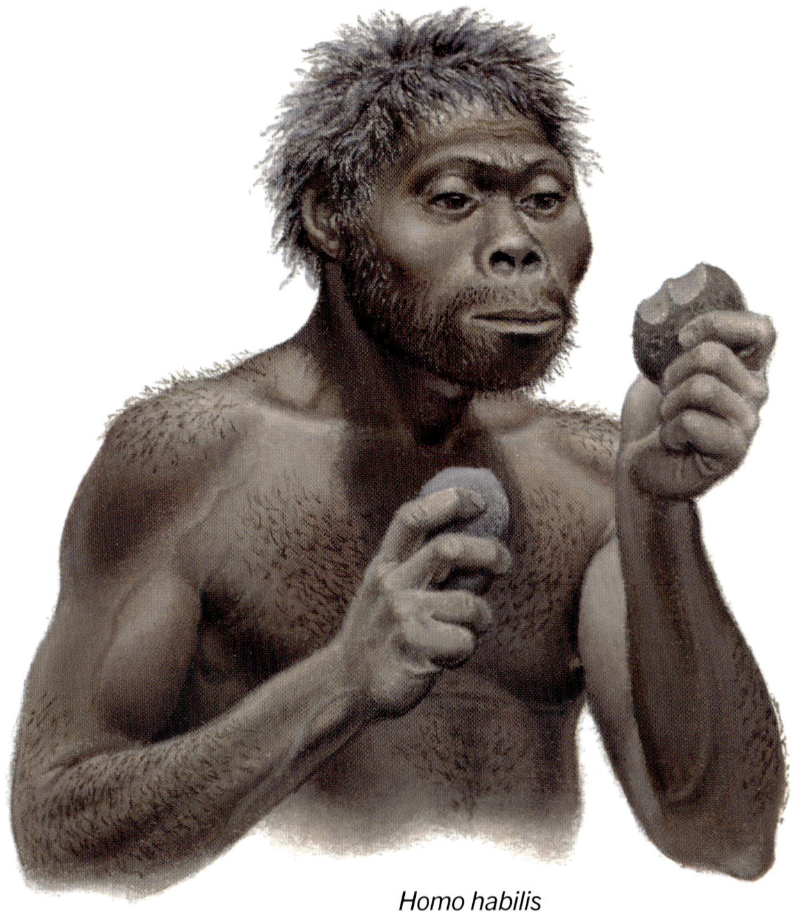

Homo habilis

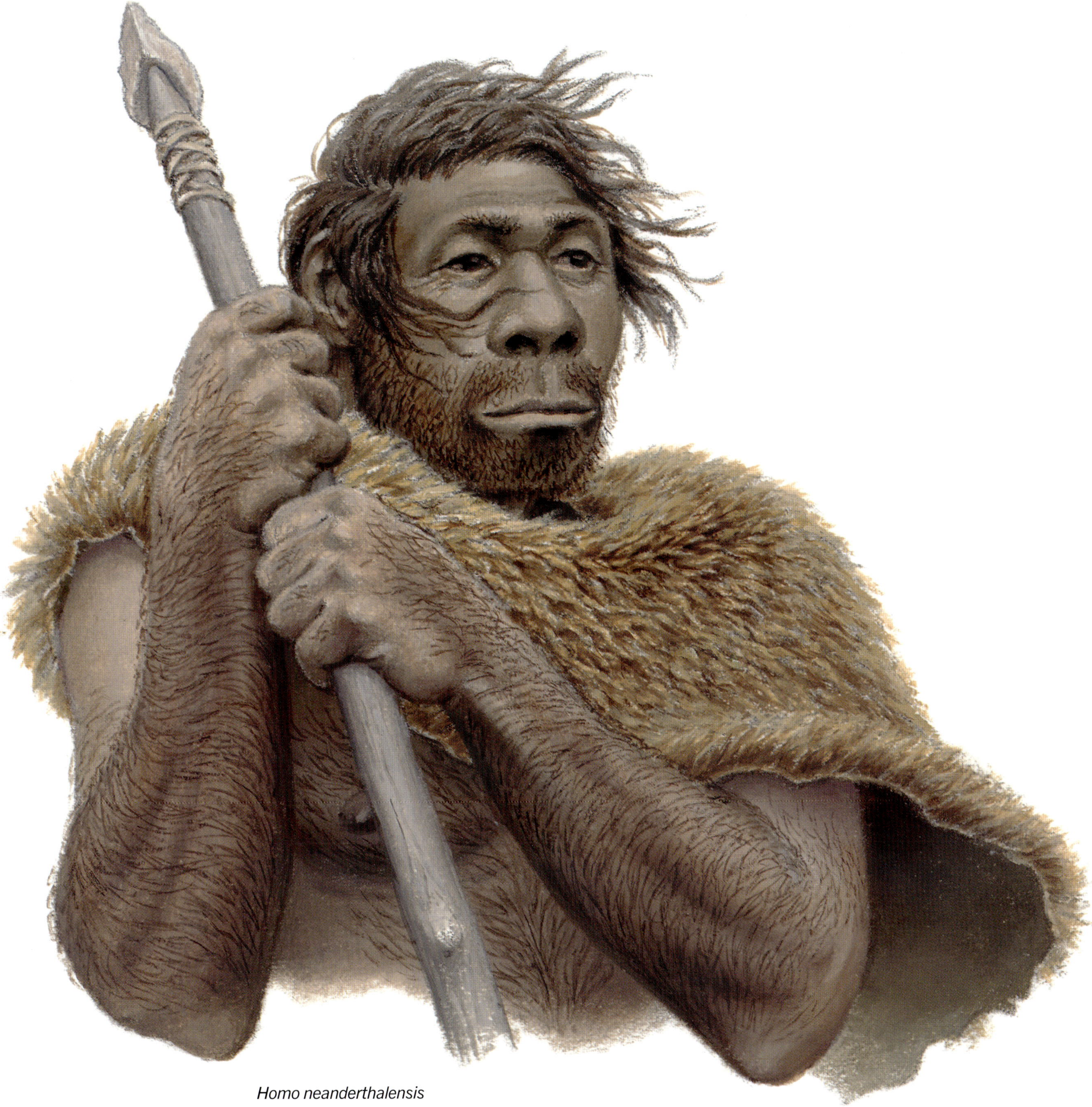

Homo neanderthalensis

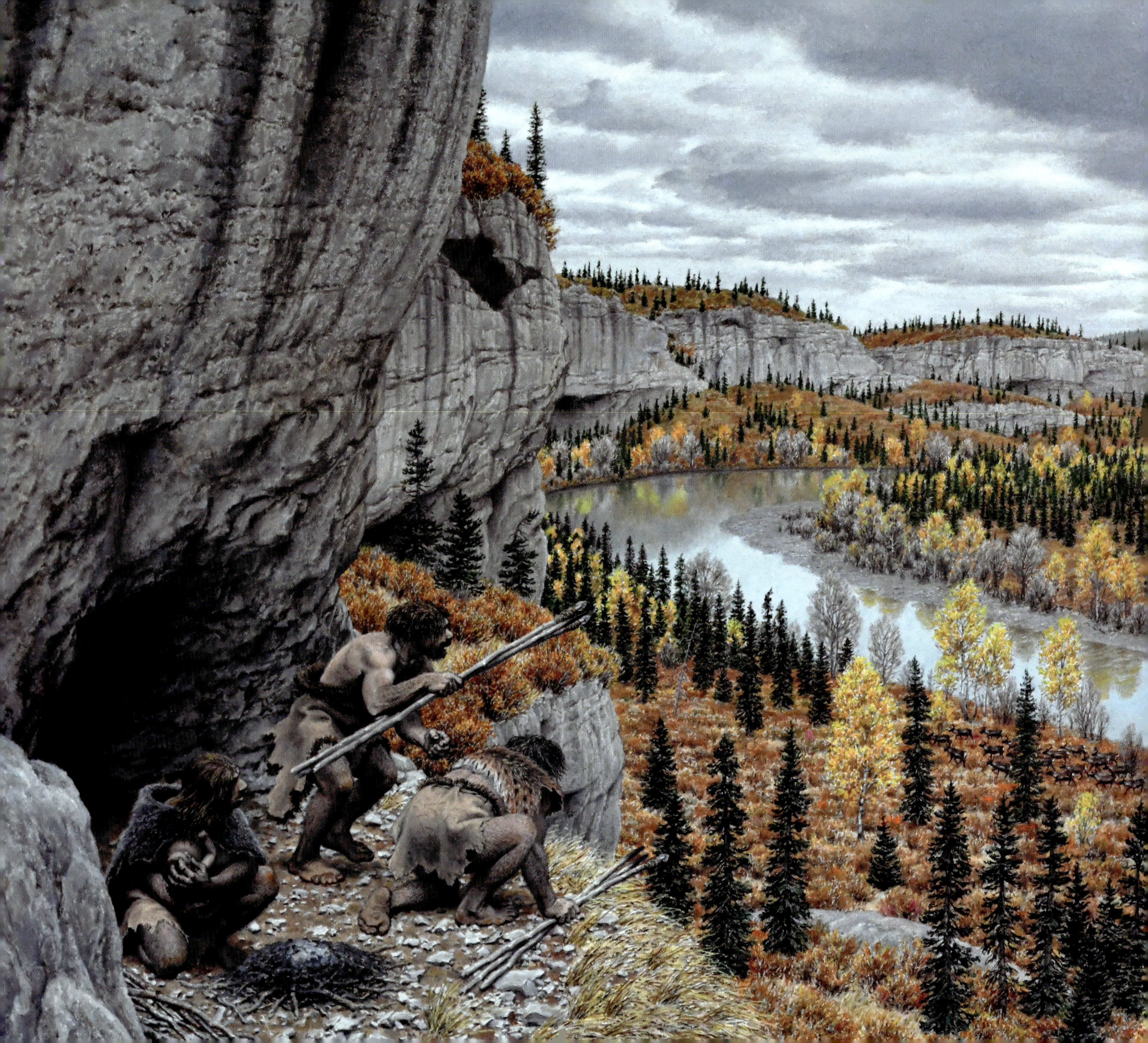

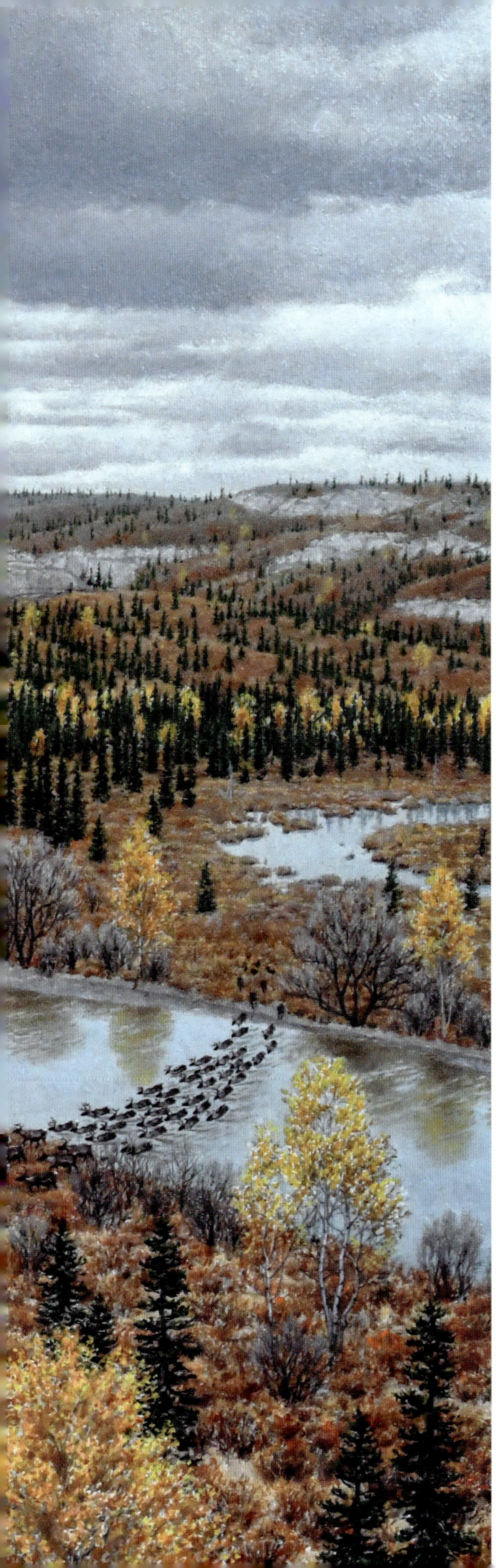

THE NEANDERTHALS

Reindeer Bonanza in the Dordogne

By the early 2000s, our ideas about *Homo neanderthalensis* had changed considerably since Jay Matternes' art had first brought this extinct hominin to life four decades earlier. During the 1960s most paleoanthropologists regarded the Neanderthals as a variation on our own theme. Since then, many have come to realize that a hominin species could be smart, resourceful, and behaviorally sophisticated without processing information in precisely the way we do.

Homo sapiens seems to be unique in possessing symbolic thought, whereby we transform our internal and exterior worlds into mental symbols that we can manipulate and rearrange to create art, lies, myths, social rules, and religion. In contrast, the Neanderthals appear never to have consistently possessed a symbolic behavioral tradition, although we can speculate that they were highly intuitive and reacted in sophisticated ways to each other and to their environments. Flourishing in harsh times, they built shelters, made tools with stone blades hafted into handles, were skillful hunters, and mastered the use of fire. But they were not us, and they almost certainly did not perceive the world as we do.

The current perspective on Neanderthals is reflected in Matternes' 2002 painting *Reindeer Bonanza in the Dordogne*. The scene is set around 50,000 years ago (before the peak of the last glaciation) and focuses on a typically small group of Neanderthal hunters in France's Vézère Valley watching a herd of reindeer ford the river. There was good hunting during periods when extensive tundra formed in the region through which the Vézère flowed, affording abundant vegetation for huge herds of reindeer and other grazers, ripe to be ambushed. In warmer times the grassy tussocks were replaced by forest, a much less propitious environment for dedicated hunters like the Neanderthals. Despite the rudimentary clothing, the frigid temperatures posed no threat to them. But their glorious pristine hunting grounds would eventually be invaded by *Homo sapiens*, spelling doom for the Neanderthal peoples.

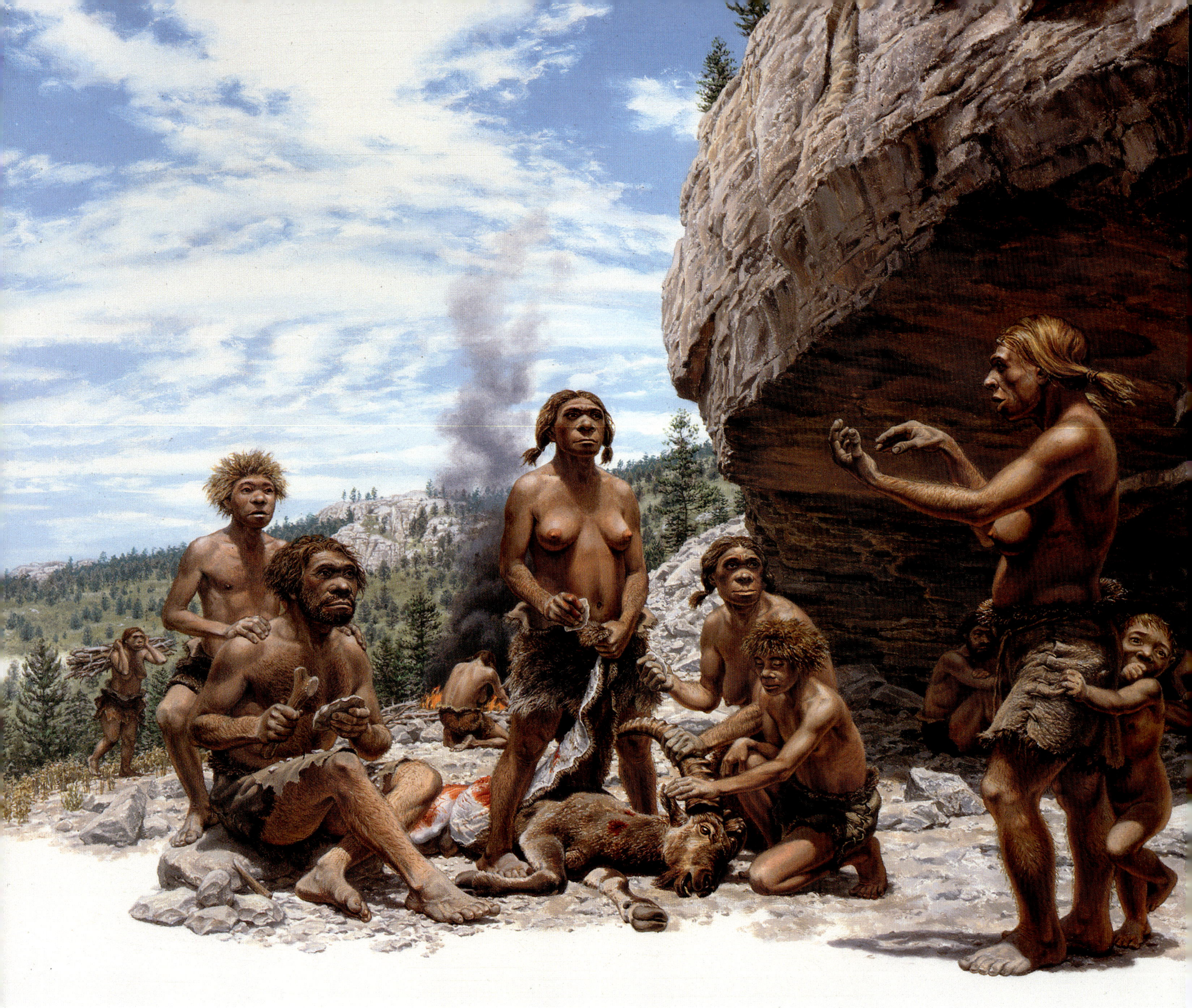

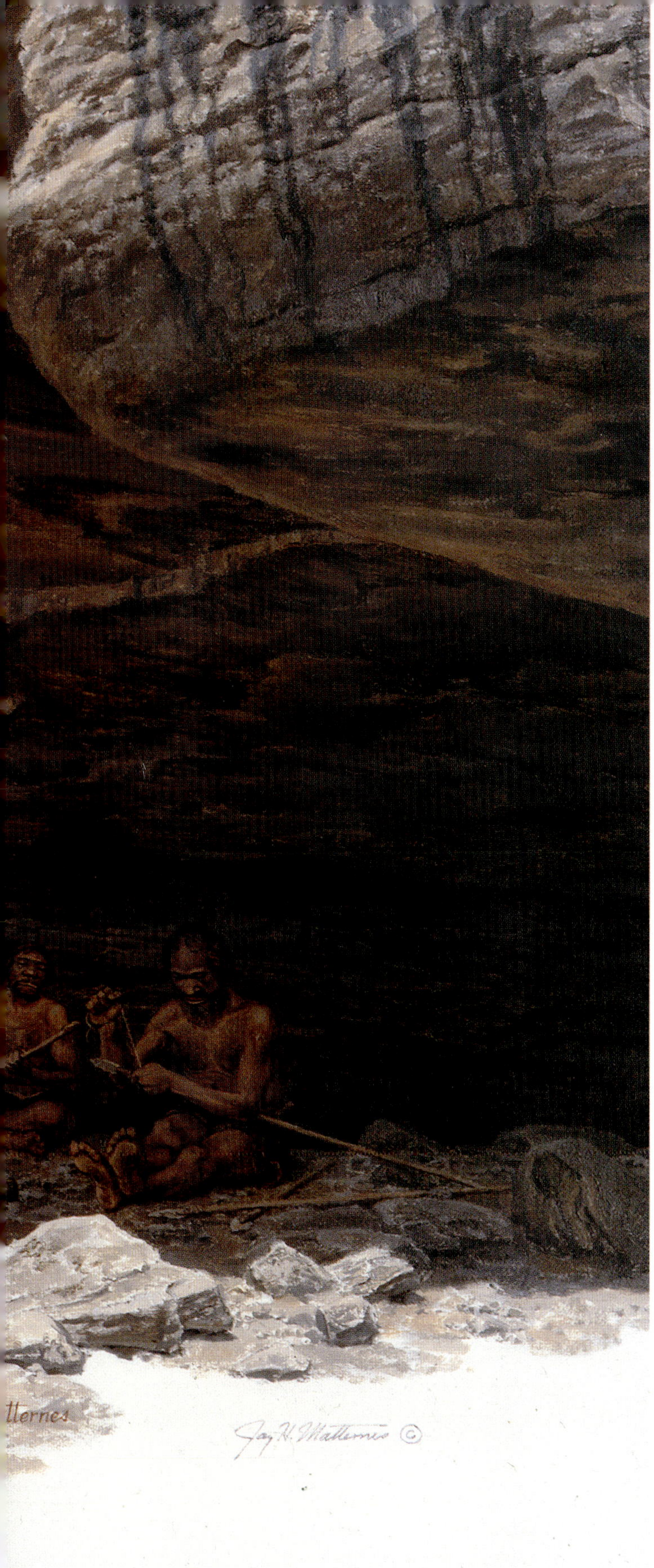

Neanderthals in the Pyrenees

The Neanderthals belonged to a Eurasian human lineage with roots going back well over 300,000 years. Their brains were as large as those of the modern humans who invaded their territory 40,000 years ago and soon drove them to extinction, but were differently shaped. The skulls of Neanderthals were longer and lower than ours are, and bore heavy brow ridges. Their rib cages flared outward to meet a broad pelvis, and their bones were thicker and heavier than our own. Archaeological evidence suggests that Neanderthals were not only sophisticated stoneworkers but smart and ecologically resourceful.

For decades after their first discovery in the second half of the nineteenth century, the Neanderthals were regularly portrayed as crude and brutish—somewhat apelike—in comparison to the early European *Homo sapiens* who appear to have been their nemesis. However, there is also evidence for some limited genetic intermingling, as seen in the "Neanderthal genes" found in modern Eurasian human populations.

During the mid-twentieth century the Neanderthals underwent a sort of rehabilitation by paleoanthropologists in the light of emerging archaeological evidence. As a result, Neanderthals were no longer viewed as a coarse hominin offshoot or ancestor, but instead as a variant of modern humans: *Homo sapiens neanderthalensis*. Matternes' 1983 portrait of a group of Neanderthals, set in the foothills of the Pyrenees, shows a neatly coiffed woman haranguing others, apparently using a spoken language. Matternes has also suggested in the picture that they cobbled animal hides together to make rough clothing. The facial reconstruction of a Neanderthal on the magazine cover dates from the same period. Matternes' treatment of facial features and expression moves the viewer to feel a connection to this ancient human, despite the low forehead and heavy brows.

Nowadays, most authorities accept that the Neanderthals were indeed on their own evolutionary trajectory and deserve recognition as their own species, *Homo neanderthalensis*. A growing archaeological record suggests that the Neanderthals interacted with the world very differently than we do, and it is very unlikely that they used language as we know it. So, rather than view them as an inferior version of ourselves, we should regard them as having a unique sensibility, and as fully deserving of their own identity.

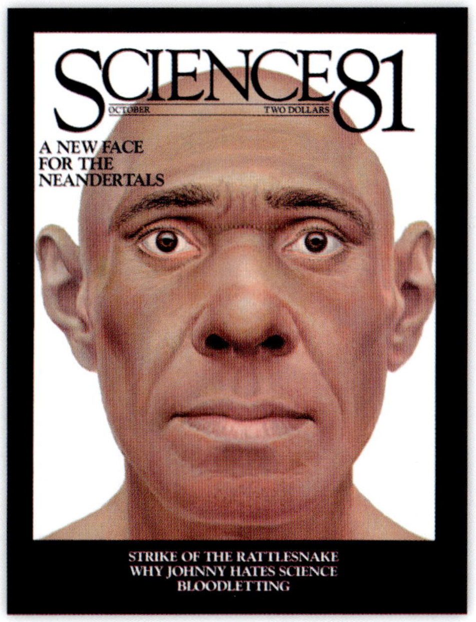

GEORGIA ON MY MIND

You wouldn't want to encounter him in a dark alley, but back in the early Pleistocene, about 1.7 million years ago, the chap in this Matternes pencil reconstruction could possibly have been mistaken for your great-great-grandfather. He is a representative of the Dmanisi hominins, whose existence was unknown and unsuspected by anthropologists until the early 1990s, when the first discovery of their remains was made.

These hominins walked fully upright but were short-statured (just under five feet) and small-brained (546–730 cc) compared with the *Homo* fossils then known from East Africa. Their faces tended to protrude, and sported distinct brow ridges. Crude stone tools used for chopping and cutting were recovered among their remains, and they may well have depended to some extent on animal proteins for survival, since nutritious vegetation was at least seasonally scarce. Odds are that these hominins must have been pretty tough to avoid being eaten themselves; in the same deposits were found remains of an assortment of powerful predators: saber-toothed cats, Etruscan wolves, and giant hyenas the size of modern lions.

During the 1990s and early 2000s, the male skull known as Dmanisi 5 and four others were discovered in the remains of an ancient trading town called Dmanisi, in the former Soviet republic of Georgia, 90 miles southwest of Tbilisi, the capital city. Located in the Caucasus, between the Black and Caspian Seas, the area was traversed for centuries via the historic Silk Road—a major artery of commerce between Asia and Europe that was famously traveled in the fourteenth century by Marco Polo.

Ruined houses in the tiny town were being excavated for their archeological interest, but proved to be full of surprises. Underneath the expected historical layers was an unsuspected prehistoric treasure trove going back almost two million years. It yielded skulls, bones, and stone tools of some of the earliest humans to cross into Asia and Europe from their African beginnings, opening a new chapter in the story of humankind: the first migration of early humans out of sub-Saharan Africa, ultimately to take over the world.

The Dmanisi discoveries have opened up new questions and debates about the variability of ancient hominins. These fossils are pretty heterogeneous, so should they all be lumped together as one evolving lineage, or do they indicate that several different hominin species existed in western Asia by shortly after two million years ago?

In drawing his reconstruction of Dmanisi 5, Matternes employed his usual three-step method, going from bones, to added muscles, to fleshed-out restoration—letting his meticulous incremental layering process lead him to the finished image without skipping any steps or allowing the intrusion of unrestrained imagination.

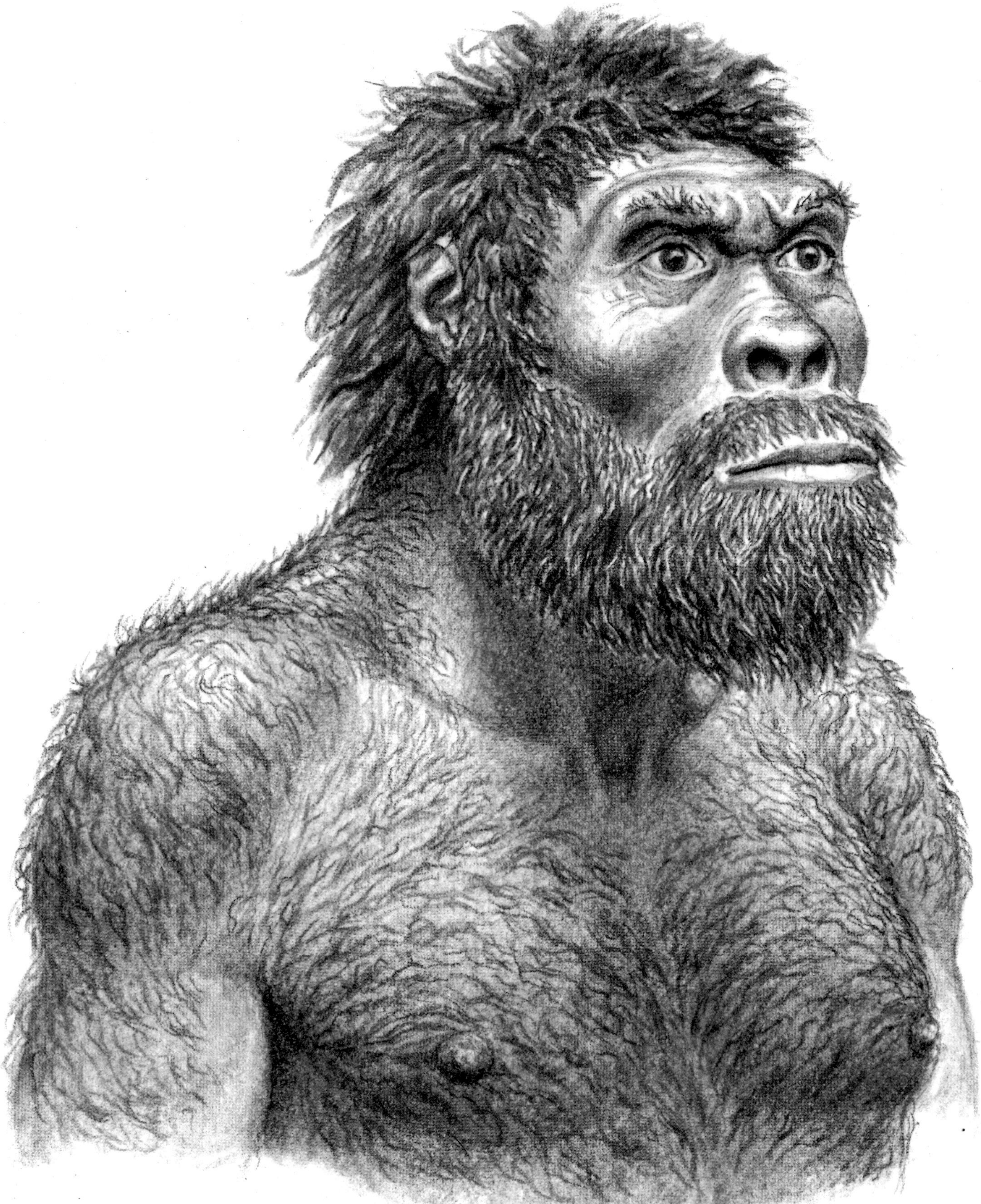

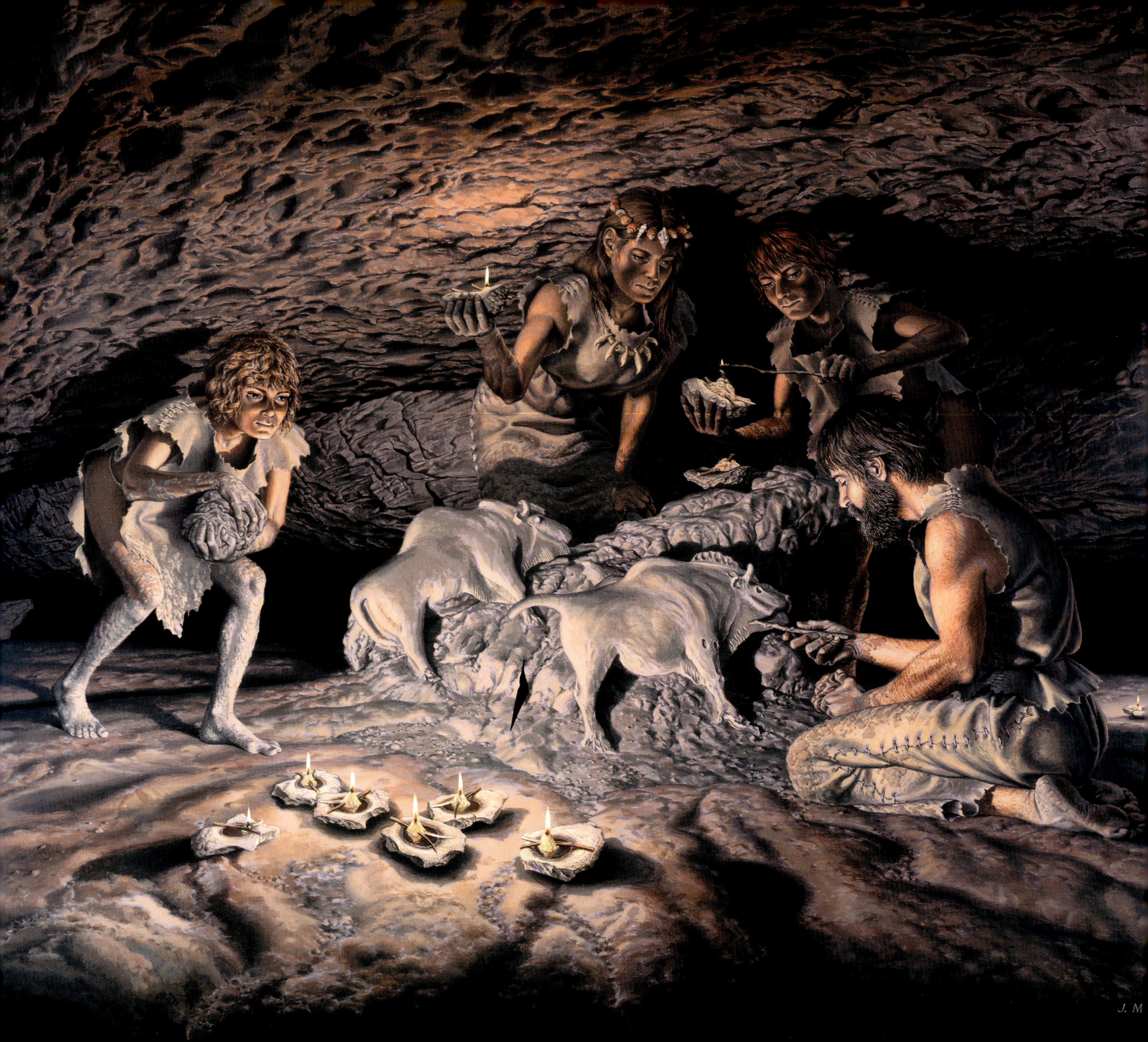
J. M

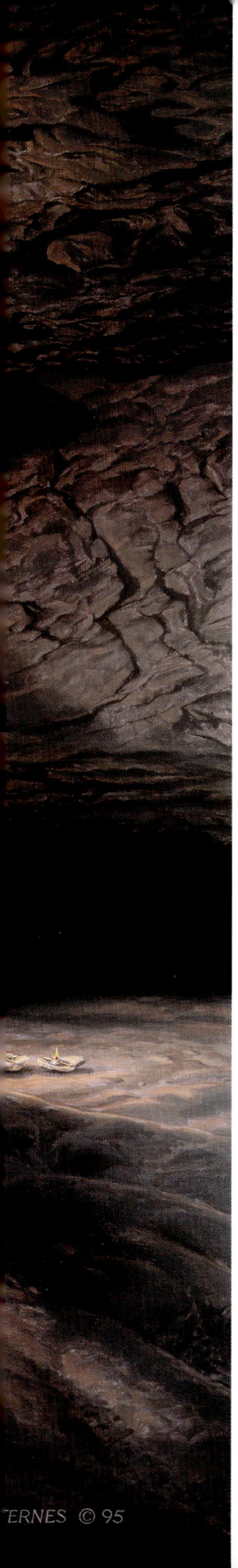

BISON SCULPTURES OF THE TUC D'AUDOUBERT

We don't know exactly who they were, those early relatives of ours. But some of them were artists who ventured into deep, labyrinthine caves to create images of the Ice Age animals that were central to their way of life. About 14,000 years ago, close to the end of the Last Glacial period in Europe, some of them walked and crawled for hours through the long, low-ceilinged, and almost inaccessible grotto now known as the Tuc d'Audoubert, in the foothills of the French Pyrenees. There, in a natural chamber deep in the Earth's bowels, they shaped wet clay into the likenesses of a pair of European bison, one of the living resources upon which they and their families depended. By the light of fat-and-marrow-burning stone lamps, they created high-relief likenesses of a bull positioned directly behind a cow with her tail raised in an invitation to mate.

We do not know exactly what was in the minds of the artists as they worked on these remarkable images. But although they must have been of great importance to their makers, it appears they were never revisited before the rediscovery of the cave in 1912, by three young boys. By that time, the prehistoric cave art tradition of southwestern France and northern Spain had already become widely known. But the Tuc d'Audoubert discovery was unprecedented, and we are fortunate that this and several other important cave art sites are located on the estate of the Bégouën family now famous for its dedication to their study and preservation.

In 1994, when Jay Matternes was hired by Japan's Gunma Museum of Natural History to reimagine and paint the Paleolithic artists at work, Count Robert Bégouën gave permission for him and anthropologist Ian Tattersall to visit the grotto to conduct research for the painting. Later, Jay confirmed that he could not have rendered the scene so convincingly had he not experienced it firsthand.

With Bégouën and his son Eric as their guides, the paleoartist and his colleague had a memorable adventure. First, the group clambered into a dinghy, hidden in the bushes near the cave entrance, which carried them inside along a subterranean stream. Since the river is too shallow for rowing, the passengers had to haul their little boat along by guide-ropes fastened to the cave walls. The only light came from small, hand-carried flashlights, making the illumination similar to that used by the Ice Age artists.

After leaving the rowboat, they squeezed through a narrow rock fissure in the cave wall that leads into a narrow atrium, from which they walked, scrambled, and crawled for hours along a path from which no visitor has been allowed to stray since the cave's discovery over a century ago.

Finally, Matternes and his companions reached a blind gallery, at the back of which a natural protrusion of rock supports the pair of clay bison, each about 18 inches high and 2 feet long. Both are elegantly limned and eternally poised to mate, their bodies hand-smoothed and completed with implements and fingernails. Near them the muddy, unshod heel prints of a couple of adults and an adolescent remain embedded in the clay floor.

Scattered about the site are skulls of cave bears that had apparently died of natural causes. Most of their canine teeth had been removed, allowing Jay to imagine that one of the human visitors could have worn a bear-tooth necklace.

Incredibly, the cave has continuously maintained its high level of humidity since the Ice Age, and its copious clay deposits have never dried out or hardened. The site is off-limits to most visitors because of its extreme fragility; even lightly touching one of the clay sculptures would leave a permanent fingerprint!

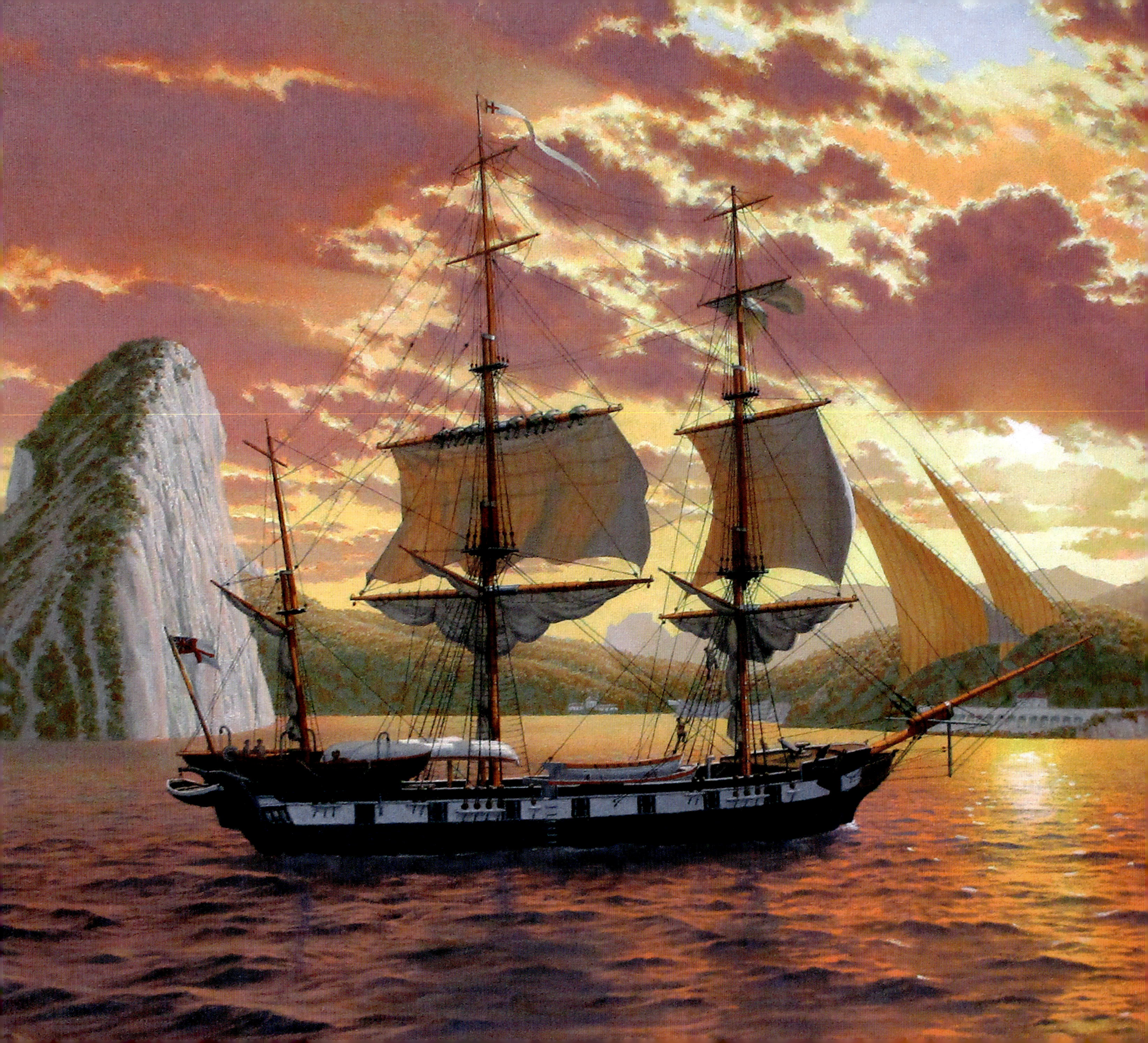

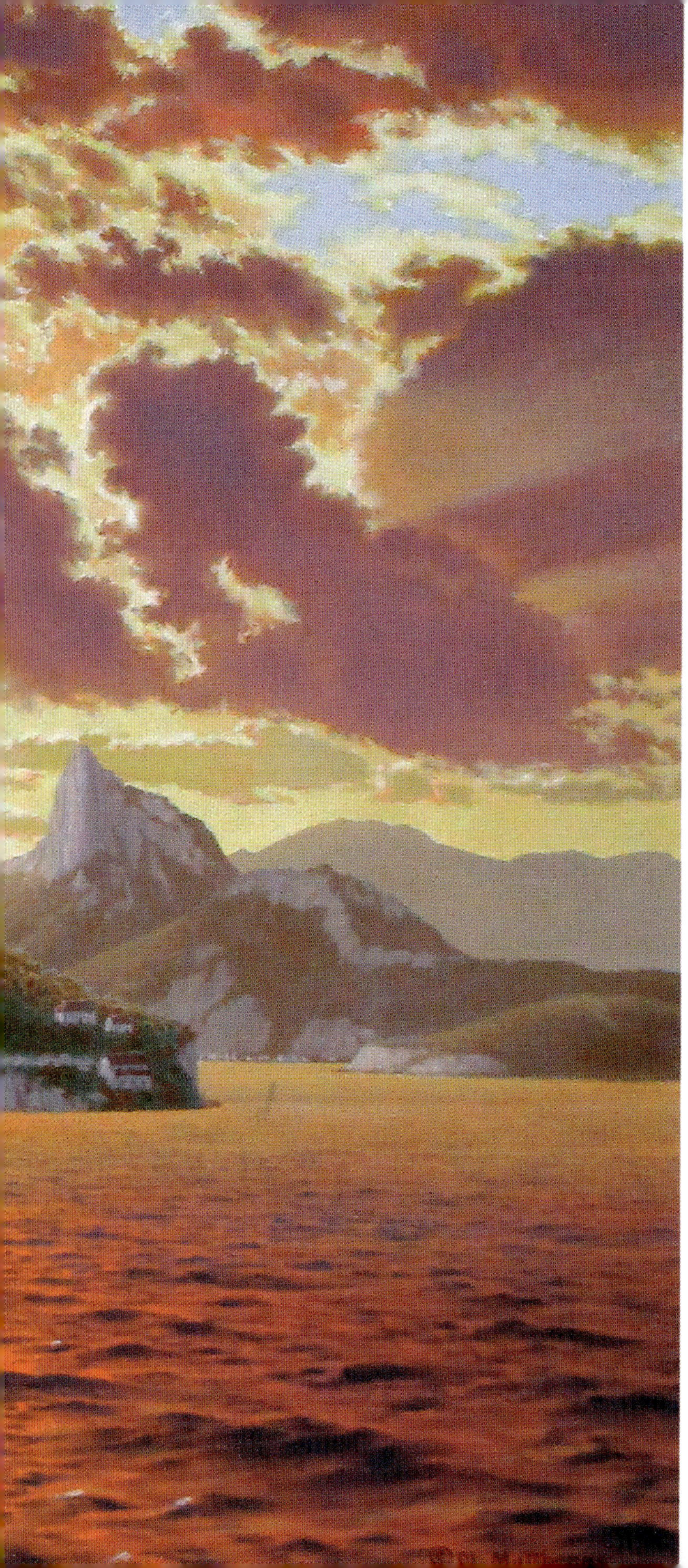

DARWIN AND HMS *BEAGLE*

HMS *Beagle* Entering Rio de Janeiro Harbor on April 3, 1832

Despite a brief stint at Edinburgh's medical school and a degree from Cambridge, Charles Darwin (1809–1882) always claimed he had acquired his "first real training or education of my mind" aboard HMS *Beagle*—when he was not being seasick.

Built at Woolwich-on-Thames in 1820, the *Beagle* was a 235-ton sloop 10-gun brig of the Royal Navy. In whimsical contrast to her formidable artillery, she was named for a friendly little dog. Other vessels in the series were named HMS *Porcupine*, HMS *Opposum*, and even HMS *Squirrel*.

Barely 90 feet long, HMS Beagle's roster included 74 men, including 8 marines and 3 young Indians who were being returned to their homeland in Tierra del Fuego.

Lieutenant Robert FitzRoy, then 26 years old, was appointed captain on a five-year voyage of discovery to make scientific measurements and to map and explore the coast of South America. After charting the coasts of Patagonia and Tierra del Fuego, the *Beagle* was to circumnavigate the globe. She sailed from Devonport, England, on December 27, 1831, and returned to Falmouth almost five years later, on October 2, 1836. Despite his youth, FitzRoy was able to navigate what sailors called "a coffin brig" through difficult, jagged channels and rough seas, accomplishing their difficult mission, and even establishing the exact longitude of Rio de Janeiro for the first time.

FitzRoy invited 22-year-old Charles Darwin, known for his "enlarged curiosity" about geology, botany, and zoology, as his "gentleman's companion" and unofficial ship's naturalist.

Young Darwin collected thousands of rocks, insects, bird and mammal skins, fossils, and fishes to send his mentor, Rev. John Henslow, back in England, where the specimens were studied by scientists, described in books, and archived in natural history museums. His observations, particularly in the Galápagos Islands off Ecuador, led to his theory of evolution by natural selection (*On the Origin of Species*, 1859), which transformed geology, botany, zoology, and all the life sciences. A younger naturalist, Alfred Russel Wallace, working alone in the jungles of Malaysia, independently came up with the same theory, announced jointly in 1858—and properly known as "the Darwin-Wallace Theory." In Matternes' painting, the reddish morning light delicately filters through the ship's sails, perhaps emblematic of a new dawn in the life sciences.

A century later (1964), the Charles Darwin Research Station in the Galápagos equipped a modern research vessel and christened her *Beagle II*. In 2003, a different sort of ship also called *Beagle 2* was launched by British astrophysicists—an unmanned spaceship shot into space to explore Mars. It reached the red planet, but crash-landed and stopped broadcasting information.

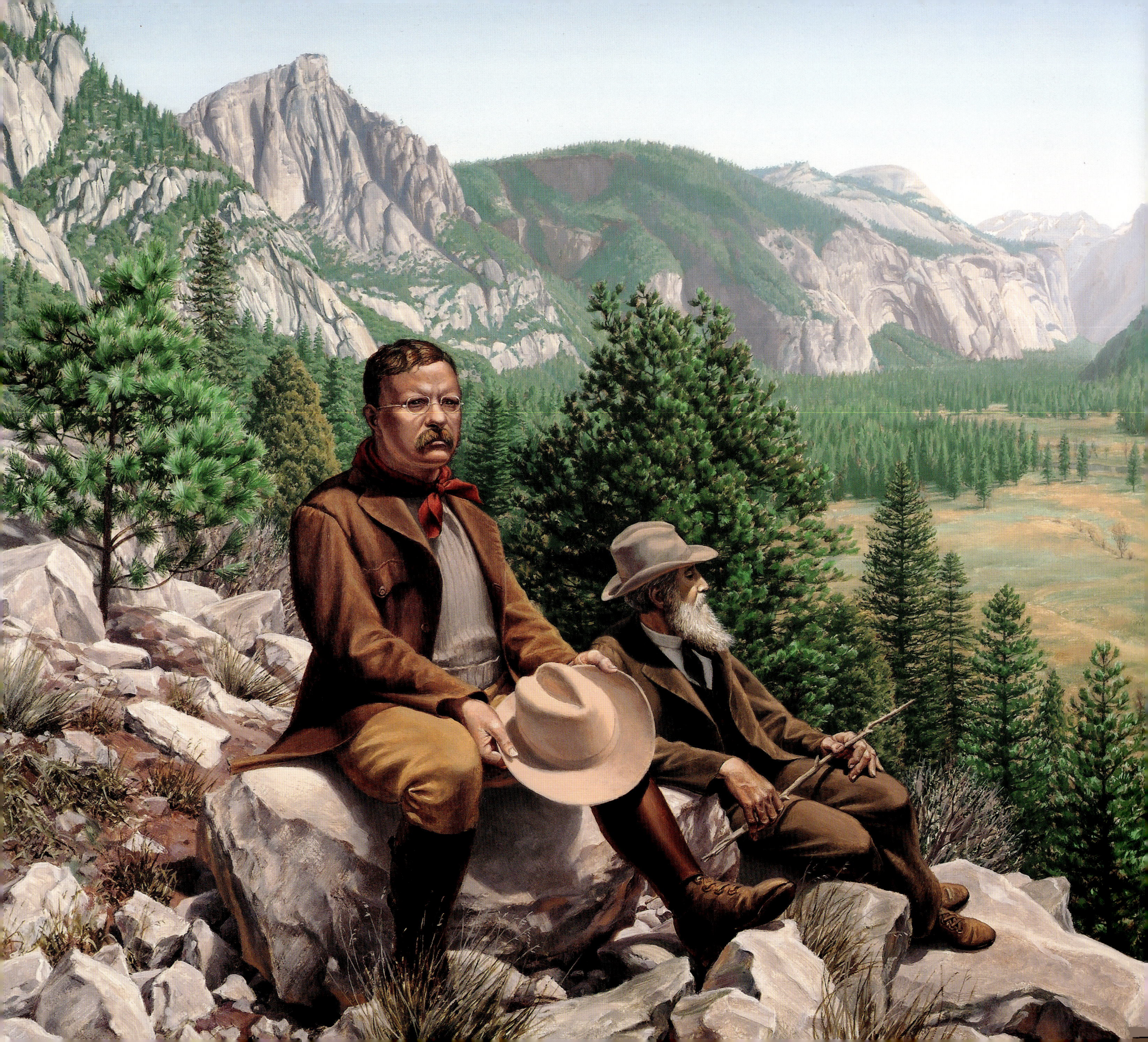

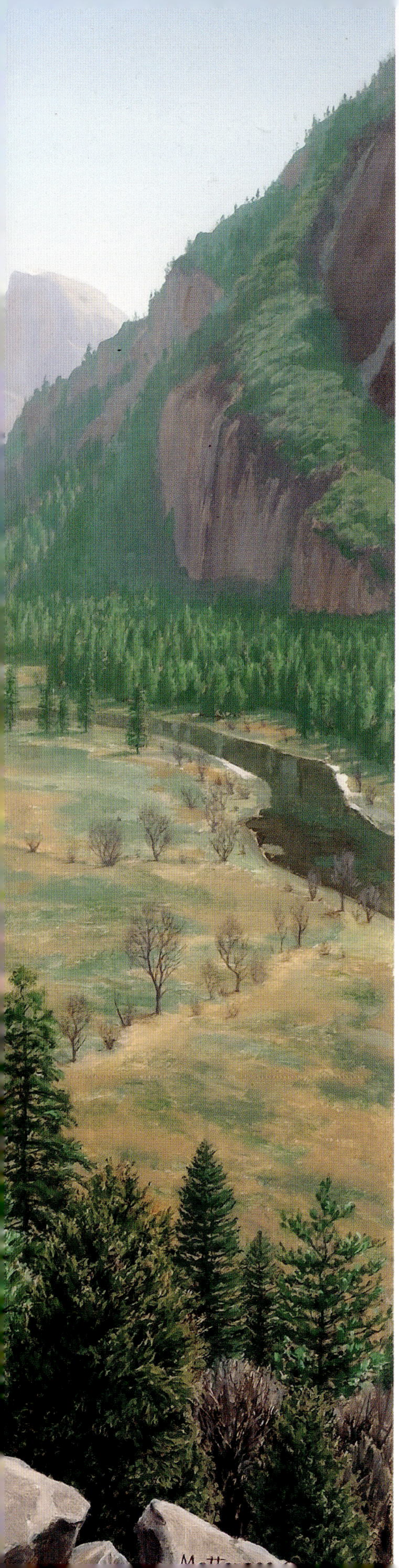

TEDDY ROOSEVELT AND JOHN MUIR AT YOSEMITE

Naturalist John Muir's three-day camping trip with President Theodore Roosevelt in the northern California wilderness in 1903 was a turning point in the history of wildlife and wilderness conservation. Traveling with pack mules and sleeping under the stars, Muir was able to persuade Roosevelt to place the Yosemite Valley and the nearby Mariposa Grove of "big trees" under permanent federal protection as part of a new national park. Muir also founded and presided over the Sierra Club, an organization that still actively promotes conservation of America's wild natural heritage. Above all, he stressed the importance of understanding the interrelatedness of all living things in shaping public policy. "When we try to pick out anything by itself [to save]," he said, "we find it hitched to everything else in the Universe."

The president was convinced: "Our people should see to it that [Yosemite and its Big Trees] are preserved for their children and their children's children forever, with their majestic beauty all unmarred." A century after that historic meeting, as part of a National Park Service art competition, Matternes painted this double portrait of Roosevelt and Muir enjoying and inspired by the magnificent vista of Yosemite Valley.

Roosevelt left a troubling and contradictory legacy. While he genuinely loved animals, he also took pleasure in shooting them for trophies in North and South America and on African safaris. But when he set off as a young man to bag a bison in Wyoming, he was appalled to witness the pathetic remnants of herds that only a few years earlier had ranged the Plains in the tens of millions. In later years, he donated thousands of carcasses from his "expeditions" to the Smithsonian Institution and the American Museum of Natural History for scientific study and public display, but still was troubled by his deep inner contradiction. After Muir urged him to "give up the gun" for good, however, he did so, and became instead the most active conservationist president in the nation's history. For starters, he established the National Bison Range, and urged ranchers to breed captive herds and return them to the wild, which succeeded in saving the species from extinction. And that was just the beginning.

By the end of his presidency, Roosevelt had made his friend John Muir proud by creating 51 federal bird sanctuaries, 18 national monuments, 5 national parks (including Yosemite), and 4 national game refuges; enlarging or creating 150 national forests; and setting aside 230 million acres of land under federal protection. Matternes found that record well worth celebrating in this affectionate double portrait of the president and the conservationist.

OPENING THE OREGON TRAIL

Fitzpatrick at Scotts Bluff

The opening of the American West commenced with the establishment of the famous Oregon Trail. Before the discovery of gold in California, Easterners thought of the West as impenetrable wilderness and "badlands," a haven only for rough trappers and mountain men. During the early nineteenth century, would-be settlers had to reach California and the Pacific coast by sailing around the entire continent of South America, as the Panama Canal was only a distant dream.

It was unimaginably difficult to follow the overland route by foot and wagon, traversing steep gullies, hills, and rivers, not to mention frequent raids by marauding Indians to steal the settlers' supplies and horses. There were also scorching deserts and the frigid Rocky Mountains to be surmounted. Eventually thousands of persevering pioneers somehow managed to survive the Oregon Trail and reach the West. Between 1843 and 1860, some 400,000 men, women, and children followed the treacherous 2,000-mile route, averaging four months to make the cross-country journey. Tens of thousands perished in the attempt.

Matternes depicts trailblazer Thomas Fitzpatrick ascending what became called Scotts Bluff to scan the landscape ahead. His picture was entered in the National Park Service's Arts in the Parks competition during the mid-1970s.

Matternes visited this area in 1967, lingering at both Chimney Rock and Scotts Bluff. The landscape greatly impressed him, not only because of its vastness and rugged beauty, but also for the historical significance of the Oregon Trail, traces of which are still abundant. Driving it alone in a rented car along U.S. Route 20, he visited many landmarks, making notes and sketches of the terrain and visualizing the days, less than 200 years earlier, when it had lured thousands to their deaths in hope of a better life, but also enabled and accelerated the settling of the American West.

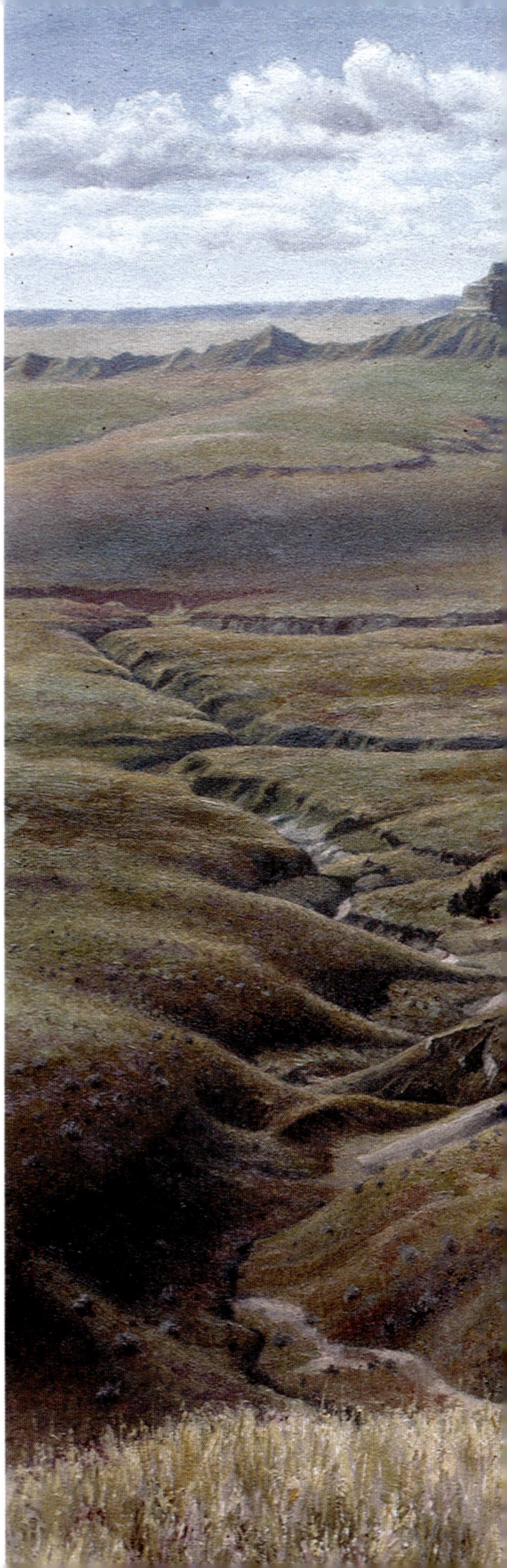

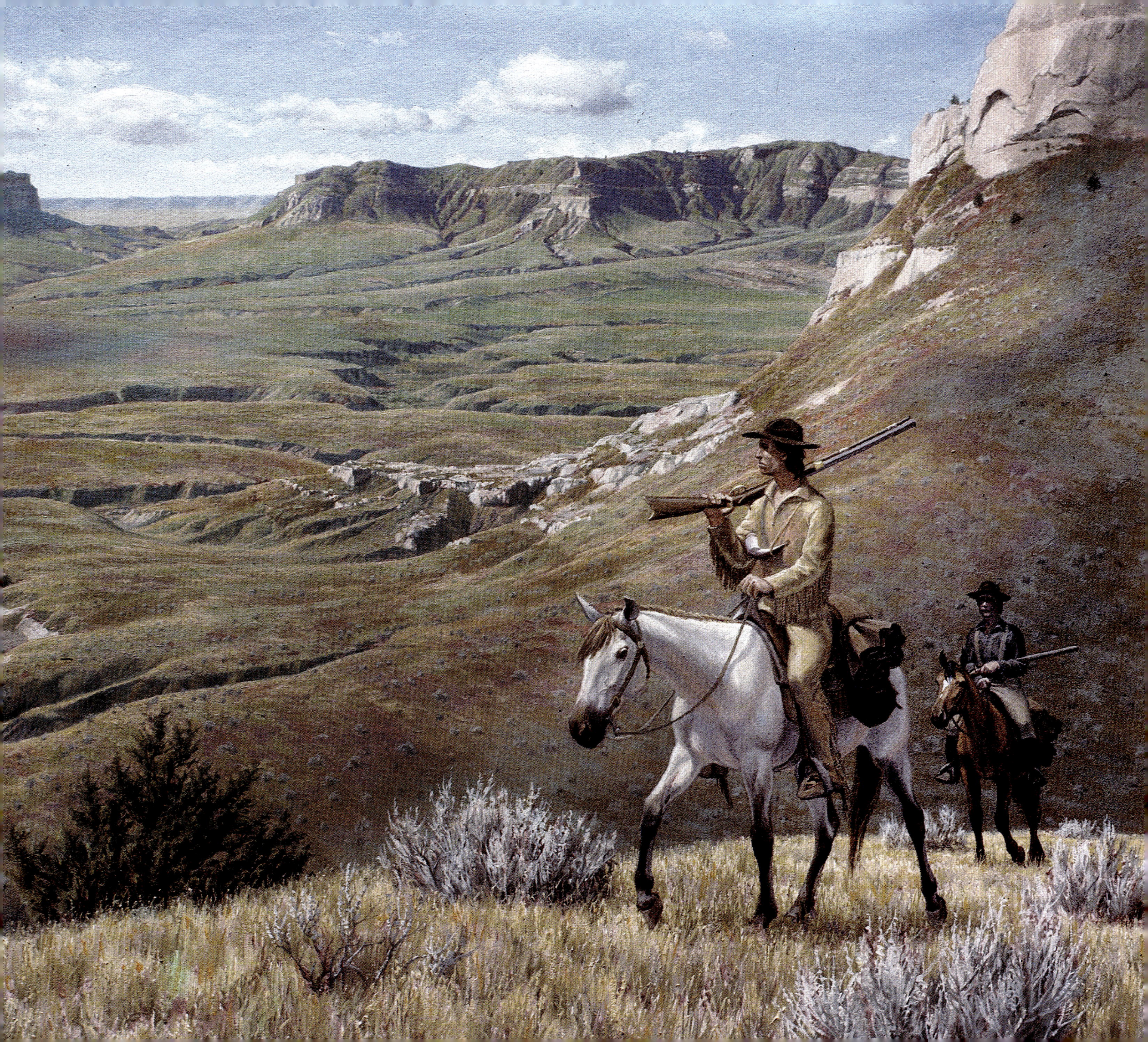

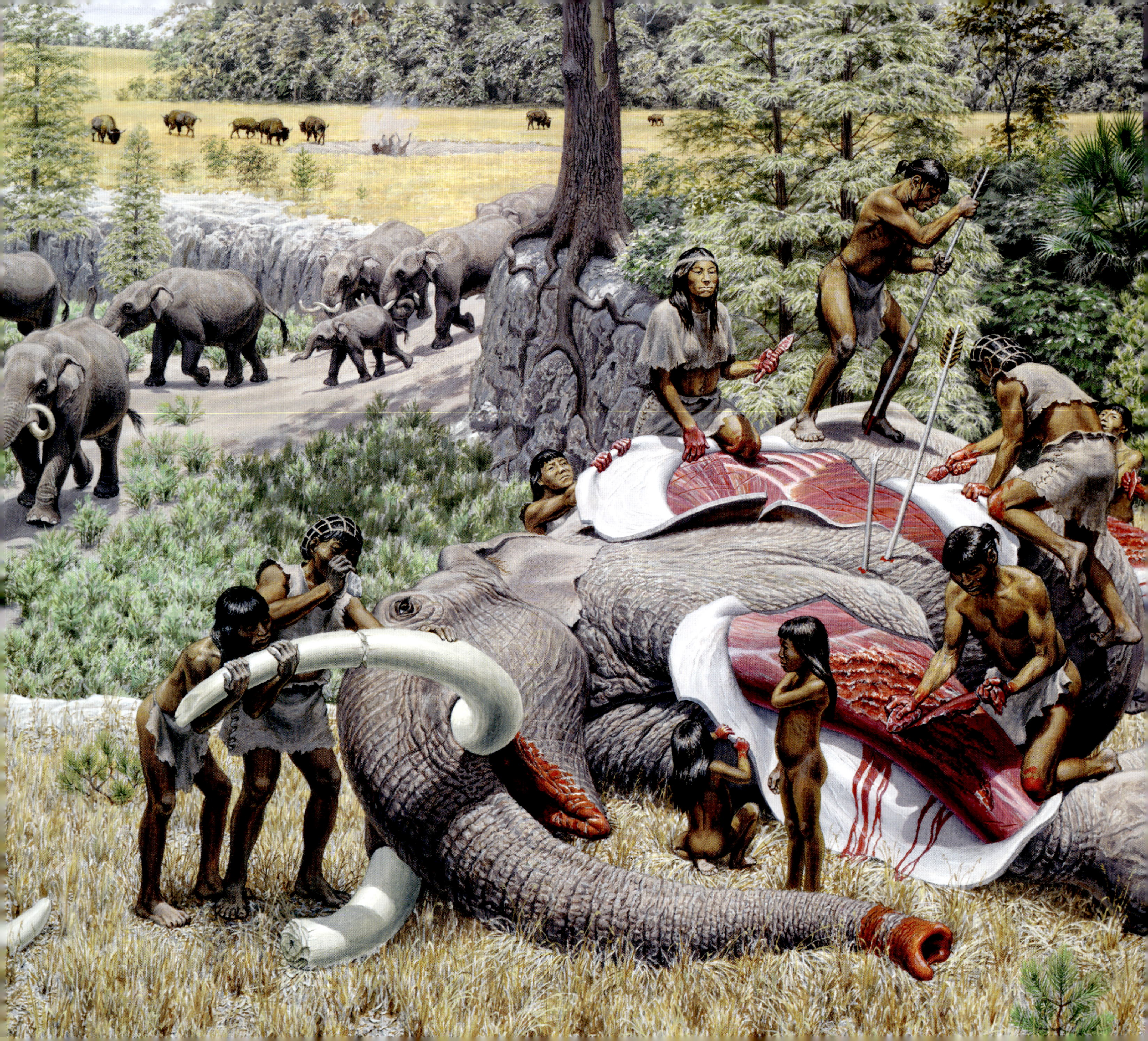

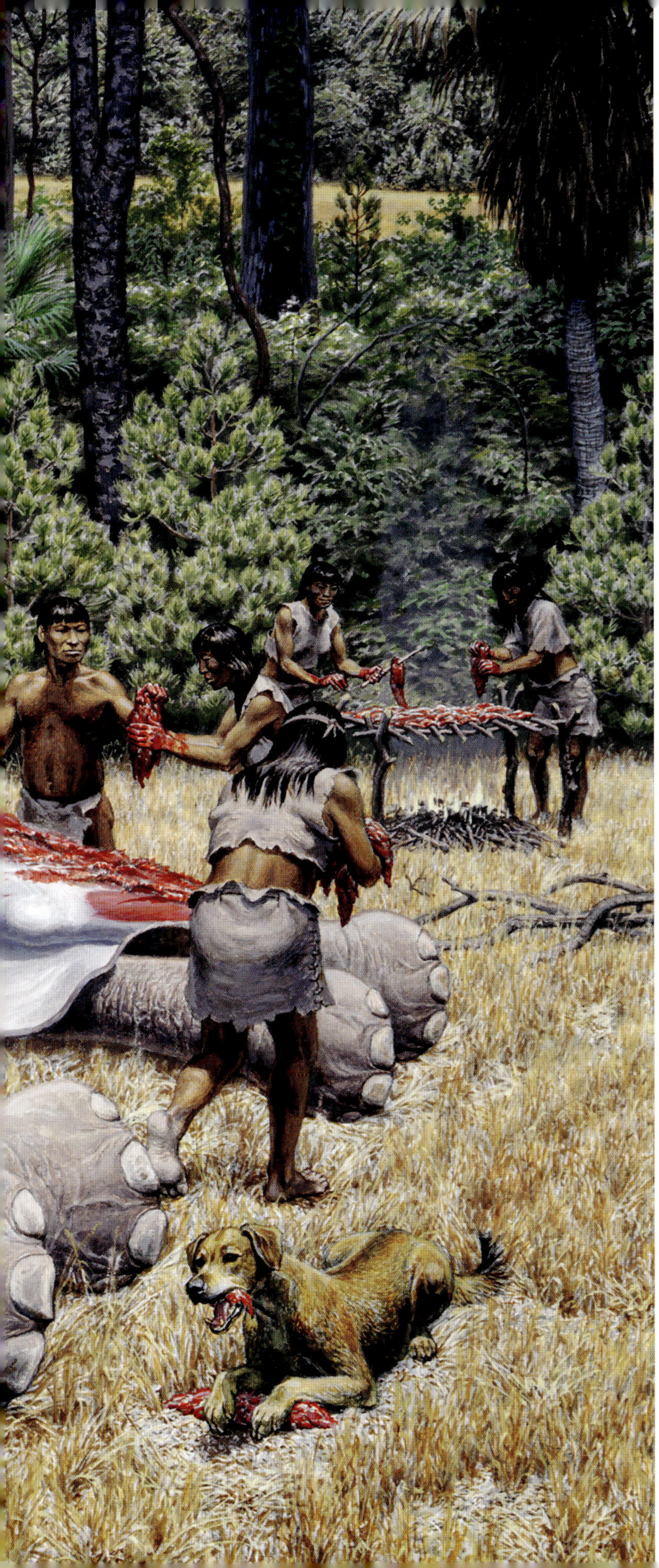

NATIVE AMERICANS

Dismantling a Mastodon Carcass: Florida Elephant Hunters of 11,000 Years Ago

Commissioned by the Florida Museum of Natural History, University of Florida, Gainesville, this painting depicts the Aucilla River archaeological site in the Florida Panhandle—a limestone sinkhole from which a great number of Late Pleistocene vertebrates have been recovered, along with sharp-edged stone tools found with remains of a slaughtered mastodon. Matternes collaborated with scientists from several disciplines in constructing this picture, and spent a week making guided forays to the site. Present are both American mastodons and Columbian mammoths, as well as tapirs, horses, llamoid camels, and bison, indicating the existence of a complex mosaic of local habitats.

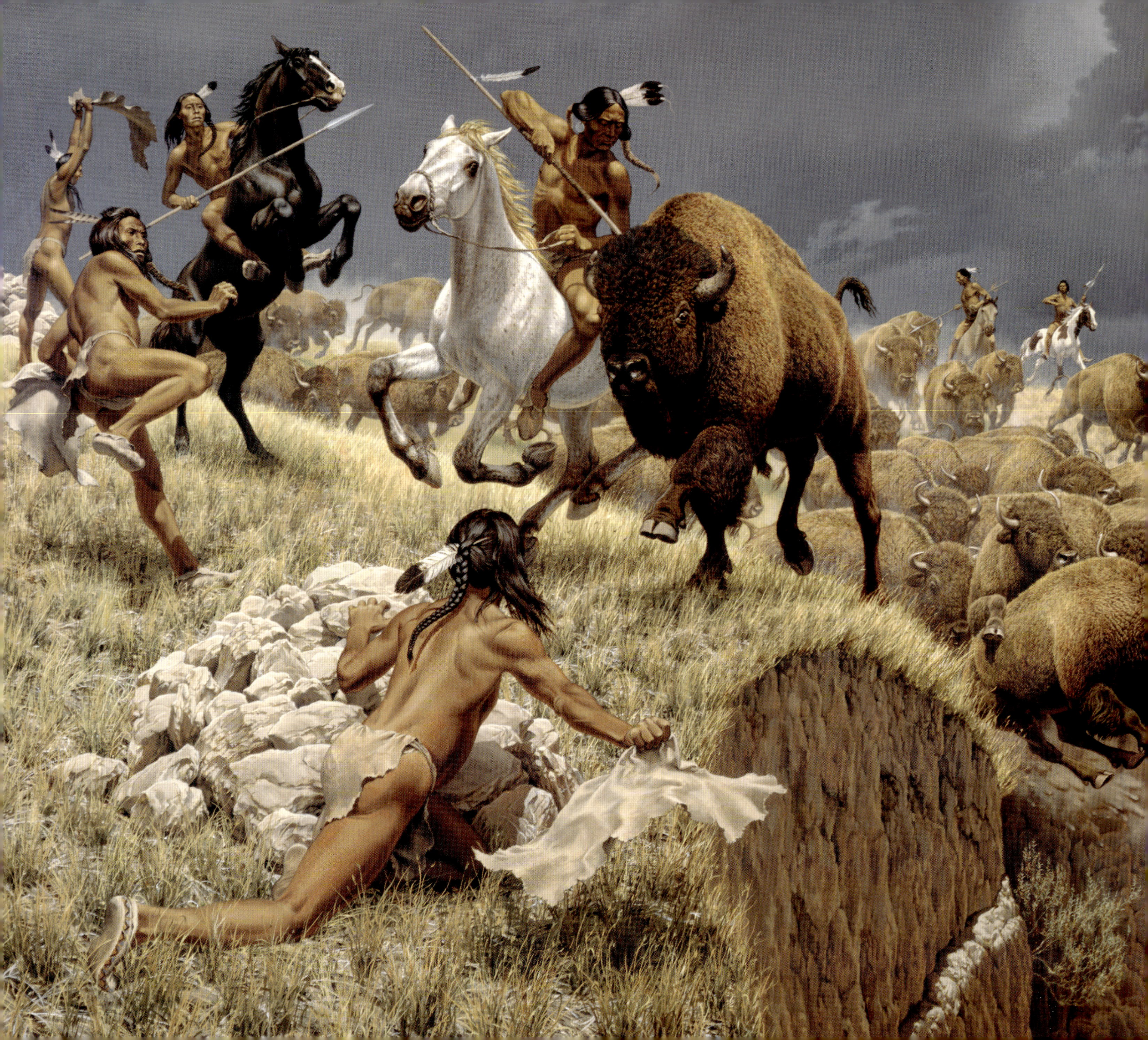

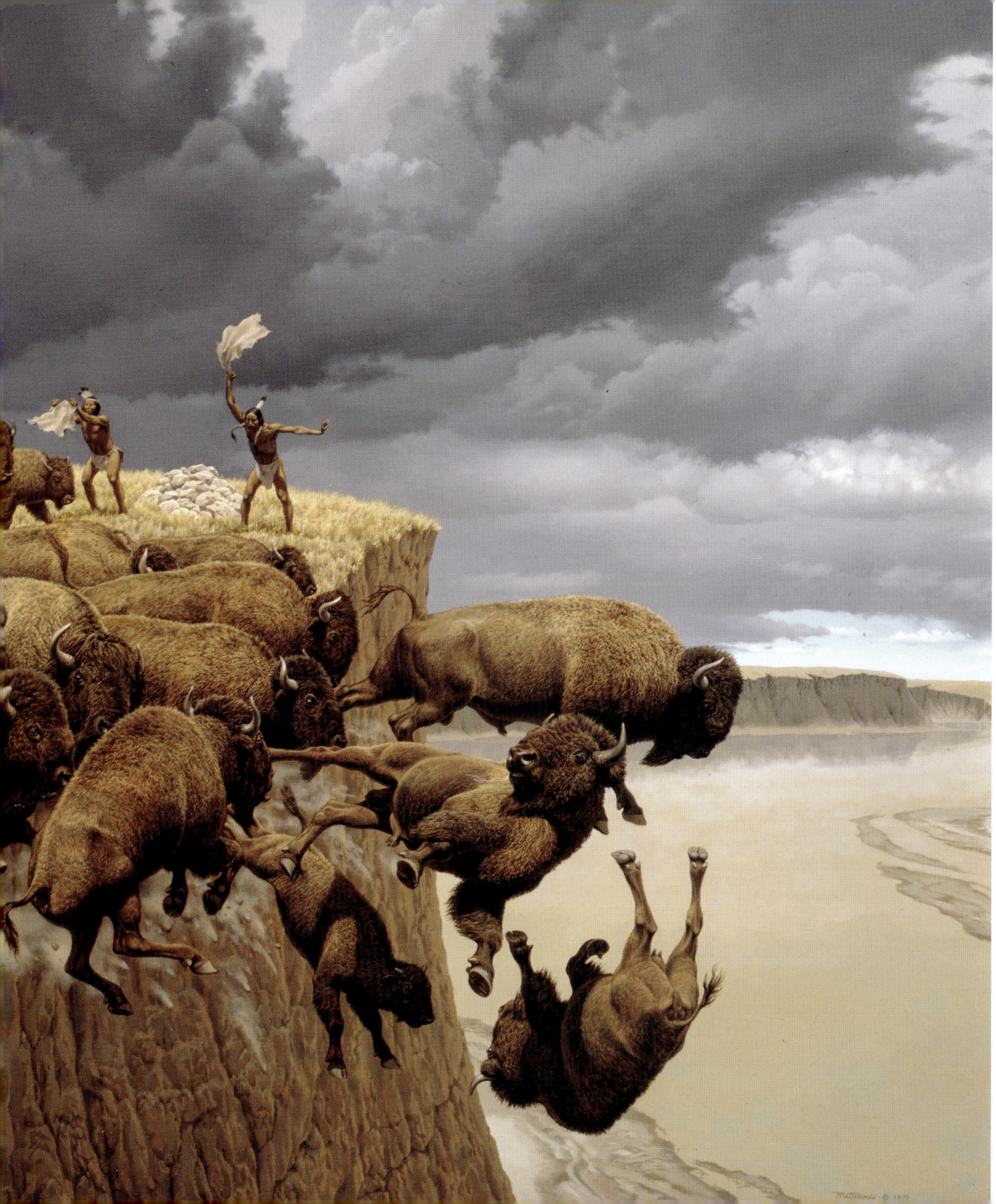

Buffalo Jump

In the early nineteenth century, Plains Indians stampede bison over a 100-foot-high cliff in central Montana, near the confluence of the Judith and Missouri Rivers. Lewis and Clark reported seeing tons of bleached bison bones at the base of such cliffs, where Indian hunters had butchered the carcasses. The bones can still be seen there.

Pulling Even (Archer Chasing Galloping Pronghorns)

Pronghorn antelopes, unique to the western American Plains, shared that vast expanse with the iconic bison until the late nineteenth century, when white hunters decimated the great buffalo herds that had numbered in the millions. Most Americans have never heard of pronghorns, except in the song "Home on the Range"—"where the deer and the antelope play." It is not a true antelope, but the last surviving member of an ancient family, the antilocaprids, that go back to the late Miocene.

Males sport twin bony horn cores, over which grow hooked outer horns that are shed annually after mating season. Like their fossil ancestors, pronghorns are built for speed and can top 55 miles per hour; only African cheetahs can run faster.

Jay says, "In *Pulling Even*, it is early summer, when the grasses are fresh and green. While I enjoyed painting an entire herd of running pronghorns, I thought the picture lacked drama and human interest, so I added the mounted hunter chasing among them as an afterthought. However, it later occurred to me that in reality no horse and rider could hope to keep pace with a herd of panicked pronghorns. The archer has just shot an arrow, but it is the viewer's guess whether it has found its mark."

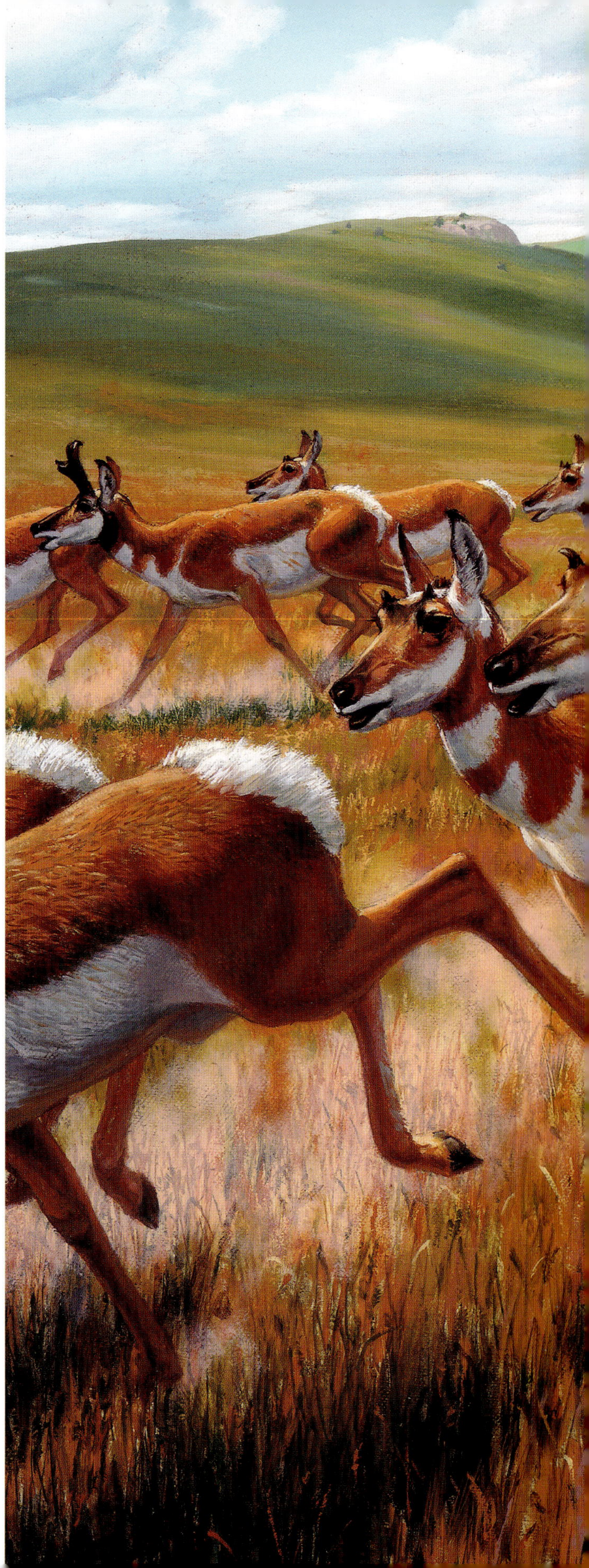

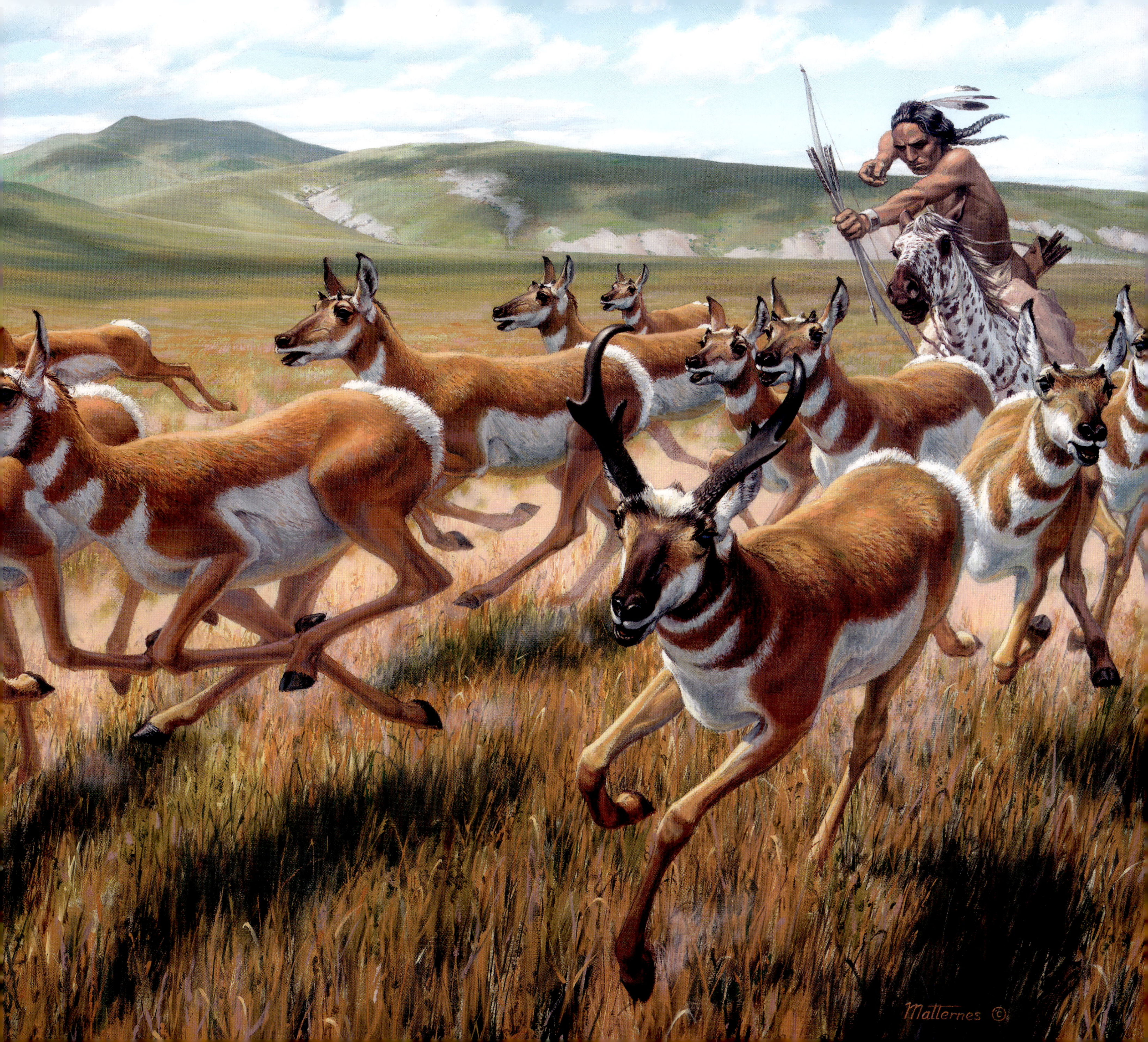
Matternes ©

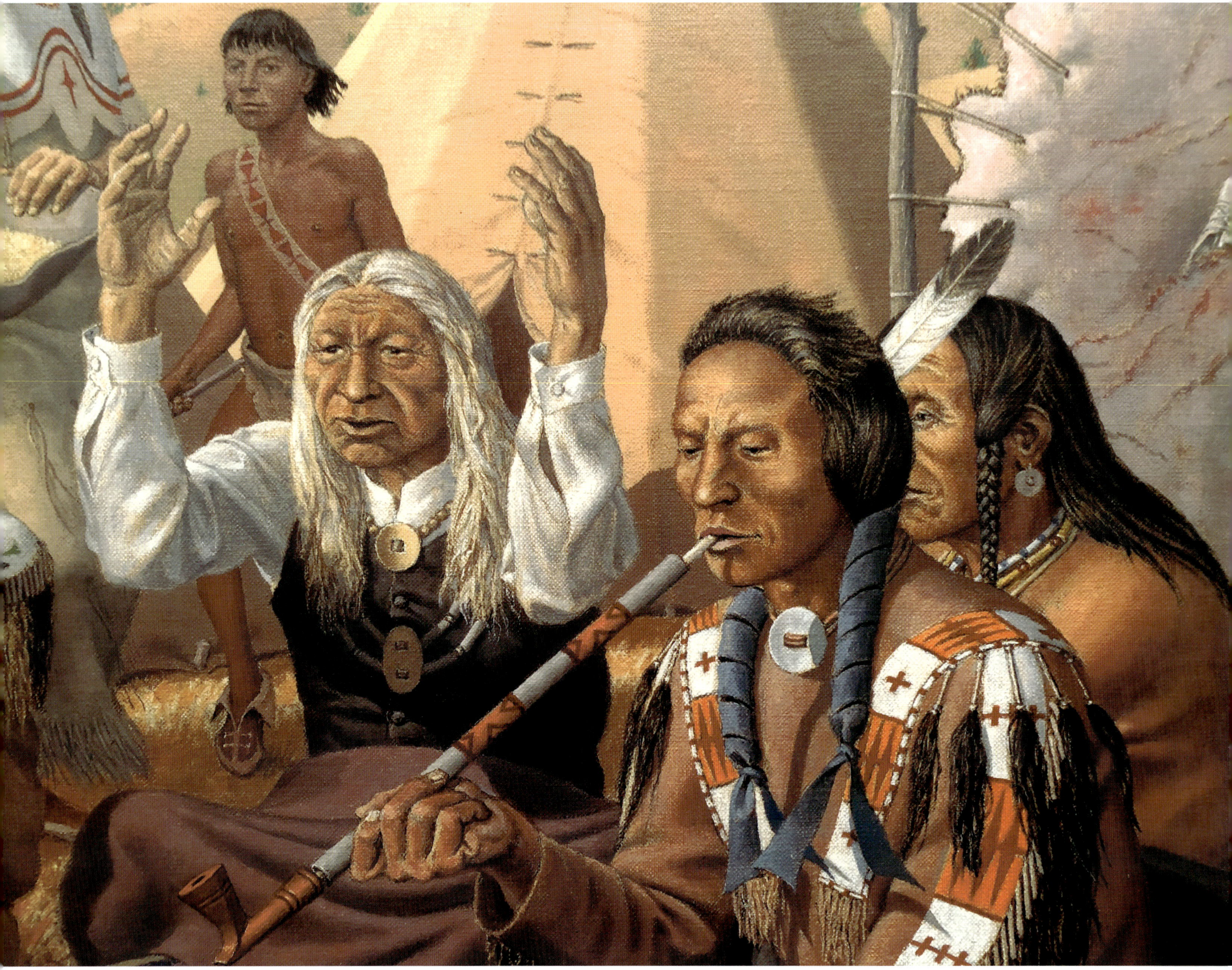

As the old storyteller passionately acts out his narrative, a thoughtful listener contemplates its meaning.

STORY TELLING

It is sometime in the 1880s, somewhere in the Dakota Territory, among a village of Lakota Sioux or other horse nomads of the Great Plains. The picture's central figure, a respected tribal elder, captivates young men and boys with his dramatic tales, stirring their imaginations and reaffirming their deep connection with the ways and traditions of their people.

The old man gesticulates passionately as he speaks and sings during his narrative of myths, events in the history of the tribe, or personal experiences to instruct, inspire, and entertain. American Indians at leisure loved to tell stories, and a skillful, truthful, and wise storyteller was accorded great affection and honor.

In this 2018 painting, Jay Matternes does not let us in on exactly what stories are being told. Could they be tales of the old man's exploits as a young hunter, his lonely vision quest for a guardian animal spirit, how he counted coups during mounted warfare with traditional enemies, or hairbreadth escapes from the murderous U.S. Cavalry? Or is he perhaps retelling ancient legends about the tribe's cosmic origins and how Grandfather Sun made the animals, or recounting how he saw with his own eyes, not that many moons ago, a time when the now-sparse buffalo herds numbered in the tens of millions, making the ground thunder when they ran in terror from the smoke-belching Iron Horses?

Jay's conception had been gestating for over thirty years, as a companion to his major piece *Buffalo Jump* (1977), a dusty action scene filled with mounted braves and stampeding bison. This time he wanted to show "a less dynamic, more quiescent picture of Plains Indians at leisure, except of course, for the women, who continue their work while the men enjoy the storytelling." Also, he decided to depict the village during a time of transition, when their material culture was being expanded by artifacts obtained through trade with the whites, but doesn't hint at the ongoing wars that were to destroy their way of life forever.

A young boy with his dog rolls a cooper's hoop, while the older lad with toy bow is fascinated by the tales of the old storyteller. One standing man holding an unlit pipe is partially wrapped in a trade blanket rather than the traditional buffalo robe, while the other sports a decorated "medicine bag," containing tobacco and perhaps a few magical charms or pebbles. The semicircle of seated men on buffalo hides listen with various degrees of attention, or stoic introspection, to the speaker, dressed in a hodgepodge of traditional elements and trade goods. All wear beaded decorations and much-prized beaded moccasins. One young man wears a deerskin shirt decorated with colored porcupine quills, which were used before trade glass beads became the new standard.

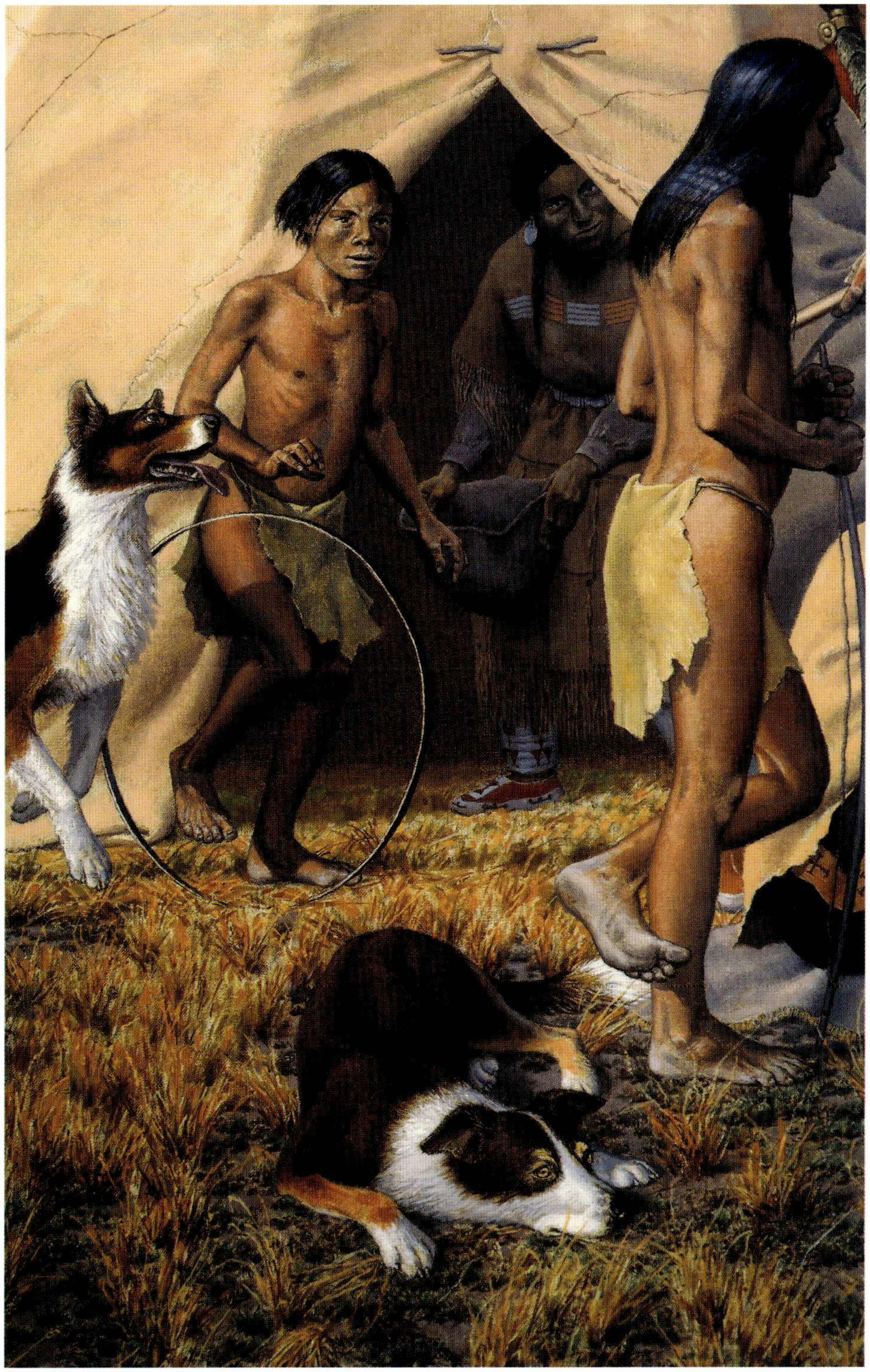

A young boy plays with an iron cooper's hoop as an older woman discards last night's hearth ashes from her lodge.

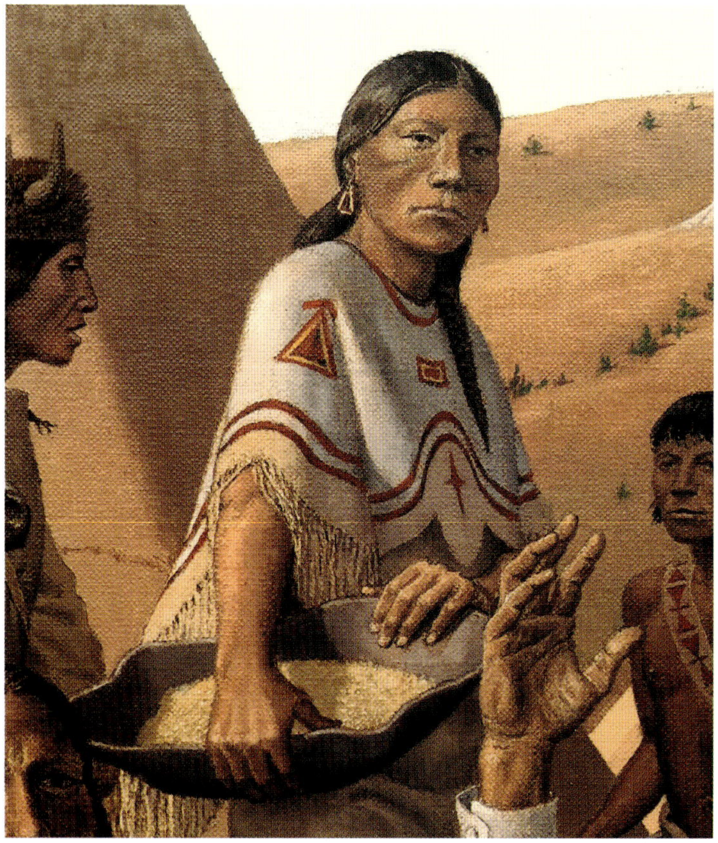

Above: A woman carries seeds she has gathered in a bowl made of dried bison skins. Right: Using an adze made of buffalo bones, another scrapes a hide to prepare it for curing. Opposite: Girls share gossip as they luxuriate on a pile of freshly skinned hides.

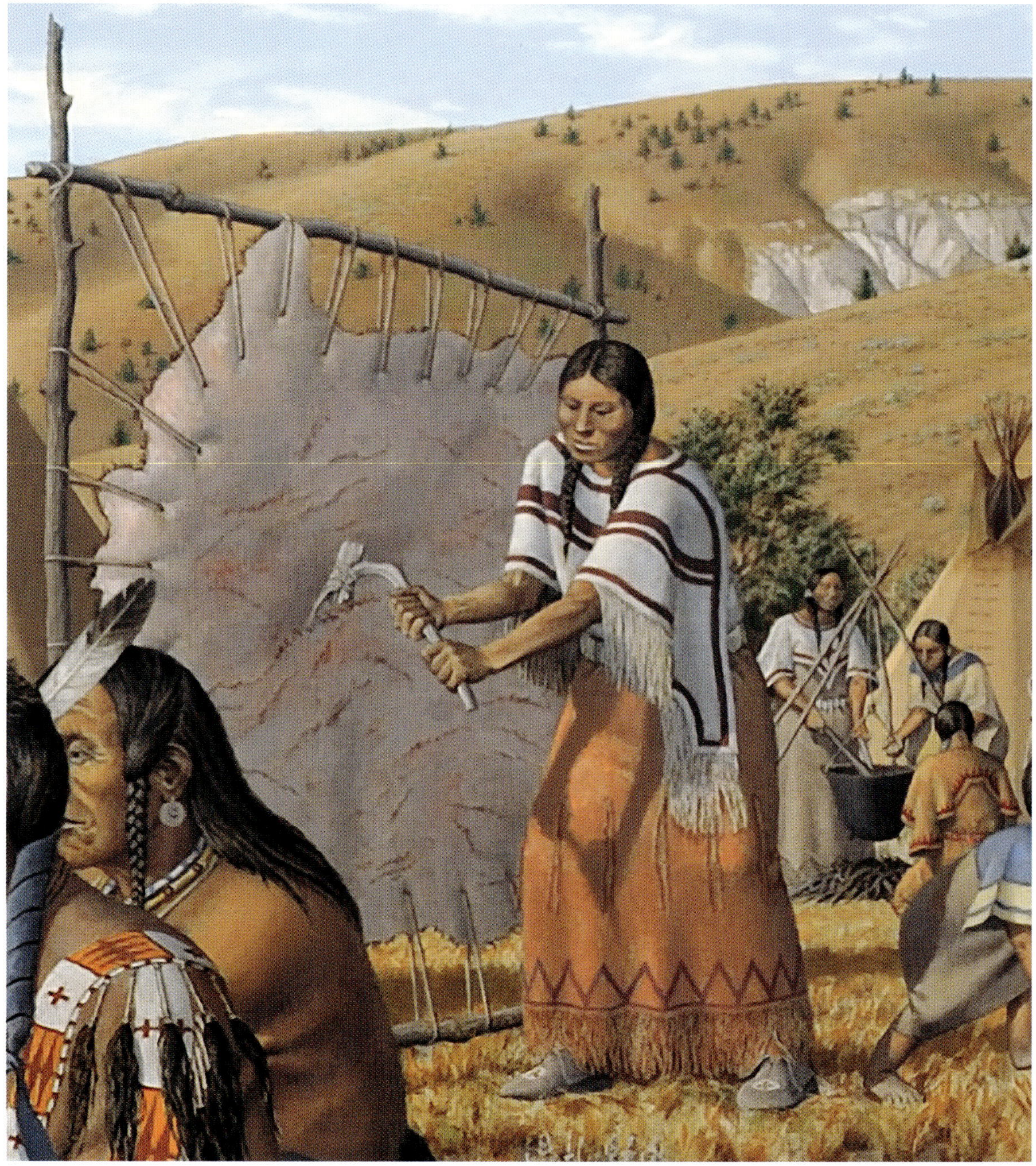

Women pursue their tasks, such as scraping a stretched buffalo hide clean before it is cured, preparing a pile of hides from a recent hunt, while two young girls lie on buffalo robes as they tease a busy woman's toddler and gossip among themselves. In the distant midground, several women cook a meaty stew in a big iron pot suspended from a tripod over a fire.

On the far right, a brave is preparing to mount one of the horses, which are kept mostly in the center of the encampment to prevent thievery by audacious rustlers from neighboring tribes, and to have their steeds near at hand in the event of surprise attack.

Many tribes decorated their buffalo-hide tepees with elaborate picto-

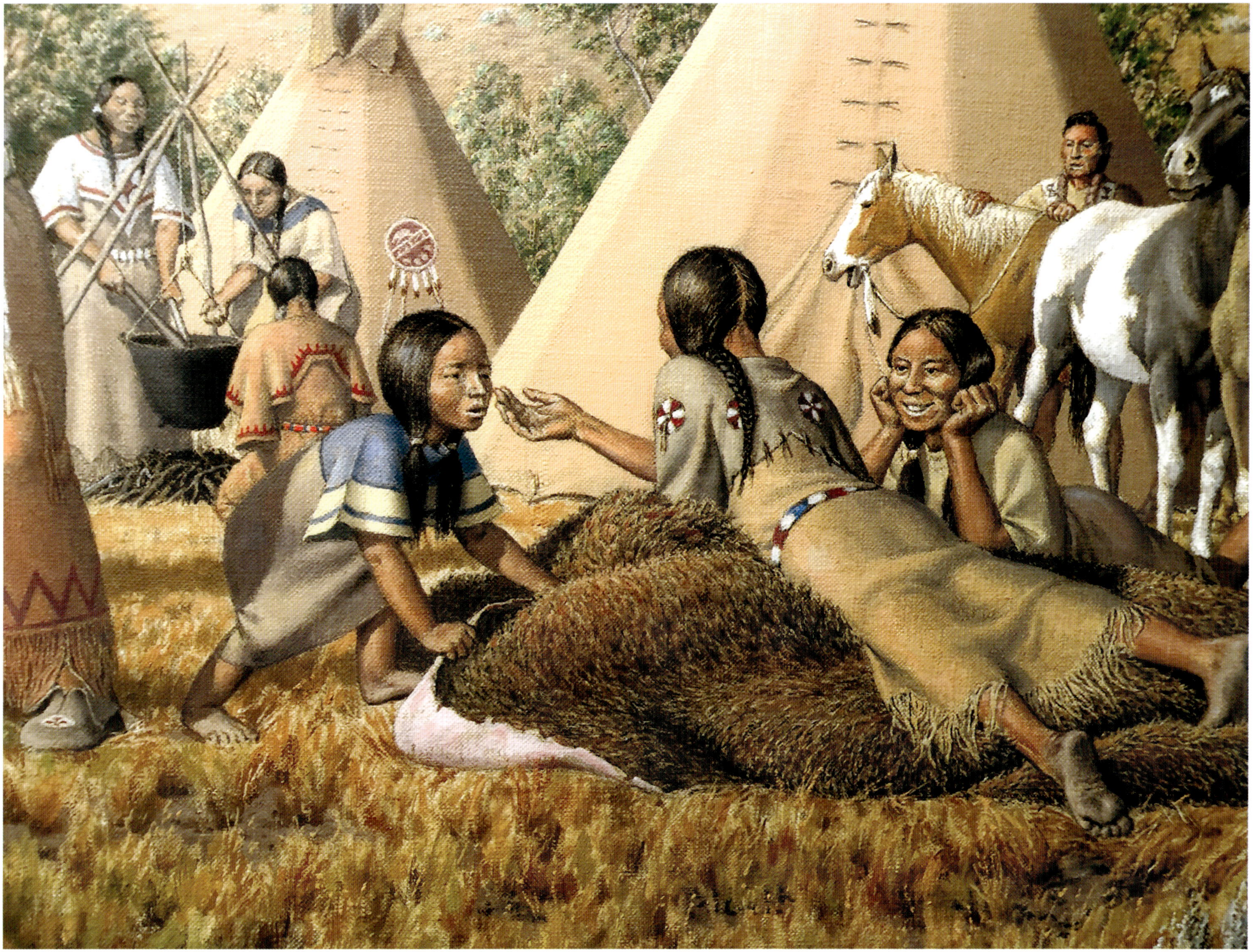

grams of abstract animals, human figures, or geometric symbols, sometimes meant to tell a story. Jay thought it remarkable that a nomadic people, never in one place for long, found the time to do that, but he chose to show the tepees without their graffiti, so as not to distract the viewer's attention from the figures.

Matternes did not intend this painting to be an ethnographic illustration, nor a photographic portrait of a particular tribe or event, but simply an artist's imaginative homage to the Plains tribes as a people and culture. "This picture was an expression of my Romantic imagination," he wrote, "a holdover from my childhood fascination with traditional Native Americans."

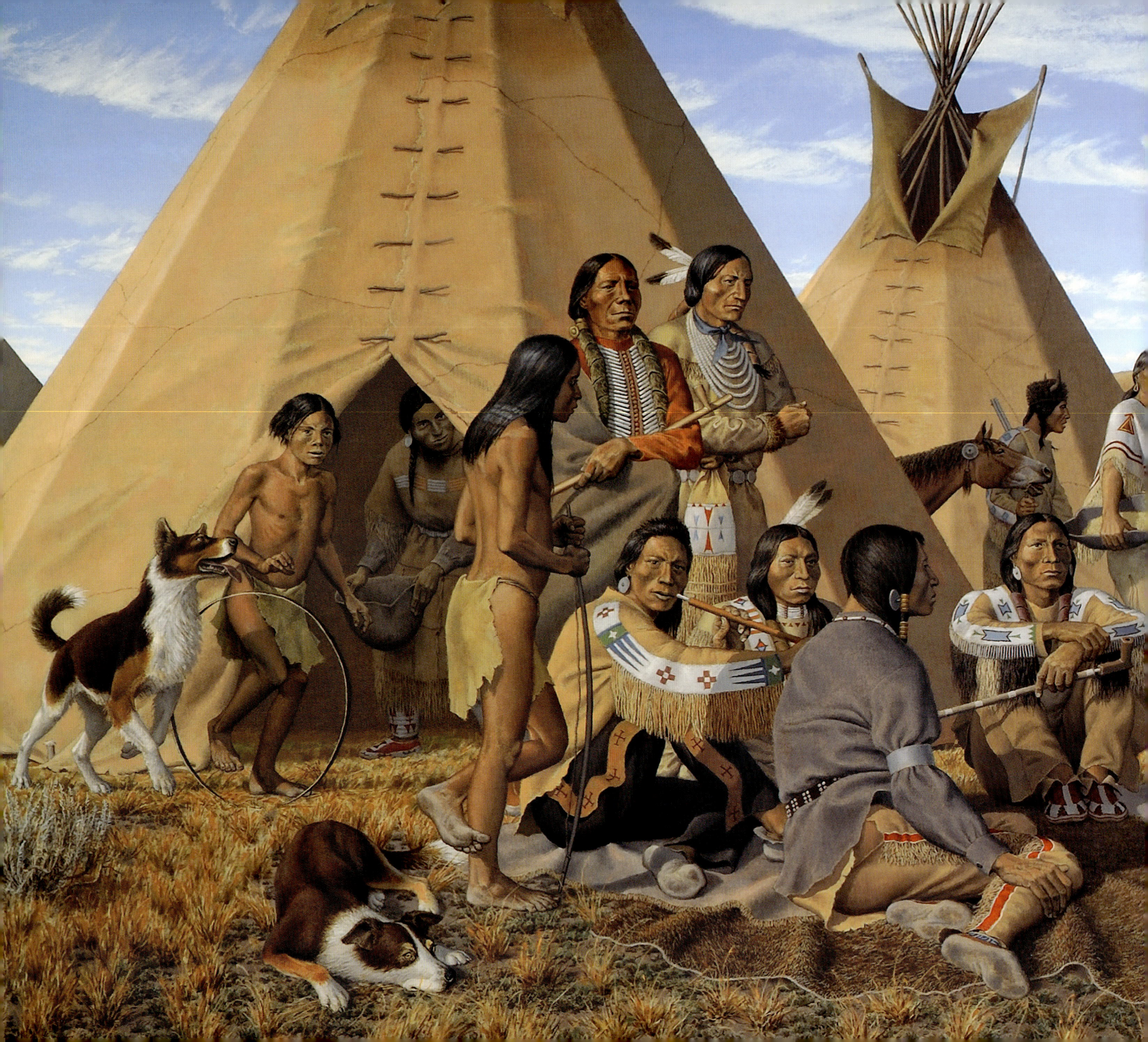

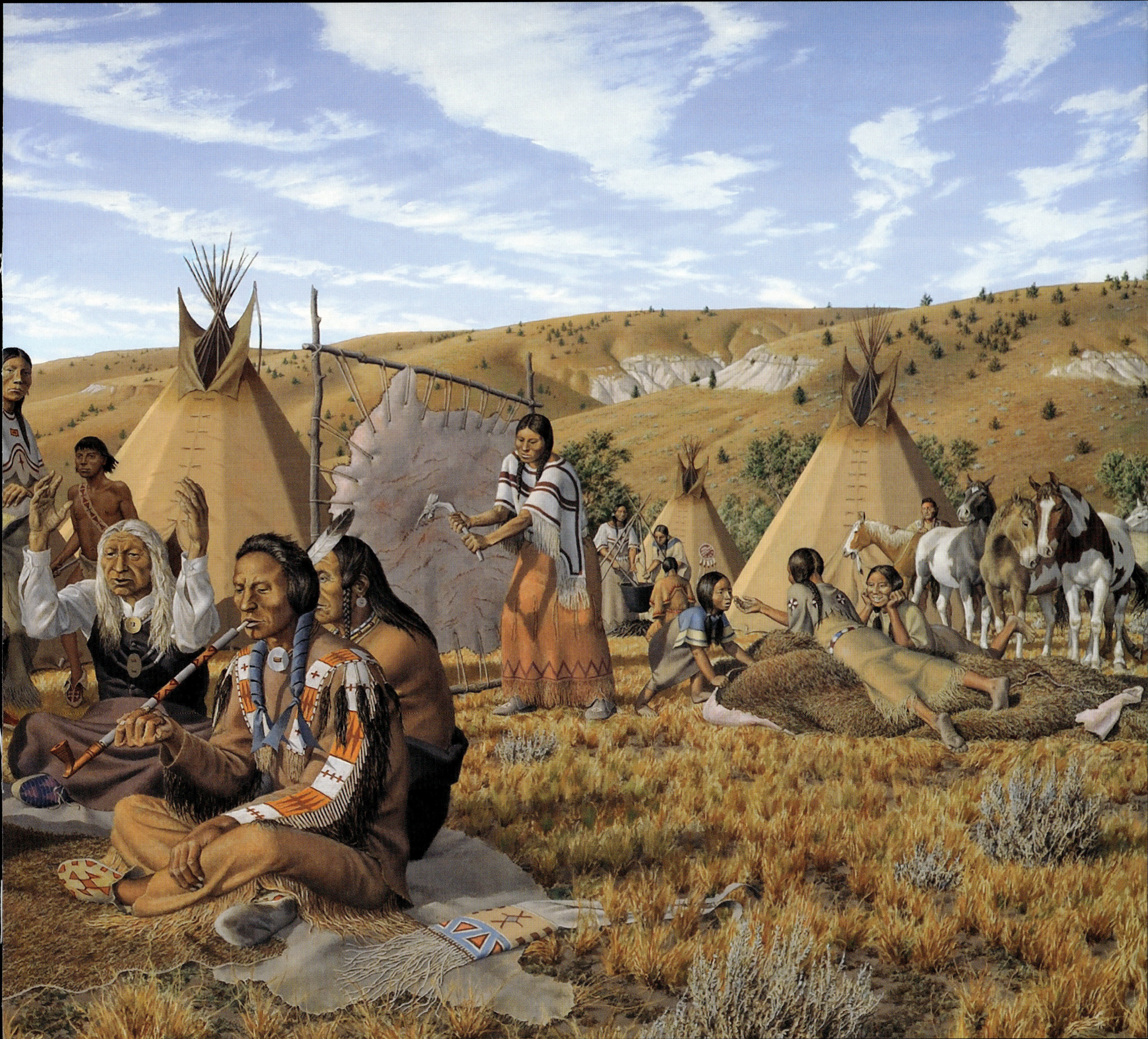

LIST OF ARTWORKS

p. 10 top left
Sketch of Anne Matternes (Sister)

p. 10 bottom left
Sketch of Lawrence Matternes (Father)

p. 10 right
Childhood crayon drawing of
Amphibious Biplane

p. 16 top
Childhood crayon drawing of
Kaydet 1425-5 Biplane

p. 16 bottom left
Childhood crayon drawing of
Cowboy with Revolver

p. 16 bottom right
Childhood crayon drawing of
Cowboys

p. 17 top
Childhood crayon drawing of
American Indian Lassoing Horse

p. 17 bottom left
Childhood crayon drawing of
Prehistoric Big Cat Hunters

p. 17 bottom right
Childhood crayon drawing of
Ape Fight

p. 18
Kudus at Waterhole
Oil on board, 12" × 14½"

p. 19
Bighorn Sheep ("First Alert"), c. 1964
Oil on board, 30" × 40"

p. 20
Bongo, 1976
Oil on canvas, 28" × 22"

p. 21 top
Sable Antelope on Beach
Oil on board, 11¼" × 13½"

p. 21, bottom
Scimitar-Horned Oryx in the Sahel
Oil sketch, 11¼" × 17"

p. 22
Zorilla vs. Gaboon Viper
Published in *Animals of East Africa* by Louis S. B. Leakey (National Geographic Books, 1969)

p. 23
Studies for *The Fight on the Pan*

pp. 24–25
The Fight on the Pan, 1998
Oil on board, 24½" × 38½"

p. 26
Cheetahs on the Athi Plains, c. 1969
Oil on canvas, 16¼" × 32¼"

p. 27
Golden Grizzlies, 1979
Oil on panel, 40" × 30"

pp. 28–29
Day and Night in the Sonoran Desert
Published in *American Deserts* by Rowe Findley (National Geographic Books, 1972)

pp. 30–31
Beneath the American Prairie
Published in *Vanishing Wildlife of North America* by Thomas B. Allen (National Geographic Books, 1974)

pp. 32–33
Monsters of the Mississippi
Published in *The Mighty Mississippi* by Bern Keating (National Geographic Books, 1971)

p. 34
The Joy of the Chase, 1974
Oil, 17½" × 21¼"

p. 35
Field Partners, 1974
Print, 16¼" × 21"

p. 36 top
First Retrieve, 1978
Oil on canvas, 23¾" × 35"

p. 36 bottom
Britts on Point, 1980
Oil on canvas, 23¾" × 35½"

p. 37
Snowy Owls with Christmas Holly, 1978
Acrylic, decal design for decorative plate

p. 38
Golden Eagles in Mountain Eyrie, 1988
Oil on canvas, 31" × 42¼"

p. 39
The Challenger, 1980–81
Acrylic on canvas, 24½" × 32"

p. 40
Oldsquaws, 1982
Acrylic, 6" × 8"
Entry, Federal Duck Stamp Contest

p. 41 top
Snow Goose (Blue Phase), 1987
Acrylic, 8" × 10"
Entry, Federal Duck Stamp Contest

p. 41 bottom
Barrow's Goldeneye, 1986
7¼" × 10"
Entry, Federal Duck Stamp Contest

p. 42
Orangutan Mother with Infant, 2023
Charcoal, 8½" × 8¾"
Collection of the artist

p. 43
Sketches of Orangutans in the Philadelphia Zoo, c. 1950
Collection of the artist

p. 45
Anatomical Dissection of Orangutan "Rango," 1972

pp. 46–47
Fig Feast, 1997
Oil on board, 11¾" × 15¾"

p. 48
Chimps Rampage in Jane Goodall's Camp
Charcoal
Collection of the artist
Published in *My Friends, the Wild Chimpanzees* by Jane Goodall (National Geographic Books, 1967)

p. 49 left
Chimp David Greybeard Steals Banana from Goodall's Tent
Charcoal
Collection of the artist
Published in *My Friends, the Wild Chimpanzees*

p. 49 right
Male Chimp Rodolf Kills a Young Baboon
Published in *My Friends, the Wild Chimpanzees*

p. 50
Sketches for *Chimp Rain Dance*

pp. 50–51
Chimp Rain Dance
Published in *My Friends, the Wild Chimpanzees*

p. 52
Chimp vs. Man
Charcoal
Jacket illustration for *The History of Noell's Ark Gorilla Show* by Anna Mae Noell (1979)

p. 53
Anatomical Dissections of Chimp "Mitzi," 1996
Graphite pencil

p. 54
Comparative Anatomy of Chimp and Human
Published in *My Friends, the Wild Chimpanzees*

p. 55
Chimp at Noell's Ark Flinging Apple
Collection of the artist

p. 56
Mid-Morning Rest, 1969
Watercolor on paper

p. 57
"Are You Looking at Me?" c. 1978
Charcoal pencil on rag drawing paper

p. 59
Massa, 1967
Oil on board, 40" × 30"

pp. 60–61
Anatomical Dissections of Gorilla "Massa," 1986
Collection of the artist

pp. 63–64
Pencil Sketches of East African Forest Vegetation, 1968

p. 66
Cattle Breeds in North America
Published in *The American Cowboy in Life and Legend* by Bart McDowell (National Geographic Books, 1972)

pp. 68–75
Sketches of Karisoke Gorillas, 1974
Collection of the artist

p. 77
Dino Jubilee!, reference drawing for miniature dioramas at the National Museum of Natural History, Smithsonian Institution
Collection of the artist

pp. 78–79
Brachiosaurus and *Camarasaurus*
Pastel drawing
Smithsonian

p. 80
Ceratosaurus
Acrylic
Smithsonian

p. 81
Trachodon
Charcoal drawing
Smithsonian

pp. 82–83
Thescelosaurus, Monoclonius, Gorgosaurus, Triceratops, 1962
Reference drawings for diorama at the National Museum of Natural History, Smithsonian Institution
Collection of the artist

p. 84
Ornitholestes, 1975
Collection of the artist

p. 85
Struthiomimus
Collection of the artist

pp. 86–87
Brachauchenius and *Elasmosaurus*
Smithsonian

pp. 88–94
Restorations of Early Eocene Mammals
American Museum of Natural History

p. 95
Earliest North American Mammals (Triconodont-like)
Pencil
Published in *Our Continent: A Natural History of North America* (National Geographic Books, 1976)

pp. 96–97
Eocene Mammals Habitat
Acrylic
Published in *Agate Fossil Beds* (National Park Handbook 107, 1980)

p. 98
Uintatherium, c. 1960
Collection of the artist

p. 99
Deinotherium, unfinished color sketch, c. 1968

p. 100
Late Oligocene–early Miocene mural, 1961–62
12' 3" × 24'
National Museum of Natural History, Smithsonian Institution

p. 101
Late Pliocene mural, 1964
12' 3" × 18' 11"
National Museum of Natural History, Smithsonian Institution

p. 102
Pelorovis
Published in *Animals of East Africa* by Louis S. B. Leakey (National Geographic Books, 1969)

p. 103
Black Rhinos
Oil on board, 12¼" × 17½"
Collection of the artist

pp. 104–5
When Olduvai Was Green
Published in *Early Man* by F. Clark Howell (Time-Life, 1965)

p. 106
Mesohippus, preliminary detail sketch for Oligocene mural, c. 1961
Smithsonian

p. 107
Desmatippus, detail of late Oligocene–early Miocene mural, 1961–62
Smithsonian

p. 108
Sketch of *Orohippus*, preliminary drawing for Eocene mural, c. 1959
Collection of the artist

p. 109 top
Desmatippus
Detail of painting published in *Agate Fossil Beds* (National Park Handbook 107, 1980)

pp. 110–11
Galloping *Merychippus* horses attacked by felid *Pseudaelurus*
Published in *Agate Fossil Beds* (National Park Handbook 107, 1980)

pp. 112–13; details p. 109 bottom, 114–15
Fairbanks Gold-Bearing Mucks, 1975
12' × 20'
National Museum of Natural History, Smithsonian Institution

pp. 116–17; details pp. 118–19, 122–23
Primates through Time mural, 1992
Acrylic on canvas, 5' × 10'
American Museum of Natural History

p. 120–21
Sketches of *Aegyptopithecus* for *Primates through Time* mural

p. 124
Golden-Crowned Sifaka, 1993
Oil on board, 31" × 23"

p. 125
Brown Lemurs, 1996
Oil on board, 16" × 20"

pp. 126–27
Giant Baboons, 1969
Published in *Animals of East Africa* by Louis S. B. Leakey (National Geographic Books, 1969)

p. 128
Anubis Baboons, 2021
Oil on board, 12¼" × 10½"
Collection of the artist

p. 129
Art Deco Baboons, 1954
Lithograph print, student exercise
Collection of the artist

pp. 130–35
Ardipithecus, *Three-Stage Reconstruction Drawings,* 1997–2009
Collection of the artist

p. 136
Ardipithecus Family Group, with Juveniles in Trees, 2012
26½" × 21"

p. 137
Ardipithecus Family Group, Two Adult Females Foreground, 2009
18" × 13½"
Collection of the artist

pp. 138–39; details pp. 140–41
Paranthropus *Family Group "Muralette,"* 1993
Acrylic on canvas, 5' × 10'
American Museum of Natural History

p. 139
Sculpture of *Paranthropus* head, 1960
Collection of the artist

p. 142; details pp. 142–45
Hippo Feast at Ileret, c. 1974
Kenya National Museum, Nairobi

p. 146
Family Feast
Oil on board, 6' 11" × 5' 3"
Published in *National Geographic* magazine, 1995

p. 147
Hominins Butchering Zebra Carcass, 1995
Acrylic on canvas, 13¾" × 20⅝"

pp. 148–49
Homo erectus *and Prairie Fire*, 1983
Alkyd oil on board, 22" × 31"
Illustration, National Geographic Books

pp. 150–51
Laetoli Footprints
Acrylic on board, 21" × 33"
Published in *National Geographic* magazine, April 1979

pp. 152–53
Once We Were Not Alone
Published in *Scientific American*, January 2000

pp. 154–55
Marathon Man
Oil on board, 14" × 42"
Collection of the artist
Published in *National Geographic* magazine, 1985

pp. 156–57
Australopiths Steal Hyena's Kill
Charcoal
Collection of the artist
Published in *Early Man* by F. Clark Howell (Time-Life, 1965)

pp. 158–59
Olduvai Lakeside Scene
Charcoal, 16¼" × 24¼"
Published in *Lucy's Child: The Discovery of a Human Ancestor* by Donald Johanson (William Morrow, 1989)

pp. 160–61
H. habilis vs. *Paranthropus*
Oil on board
Collection of the artist
Published in *Early Man* by F. Clark Howell (Time-Life, 1965)

pp. 162–63
Family Album
Collection of the artist
Published in *Scientific American*, January 2000

pp. 164–65
Reindeer Bonanza in the Dordogne (Neanderthals), 2014
Oil on board, 18" × 24"

pp. 166–67
Neanderthals in the Pyrenees, 1982
Alkyd oil, 20" × 33"
Published in *National Geographic* magazine, November 1985

p. 169
Dmanisi Male
Charcoal
Collection of the artist

pp. 170–71
Bison Sculptures of Tuc d'Audoubert, 1994
Acrylic on canvas, 5' 3" × 6' 11"
Gunma Prefecture Museum, Japan

pp. 172–73
HMS Beagle *Entering Rio de Janeiro Harbor on April 3, 1832*, 1999
Oil on board, 1' 5" × 2' 3"
Collection of the artist

pp. 174–75
Teddy Roosevelt and John Muir at Yosemite, 1987
Oil on board, 18½" × 25½"
Collection of the artist

pp. 176–77
Opening the Oregon Trail, 1988
Oil on canvas, 18½" × 25½"

pp. 178–79
PaleoIndians Dismantle a Mastodon, 2003
Acrylic on canvas, 32" × 64"
Florida Museum of Natural History, University of Florida, Gainesville

pp. 180–81
Buffalo Jump, 1976
Oil on canvas, 36" × 72"

pp. 182–83
Pulling Even (Chasing Pronghorns), c. 1978
Oil on board, 17½" × 25"
Collection of the artist

pp. 188–89; details pp. 184–87
Plains Indian Village (Story Telling), 2023
Oil on board, 36" × 72"

INDEX

Page numbers in *italics* refer to illustrations.

ACKNOWLEDGMENTS

Special thanks to
Polly Cleveland and Tom Haines, Ernest Farino, David Cox, Carol Story and David Parkinson, Denis Finnin, James Farlow, Matthew Carrano, Kirk Johnson, Alan Jay Kaufman, RJ Coté, Will Harcourt-Smith, James Van Abbema, Judith Jennings, Peter William Von Sholly and Junior Mint, Errol Fuller, Catherine Wallis, Rachel Baron and staff (Old Town Editions, Alexandria, Virginia), Frederick Manthi and Photo Staff at Kenya National Museum, Nairobi; Edward ("Ned") Barnard, Roy Pinney, Melvin Van Peebles, Gary Sawyer, Jeannine Frank and Michael J. Levin, Larry Luckman, John and Hazel Marsden, Barbara and Michael MacRoberts, Lindsay Fulcher, Grenville Lucas, Enrico Banducci, Bill and Linda Susman, June Bruno, Henry and Gloria Jarecki, Robert Carneiro, Paula Henderson, John Terry, Donald Stone Sade, Earl W. Count, George Beccaloni, Stephen Jay Gould, David Bergman, James Demes, Sheldon Harnick, Jackie Mason, Michael Shermer, Donald Prothero, Gary and Michelle Hallgren, Shoko Ozeki-Gardner, Randal Keynes, Susan and Bob Wilder, Gerry Ohrstrom, Norman Shaifer, Bob Adelman, C. R. Tripp, Ray Troll, Mary Lynn Price, Patti Hinsdale, Gina Douglas, James Healey, Bob Silverstein, Judy Twersky, Marian Brickner, Diana Reiss, Paul Wason, Samson and Jackie Pollen, Wong-Baxter, Ivory Madison and Abraham Mertens, Gracie and Archer, Jacques Semmelman, Annie Gottlieb, Joyce Johnson, Sophie Weiler, Vaughan R. Southgate

PHOTO CREDITS

p. 6: Jay in studio, photo by Willard Whitson and Susan Arthur-Whitson, October 21, 2012; p. 8: Jay with skull, photo by Bob Campbell; p. 9: young Jay with mom and brother, artist's collection; p. 11: American Museum of Natural History; p. 13: young Jay sketching Fairbanks mural, photo by Victor R. Boswell, *National Geographic* magazine, March 1972; p. 14: Jay with car, photo by Richard Milner, 2015, Vienna, Virginia; opening of "Deep Time" Hall, Smithsonian, 2019; p. 15: Jay measuring skull, photo by Linda Bartlett, *Science 81*; pp. 16–17: childhood drawings, photos by Greg Staley; p. 22: "Zorilla vs. Gaboon Viper" from *Animals of East Africa* by Louis S. B. Leakey, National Geographic Books, 1969; p. 23: studies for *The Fight on the Pan*, photos by Greg Staley; pp. 28–29: "Sonoran Desert Day and Night" from *American Deserts* by Rowe Findley, National Geographic Books, 1972; pp. 30–31: "Beneath the American Prairie" from *Vanishing Wildlife of North America* by Thomas B. Allen, National Geographic Books, 1974; pp. 32–33: "Monsters of the Mississippi" from *The Mighty Mississippi* by Bern Keating, National Geographic Books, 1971; p. 44: Jay dissects large male orang, photo by Chip Clark, 1976, courtesy Smithsonian; pp. 46–47: *Fig Feast* (orangutan and hornbills), 1997, artist's collection; pp. 50–51: "Chimp Rain Dance" from *My Friends the Wild Chimpanzees* by Jane Goodall, National Geographic Books, 1967; p. 52: "Chimp vs. Man," cover illustration for *The History of Noell's Ark Gorilla Show* by Anna Mae Noell, 1979; p. 53: chimpanzee dissections, 1996; p. 54: comparative anatomy of chimp and *Homo sapiens* from *My Friends the Wild Chimpanzees* by Jane Goodall, National Geographic Books, 1967; p. 55: sketches of chimp at Noell's Ark flinging apple, artist's collection; p. 65: Dian Fossey with tracker Nmeye, photo by Jay Matternes; pp. 66–67: "Breeds of Cattle" from *The American Cowboy in Life and Legend* by Bart McDowell, National Geographic Books, 1972; p. 72: "Tiger," juvenile gorilla, photo by Jay Matternes, 1974; p. 76: *King Kong* publicity still, inscribed by Dian Fossey, artist's collection; p. 79: *Brachiosaurus* and *Camarasaurus*, pastel drawing, Smithsonian; p. 80: *Ceratosaurus*, Smithsonian Books, 2019; p. 81: *Trachodon*, Smithsonian Books, 2019; pp. 82–83: *Thescelosaurus*, 1962, Smithsonian; p. 84: *Ornitholestes*, 1976, Smithsonian; p. 85: *Struthiomimus*, Smithsonian; pp. 86–87: *Brachauchenius* and *Elasmosaurus*, Smithsonian; pp. 88–94: "Some Early Mammals Restored," all photos by Denis Finnin, Photographer Emeritus, American Museum of Natural History; p. 95: "Earliest Mammals (Triconodont-like)" from *Our Continent*, National Geographic Books, 1976; pp. 96–97: "Early Mammals" from *Agate Fossil Beds*, National Park Handbook 107, 1980, photo by Matternes from original; p. 98: *Uintatherium* sketch, c. 1960, Smithsonian; p. 100: early Miocene mural, National Museum of Natural History (Smithsonian); p. 101: later Pliocene mural, National Museum of Natural History (Smithsonian); p. 102: *Pelorovis* from *Animals of East Africa* by Louis S. B. Leakey, National Geographic Books, 1969; pp. 104–5: "When Olduvai Was Green" from *Early Man* by F. Clark Howell, Time/Life Books, 1965, photo courtesy Smithsonian; p. 107: *Desmatippus* (detail of early Miocene mural), Smithsonian; p. 108: herd of *Orohippus* (mid-Eocene horses), c. 1959, Smithsonian; p. 109: early Miocene horses, detail from *Agate Fossil Beds*, National Park Handbook 107, 1980; pp. 110–11: galloping *Merychippus* and felid *Pseudaelerus* from *Agate Fossil Beds*, National Park Handbook 107, 1980, photo by Matternes from the original; pp. 112–15: Fairbanks Gold-Bearing Mucks mural and details, 1975, National Museum of Natural History (Smithsonian); pp. 116–19: *Primates through Time* mural and details, 1992, American Museum of Natural History; pp. 126–27: Giant baboons (*Theropithecus* sp.) from *Animals of East Africa* by Louis S. B. Leakey, National Geographic Books, 1969; p. 129: *Art Deco Baboons*, lithographic print, 1954, artist's collection; pp. 130–35: all photos in *Ardipithecus* chapter by Alan Jay Kaufman; pp. 138–41: *Paranthropus* family habitation "muralette" and details, American Museum of Natural History; sculpture of *Paranthropus* head by Matternes; pp. 142–45: *Ileret Hippo Feast* mural and details, photos courtesy Frederick Kyolo Manthi, Kenya National Museum, Nairobi, c. 2020; p. 146: *Family Feast*, *National Geographic* magazine, 1995; p. 147: hominins butchering a zebra carcass, 1995, Gunma Prefecture Museum, Japan; pp. 148–49: *Homo erectus* and fire, National Geographic Books, 1983. pp. 150–51: Laetoli footprints, *National Geographic* magazine, April 1979; pp. 152–53: *Once We Were Not Alone*, *Scientific American*, January 2000; pp. 154–55: *Marathon Man*, *National Geographic* magazine, 1985; pp. 156–57: "Australopiths Steal Hyena's Kill" from *Early Man* by F. Clark Howell, Time/Life Books, 1965; pp. 158–59: "Olduvai Lakeside Scene" from *Lucy's Child: The Discovery of a Human Ancestor* by Donald Johanson and James Shreeve, William Morrow, 1989; pp. 160–61: "*H. habilis* vs. *Paranthropus*" from *Early Man*, Time/Life Books by F. Clark Howell, 1965; pp. 162–63: *Family Album* (hominins), *Scientific American*, January 2000; pp. 164–65: *Reindeer Bonanza* (Neanderthals), photo by Will Harcourt-Smith, 2020; pp. 166–67: Neanderthals in Pyrenees, *National Geographic* magazine, November 1985; pp. 170–71: Ice Age bison sculptures, Gunma Prefecture Museum, Japan; pp. 178–79: PaleoIndians dismantle a mastodon, Florida Museum of Natural History, University of Florida, Gainesville; pp. 184–89: *Story Telling* (Plains Indian village), all photos by Greg Staley and Alan Jay Kaufman; p. 190: author photos, Matternes by Alan Jay Kaufman, Milner by Roy Pinney, Tattersall by Malcolm McKenna, Maurico Antón courtesy himself; jacket flap: author photo by Alan Jay Kaufman.

FOR JUDE: LOVE BEYOND WORDS! —RM

FOR DELPHINE: FOREVER! —JM

Designer: Rick DeMonico

First edition

10 9 8 7 6 5 4 3 2 1

Library of Congress Cataloging-in-Publication Data
Names: Milner, Richard, 1941- author. | Tattersall, Ian, author. | Antón, Mauricio, writer of foreword.
Title: Jay Matternes : paleoartist and wildlife painter / by Richard Milner with Ian Tattersall ; foreword by Mauricio Anton.
Description: [First edition]. | New York : Abbeville Press, [2024] | Includes index. | Summary: "A retrospective of Jay Matternes' career"—Provided by publisher.
Identifiers: LCCN 2024007258 | ISBN 9780789214805 (hardcover)
Subjects: LCSH: Matternes, Jay H.—Criticism and interpretation. | Paleoart—United States. | Wildlife painting, American.
Classification: LCC ND237.M338 M55 2024 | DDC 759.13—dc23/eng/20240314
LC record available at https://lccn.loc.gov/2024007258

Jacket front: *The Fight on the Pan*. See pp. 24–25.
Jacket back: *Hominin Confrontation*. See pp. 160–61.
Binding: Mural of the Fairbanks Gold-Bearing Mucks. See pp. 112–13.
Endpapers: *Day and Night in the Sonoran Desert*. See pp. 28–29.
Frontispiece: Late Pliocene horses and mastodons. Detail of Idaho Riverbank mural. See page 101.

Permissions

Permission to publish letters from the Fossey-Matternes Correspondence, courtesy of the Mountain Gorilla Conservation Fund, proprietors of the Dian Fossey Estate, and Russell & Volkening

Song Lyric, p. 76: "I Had the Craziest Dream Last Night" by Warren/Gordon, © 1943, WB Music Corp., Four Jays Music Publishing Company

For bulk and premium sales and for text adoption procedures, write to Customer Service Manager, Abbeville Press, 655 Third Avenue, New York, NY 10017, or call 1-800-ARTBOOK.

Visit Abbeville Press online at www.abbeville.com.

AUTHORS

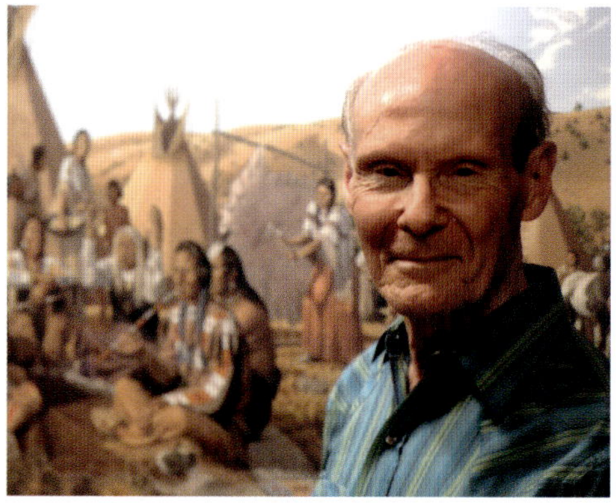

JAY H. MATTERNES is a wildlife painter and sculptor, born on Corregidor Island, Philippines, and raised in Pennsylvania as well as at several army bases to which his surgeon father was assigned. On completion of high school, he received a full four-year scholarship to study art at the Carnegie Institute of Technology (now Carnegie Mellon University) in Pittsburgh. After graduation he was hired at the Cleveland Museum of Natural History, where he painted backgrounds for dioramas. Eventually, his childhood passion for drawing wildlife evolved into expertise as a paleoartist specializing in fossil primates and hominins. During a long career as a freelance museum artist, Matternes created classic murals on the Age of Mammals at the Smithsonian Institution and primate evolution at the American Museum of Natural History. He also illustrated popular Time-Life books, including the best seller *Early Man*, numerous articles in *National Geographic* and *Scientific American*, and Jane Goodall's first book on chimpanzees. He has conducted field research in the wilderness of Alaska, Japan, France, Colombia, and Africa. During his eighties, he painted a bustling 1880s Sioux Indian Village that had been gestating in his mind for thirty years.

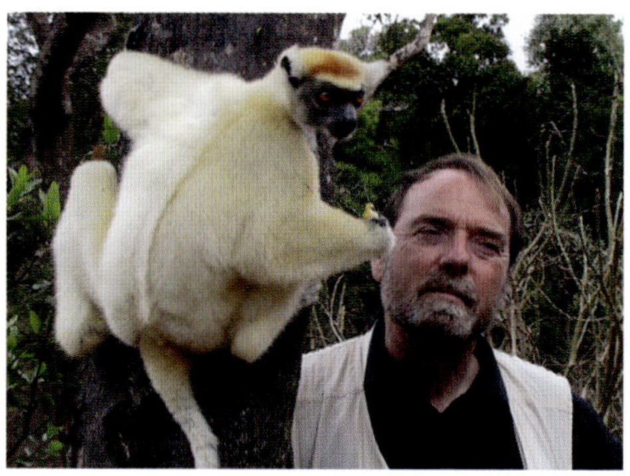

IAN TATTERSALL and Jay H. Matternes have worked together since the early 1990s, notably on the Hall of Human Biology and Evolution at the American Museum of Natural History in New York City, where Tattersall is Curator Emeritus in the Division of Anthropology, and at Japan's Gunma Museum. Tattersall studied at Christ's College, Cambridge, and completed his graduate work at Yale. Among his many books are *The Human Odyssey: Four Million Years of Human Evolution* (Macmillan, 1993), *Masters of the Planet: The Search for Our Human Origins* (Palgrave Macmillan, 2012), *The Strange Tale of the Rickety Cossack and Other Cautionary Tales from Human Evolution* (Palgrave Macmillan, 2015), and *Understanding Human Evolution* (Cambridge University Press, 2022). He began this retrospective of Matternes' life's work and invited Milner to complete it. A lemur expert, Tattersall was exploring a forest in northern Madagascar when a companion photographed him receiving a spontaneous welcome from a golden-crowned sifaka, a species he discovered and that is named for him (*Propithecus tattersalli*).

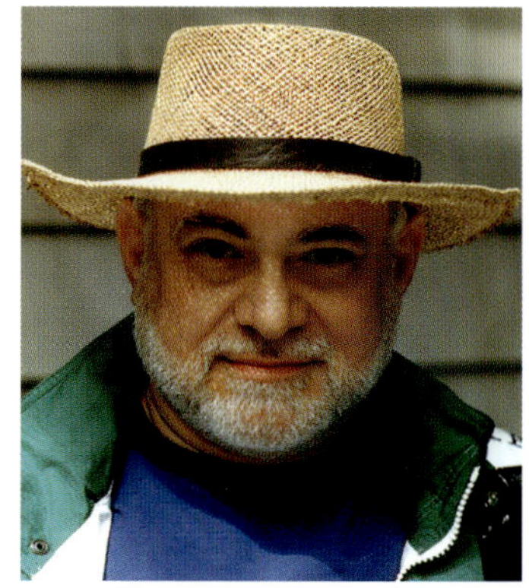

RICHARD MILNER, anthropologist, historian of science, and songwriter, was born in Brooklyn, New York, and attended Hamilton College, Queens College, UCLA, and UC Berkeley. He served as Senior Editor of *Natural History* magazine at the American Museum of Natural History, where he is an Associate in the Division of Anthropology. In 2013 he helmed the Alfred Russel Wallace Centenary Celebration funded by the John Templeton Foundation (with a keynote lecture by Sir David Attenborough) at the AMNH and UCLA. Milner's articles on Darwin, Wallace, and evolution have appeared in *Scientific American, Natural History*, and the *Journal of the Linnean Society of London*. Books include *Black Players* with Christina Milner (Little, Brown, 1971), *The Last Human*, edited by Gary Sawyer (Yale University Press, 2007), *The Encyclopedia of Evolution* (Facts on File, 1990), *Darwin's Universe: Evolution from A to Z* (University of California Press, 2009), *Charles R. Knight: The Artist Who Saw through Time* (Abrams, 2012), and *Alfred Russel Wallace in Paradise: Odyssey of a Victorian Evolutionist* (Princeton University Press, in press). He has performed his one-man musical *Charles Darwin: Live & In Concert* in London, Edinburgh, and Australia and on Darwin's beloved Galápagos Islands.

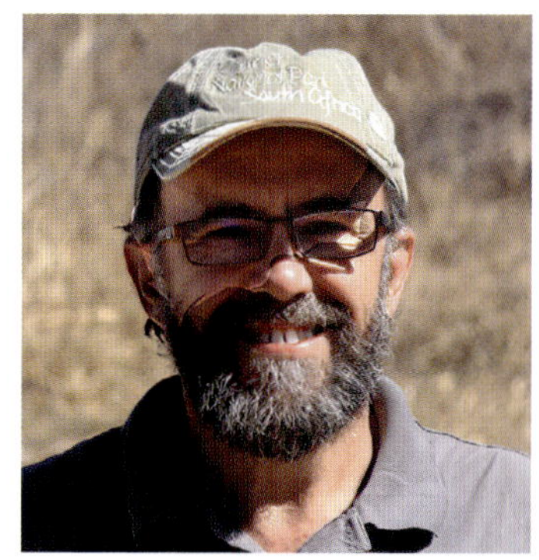

MAURICIO ANTÓN was born in Bilbao, Spain, and studied fine arts in Caracas, where he lived for several years. As a boy he was inspired by the paintings of extinct animals by Jay Matternes. Now based in Madrid, he has been a full-time paleoartist since 1987, specializing in the reconstruction of fossil mammals. He has created artwork for museum exhibits worldwide, including among others Madrid's Museo Nacional de Ciencias Naturales and the American Museum of Natural History. He has written and illustrated *El Secreto de los Fósiles* (2007) and *Sabertooth* (2013), and has coauthored and illustrated many others, including *The Big Cats and Their Fossil Relatives* and *The National Geographic Book of Prehistoric Mammals.* Antón's artwork is inseparable from his research on the anatomy and adaptations of fossil mammals, and the results have been published in the *Journal of Human Evolution* and the *Zoological Journal of the Linnean Society*. In 2020 he held the Basler Chair of Excellence at East Tennessee State University. Since 2013 he has led an annual art safari, Drawing the Big Cats, in northern Botswana, where he teaches the anatomy and adaptations of felines and the techniques to draw and paint them.

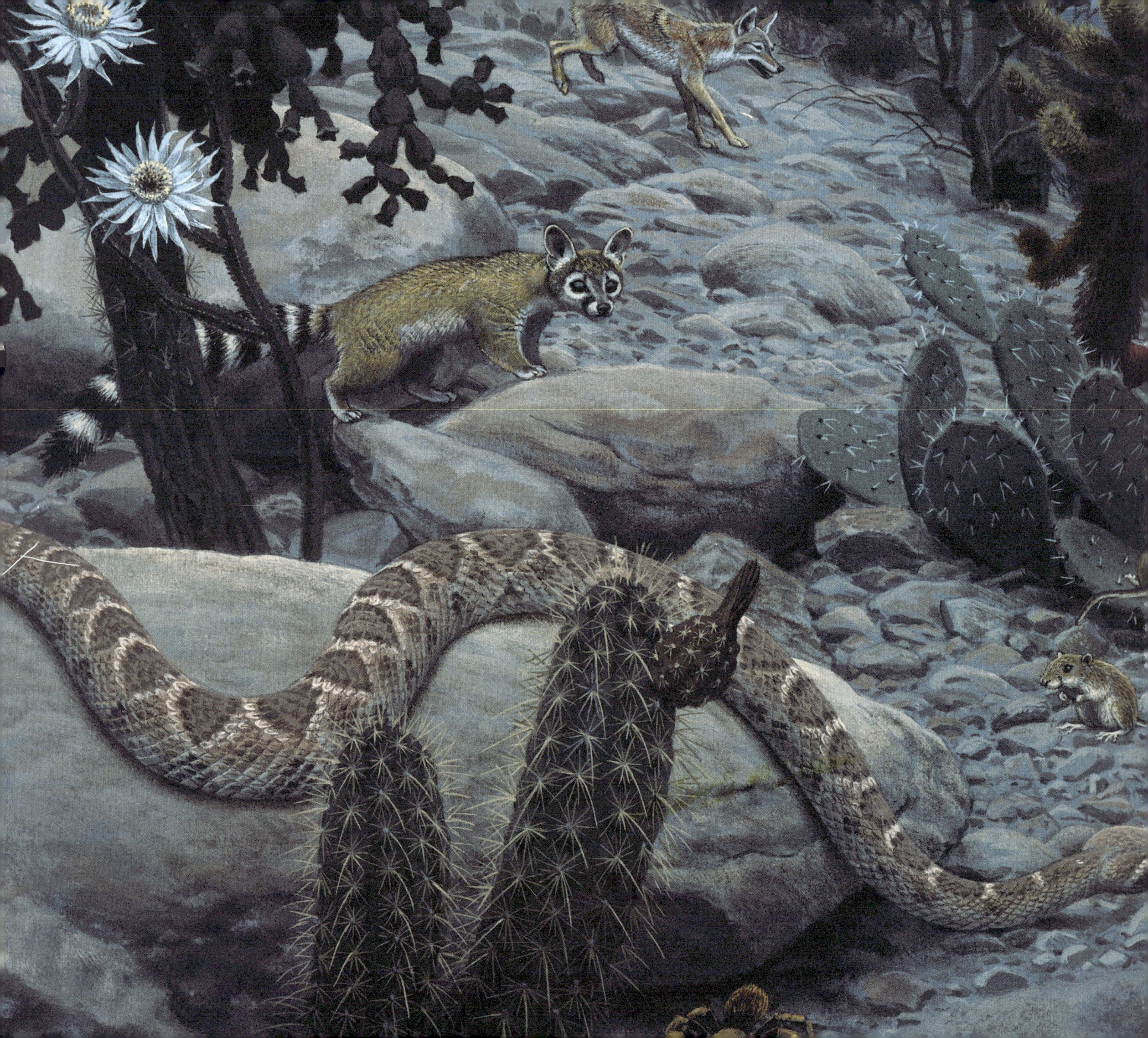